# LASER AND PHOTONIC SYSTEMS

## Design and Integration

# INDUSTRIAL AND SYSTEMS ENGINEERING SERIES

# LASER AND PHOTONIC SYSTEMS

## Design and Integration

Edited by
Shimon Y. Nof
Andrew M. Weiner
Gary J. Cheng

CRC Press
Taylor & Francis Group
Boca Raton   London   New York

CRC Press is an imprint of the
Taylor & Francis Group, an **informa** business

CRC Press
Taylor & Francis Group
6000 Broken Sound Parkway NW, Suite 300
Boca Raton, FL 33487-2742

First issued in paperback 2017

© 2014 by Taylor & Francis Group, LLC
CRC Press is an imprint of Taylor & Francis Group, an Informa business

No claim to original U.S. Government works

Version Date: 20140311

ISBN 13: 978-1-138-07629-7 (pbk)
ISBN 13: 978-1-4665-6950-8 (hbk)

---

**Library of Congress Cataloging-in-Publication Data**

---

Laser and photonic systems : design and integration / editors, Shimon Y. Nof, Andrew M. Weiner, Gary J. Cheng.
    pages cm. -- (Industrial and systems engineering)
    Summary: "Laser and photonic technologies and solutions influenced many aspects of everyday life. With new and significant recent scientific discoveries in their fields, systems perspectives and integrated design approaches can improve even further the impact in critical areas of challenge. Yet this knowledge is dispersed across several disciplines and research arenas. This book brings together a multidisciplinary group of experts in many of these areas to foster increased understanding of the ways in which systems perspectives may influence laser and photonic innovations and application integration"-- Provided by publisher.
    Includes bibliographical references and index.
    ISBN 978-1-4665-6950-8 (hardback)
    1. Lasers. I. Nof, Shimon Y., 1946-

TA1675.L334 2014
621.36'6--dc23                                                                              2014001341

---

**Visit the Taylor & Francis Web site at**
**http://www.taylorandfrancis.com**

**and the CRC Press Web site at**
**http://www.crcpress.com**

*This book is dedicated to the scientists, engineers, explorers, and inventors of light,*

*laser, and photonics throughout human history. May we always be enlightened.*

# Contents

# *Foreword*

## Laser and Photonic Systems: Dreams and Visions that Come True

 As someone with a nearly 60-year career in materials research, I welcome the opportunity to write this foreword for a book on a visionary subject that is dear to me. I have been blessed by knowing many of the visionaries in this field who have propelled it forward and made it come true: Charles Townes, discoverer of stimulated emission from masers and lasers; Theodore Maiman, inventor of the first laser; Amnon Yariv, pioneer in lightwave communications and nonlinear and phase conjugate optics; Zhores Alferov, developer of semiconductor heterojunctions; Nick Holonyak, inventor of the practical LED; and Jerry Woodall, developer of compound heterojunctions and inventor of the red LED, among others.

Lasers and photonics also have a 60-plus year history starting from 1951 when Charles Townes conceived of the idea for the maser. Our lives before the many transformations in all aspects of our life that have resulted from these discoveries must be bewildering to most young students who today take these transformations for granted.

Our role as scientist-engineers and innovators is to integrate new concepts to find new solutions that will improve our security and quality of life. Experience has taught us that innovations usually derive from ideas that come from challenging old paradigms and probing into how best to do things differently. Contributors to this book have done just that and have asked not just *why* and *how* but *why not* in looking beyond the technological frontiers of our known world.

Consequently, this book will appeal to members of the STEM community of all ages in exploring how critical *grand challenge* problems of the twenty-first century can be addressed by lasers and photonics.

This book addresses such transformative applications of laser and photonic systems as in medicine, communications, security, sustainable energy, advanced manufacturing, and the synthesis and modification of new materials, especially the nanoprocessing of nanomaterials. Contributors to the book describe a few well-chosen samples of the many discoveries and developments currently underway that will have a substantial impact on our lives in the remainder of the twenty-first century.

Let me offer a walk through my own history in this field since the 1970s involving many distinguished colleagues.

In 1976, I joined DARPA (Defense Advanced Research Projects Agency) as head of the Materials Research Office, where along with Richard Reynolds, Ed van Reuth, Harry Windsor, and Michael Buckley, DARPA funded pioneering research at both universities and industries in compound semiconductor heterojunctions, rapidly solidified metals, laser-assisted machining of ceramics, integrated optic flight systems, ring laser gyros, optical fiber sensors, nonlinear optic frequency doubling and phase conjugation, LIDAR (light detection and ranging, or laser imaging detection and ranging) systems, and hardening against directed high-energy weapons (to include the effects of laser-induced plasmas on seeker performance). Most of this effort at the time was early-stage 6.1 (basic) and 6.2 (applied) research according the Department of Defense research budget categories; however, it preceded the early acceleration of nanotechnology.

In the 1990s, we worked on laser beam lithography of metal oxide electrodes for PZT memory applications. Traditional metal/metal oxide patterning techniques typically involved photolithography, which is a relatively expensive and time-consuming process. As an improvement, the fabrication of metal oxide upper electrodes by metallorganic decomposition (MOD) was integrated with laser beam patterning. A variety of patterns were prepared, which demonstrates the suitability of laser patterning for many applications.

During my directorship at NIST (National Institute of Standards and Technology), I enjoyed discussing laser quantum physics with Bill Phillips, Eric Cornell, Jan Hall, Dave Wineland, Deborah Jin, and Jun Ye. Bill Phillips shared the Nobel Prize in Physics in 1997 for "developing methods to cool and trap atoms with laser light." Eric Cornell shared the Nobel Prize in Physics in 2001 for synthesizing the first Bose–Einstein condensate in 1995. Jan Hall shared the Nobel Prize in Physics in 2005 for his work in precision laser spectroscopy and the optical frequency comb technique. David Wineland shared the Nobel Prize in Physics in 2012 for groundbreaking experimental methods on manipulating quantum systems that form the building blocks for a practical quantum computer.

The work accomplishments of Deborah Jin and Jun Ye are likewise groundbreaking. Deborah Jin and her group have achieved a fermionic condensate and discovered fermionic condensates of Cooper pairs at the BCS (Bardeen–Cooper–Schrieffer) and BEC (Bose–Einstein condensate) crossover regime. Jun Ye and his group have demonstrated precision manipulation of atomic states in ultracold strontium, resulting in the most accurate atomic clock in the world (will neither gain nor lose a second in 200 million years).

These discoveries of new states of condensed matter and their quantum behavior will lead to transformative quantum information and communications technologies in the twenty-first century. In addition, they will provide precise metrology techniques that will permit the engineering of materials at the atomic level with novel properties through the manipulation of their bonding states and band structures.

During my years with NIST, an important effort was to overcome the problems associated with LADAR (laser radar) systems. The NIST Construction Metrology and Automation Group (CMAG), cooperating with the NIST Intelligent Systems Division (ISD), developed performance metrics and researched issues related to the design and delivery of a next-generation LADAR sensor that will enable general automation in structured and unstructured environments. The physics and implementation of various LADAR technologies were studied. Worldwide state-of-the-art research was compared, and the results pointed to the general trends in advanced LADAR sensor research. Interestingly, their evaluation of likely impact on manufacturing, autonomous vehicle mobility, and on construction automation has been largely materialized with LADAR and also with LIDAR.

From my memories at the National Science Foundation (NSF), I recall our efforts to sustain and advance the Laser Interferometer Gravitational Wave Observatory (LIGO). LIGO is a large-scale physics experiment to directly detect and observe gravitational waves of cosmic origin. It is a joint project between scientists at many colleges and universities and is the largest and most ambitious project ever funded by the NSF. The international LIGO Scientific Collaboration (LSC) is a growing group of researchers, over 800 individuals at roughly 50 institutions, working to analyze the data from LIGO and other detectors. Beyond its immense scientific role, LIGO is a symbol of global scientific collaboration, which is both educational and inspirational. Teachers and educators are at the frontline of generating a high-quality workforce for any nation. Many students learning at universities and sophisticated labs participate in hands-on research experiments and internships made possible by federal grants in facilities such as the LIGO. These experiences are an effective way to excite, inspire, and train future generations of scientists, engineers, and technologists who will continue innovating. An important role of this book on design and integration of laser and photonic systems is to inspire teachers, educators, and students, as well as entrepreneurs and managers, to learn and innovate.

There are other public policies that will require input from scientists and engineers to policy makers because of their complexity and contextual nature (to include technical, social, and economic factors and their interrelationships). The relevant policy alternatives will require extensive analysis and research to identify the relative advantages and disadvantages of each alternative. Examples of challenging policy issues include the following:

- Crafting certification requirements for medical practitioners and manufacturing specialists involved in the safe operation of laser and photonics systems

- Working with OSHA and EPA in setting standards and regulations for safe, environment-friendly applications of additive manufacturing and surface modifications involving nanomaterials
- Establishing physical standards (with NIST research) through ANSI and ISO for new technologies entering commercialization and use
- Determining equitable standards for broadband spectrum auctions and allocations for laser and photonics communications (especially in the quantum regime)
- Establishing privacy and equity standards for the use of cloud computing, especially where open systems are desired using HubZero capabilities

The scientific and engineering communities in the field of laser and photonics systems must take a responsible stance in working with policy makers to assure that new policies are workable and meet fairness and equity principles in the context of their intended application. This view is increasingly important as complexity escalates when we progress from local and regional to national and global regimes.

Looking to the future, one can expect additional transformative developments in how research and education in laser and photonics will be conducted, as discussed in this book. This vision provides hope that as distances continue to shrink worldwide and virtual engagement is further facilitated, scientists and engineers will collaboratively find solutions to the daunting global challenges confronting the world today and lead us to a better quality of life.

**Arden L. Bement Jr.**
*Director (2004–2010)*
*National Science Foundation*
*Director (2001–2004)*
*National Institute of Standards and Technology*
*Distinguished Professor (1992–2012)*
*Nuclear Engineering, Materials, Electrical, Computer Engineering,*
*and Industrial Engineering*
*Purdue University*

# *Preface*

Laser and photonic systems and innovations are increasingly proliferating, resulting in significant advancements in many areas. Often, they offer better approaches in addressing important and difficult problems. It is necessary to understand their emerging discoveries and systems perspectives and learn how they can improve the quality of our life by overcoming serious obstacles.

This book brings together multiple perspectives on the topic of laser and photonic systems and innovations. Such systems are of growing interest to many organizations and individuals, given their promise and potential solutions of grand societal challenges.

The book contains insights from leading researchers, inventors, and innovators. We discuss and explain a variety of techniques, models, technologies, and proven experience with laser and photonic systems, as well as their development, design, and integration.

We realize that laser and photonic technologies have already impacted solutions across many critical sectors. In this book, we are seeking to answer several questions. First, how can industrial and systems engineering knowledge (systems improvement and integration, optimization, process simulation, cyber-supported collaboration, and human supervisory control, among others) accelerate the systematization of laser and photonic systems and innovations? Can we inspire new concepts, models, scale-up methods, and significant quality improvements? Included are implications of cutting-edge laser and photonics technologies for industry and academia, scientists, engineers, students, and industry leaders. Materials are presented in terms of near-term and long-term methodological innovations, research and action agenda, and educational programs.

By bringing together chapters from multidisciplinary leading scientists and technologists as well as industrial and systems engineers and managers, we aim to stimulate new thinking that would bring a systems, networks, and system-of-systems perspective to bear on laser and photonic systems applications. Throughout the chapters, we challenge ourselves to explore opportunities for revolutionary and broader advancements. A particular emphasis throughout the book is the identification of emerging research and application frontiers where there are promising contributions to lasers, optics, and photonics applications in fields such as manufacturing, healthcare, security, and communications. The last chapter provides a summary of the subjects discussed in the previous chapters, pointing out important developments and implications to education.

The coeditors and coauthors appreciate and acknowledge the contributions of many individuals and organizations that took part in this effort and helped us bring this book to fruition. We hope that you, the readers, will benefit from this book and leverage the knowledge to exciting new frontiers of successful solutions.

# Editors

**Shimon Y. Nof**, PhD, DHC, is professor of industrial engineering, Purdue University, and has held visiting positions at MIT and at universities in Chile, the European Union, Hong Kong, Israel, Japan, Mexico, and Taiwan. His research interests include collaborative control theory and collaborative robotics. He is the director of the NSF and the industry-supported PRISM Center (Production, Robotics and Integration Software for Manufacturing and Management) linked with PGRN (PRISM Global Research Network); recent chair of the IFAC (International Federation of Automatic Control) Coordinating Committee, Manufacturing & Logistics Systems; recent president and current board member of IFPR (International Foundation of Production Research); fellow of the Institute of Industrial Engineers; fellow of the IFPR; and inaugural member of Purdue's *Book of Great Teachers*. He teaches courses on integrated production systems design, computing in Industrial Engineering (IE), design of e-work and e-service systems, industrial robotics and assembly, and research in computer and communication methods for supply networks. He is coinventor of four patents; has published over 500 articles on production and manufacturing systems, robotics, and automation; and is the author, coauthor, and editor of 12 books, including the *Handbook of Industrial Robotics* (first and second editions) and the *International Encyclopedia of Robotics*, both winners of the Most Outstanding Book in Science and Engineering Award, as well as *Information and Collaboration Models of Integration*, *Industrial Assembly*, and the *Springer Handbook of Automation*. Dr. Nof was a cochair of the inaugural Gavriel Salvendy International Symposium on Frontiers in Industrial Engineering in 2010, on the topic "Cultural Factors in Decision Making and Action," which resulted in the publication of the CRC Press book entitled *Cultural Factors in Systems Design* (2012).

**Andrew M. Weiner** earned an ScD in electrical engineering from MIT, worked at Bellcore and joined Purdue University in 1992.

He is the Scifres Family Distinguished Professor of Electrical and Computer Engineering. His research focuses on ultrafast optics signal processing and applications to high-speed optical communications and ultrawideband wireless communication. He is especially well known for his pioneering work on programmable generation of arbitrary ultrashort pulse waveforms, which has found application both in fiber-optic networks and in ultrafast optical science laboratories around the world. He authored the textbook *Ultrafast Optics* (Wiley, 2009), eight book chapters, and approximately 270 journal articles and is the inventor of 15 US patents. Professor Weiner is a fellow the Optical Society of America and of the Institute of Electrical and Electronics Engineers (IEEE) and is a member of the US National Academy of Engineering. He has won numerous awards for his research, including the Hertz Foundation Doctoral Thesis Prize (1984), the Adolph Lomb Medal of the Optical Society of America (1990), the Curtis McGraw Research Award of the American Society of Engineering Education (1997), the International Commission on Optics Prize (1997), the Alexander von Humboldt Foundation Research Award for Senior US Scientists (2000), and the IEEE Photonics Society Quantum Electronics Award (2011). He is joint recipient, with J.P. Heritage, of the IEEE LEOS William Streifer Scientific Achievement Award (1999) and the OSA R.W. Wood Prize (2008) and has been recognized by Purdue University with the inaugural Research Excellence Award from the Schools of Engineering (2003), with the Provost's Outstanding Graduate Student Mentor Award (2008), and with the Herbert Newby McCoy Award for outstanding contributions to the natural sciences (2013). In 2009, Professor Weiner was named a US Department of Defense National Security Science and Engineering Faculty Fellow. Dr. Weiner recently served a three-year term as chair of the National Academy's US Frontiers of Engineering Meeting. He currently serves as the editor-in-chief of *Optics Express*, an all-electronic, open-access journal that publishes more than 3000 papers a year, emphasizing innovations in all aspects of optics and photonics.

**Gary J. Cheng** earned a PhD in mechanical engineering from Columbia University (2002), is an associate professor of industrial engineering and mechanical engineering (by courtesy), Purdue University. His research area is advanced functional materials, scalable micro/nanomanufacturing processes, laser–matter interaction, and mechanical/

physical property enhancement of materials. His research work is highly interdisciplinary, covering nanotechnology, materials science, biomedical, solid mechanics, and manufacturing. His research has been awarded an NSF CAREER award (2006), an ONR Young Investigator Award (2007), an SME K.K. Wang Outstanding Young Manufacturing Engineer Award (2007), and an ASME Chao and Trigger Young Manufacturing Engineer Award (2012). He has also been selected to serve as a faculty scholar at Purdue University (2013–2018), and was elected a fellow of the ASME in 2013. He has published more than 120 peer-reviewed papers, including more than 70 journal papers, and has 6 US patents awarded, with 9 other US patents filed.

# Contributors

**Avital Bechar**
The Institute of Agricultural
    Engineering
Volcani Center
Bet-Dagan, Israel

**Peter Bermel**
Birck Nanotechnology Center
and
School of Electrical and Computer
    Engineering
Purdue University
West Lafayette, Indiana

**Satish T.S. Bukkapatnam**
School of Industrial Engineering
    and Management
Oklahoma State University
Stillwater, Oklahoma

**Changqing Chang**
School of Industrial Engineering
    and Management
Oklahoma State University
Stillwater, Oklahoma

**Xin W. Chen**
Department of Industrial and
    Manufacturing Engineering
Southern Illinois
    University–Edwardsville
Edwardsville, Illinois

**Gary J. Cheng**
Birck Nanotechnology Center
and
School of Industrial Engineering
Purdue University
West Lafayette, Indiana

**Nancy Huang**
School of Industrial Engineering
Purdue University
West Lafayette, Indiana

**Ranga Komanduri**
School of Industrial Engineering
    and Management
Oklahoma State University
Stillwater, Oklahoma

**Prashant Kumar**
Birck Nanotechnology Center
and
School of Industrial Engineering
Purdue University
West Lafayette, Indiana

**Seokcheon Lee**
School of Industrial Engineering
Purdue University
West Lafayette, Indiana

**Mark R. Lehto**
School of Industrial Engineering
Purdue University
West Lafayette, Indiana

**Peter  Lorraine**
GE Global Research Center
Niskayuna, New York

**Gaurav Nanda**
School of Industrial Engineering
Purdue University
West Lafayette, Indiana

**Shimon Y. Nof**
Production, Robotics, and
    Integration Software for
    Manufacturing and Management
    Center
School of Industrial Engineering
Purdue University
West Lafayette, Indiana

**David D. Nolte**
Department of Physics
Purdue University
West Lafayette, Indiana

**Cesar Reynaga**
School of Industrial Engineering
Purdue University
West Lafayette, Indiana

**Leyuan Shi**
Department of Industrial and
    Systems Engineering
University of Wisconsin–Madison
Madison, Wisconsin

**Juan D. Velasquez**
Production, Robotics, and
    Integration Software for
    Manufacturing and Management
    Center
College of Engineering
Purdue University
West Lafayette, Indiana

**Juan P. Wachs**
School of Industrial Engineering
Purdue University
West Lafayette, Indiana

**Longfei Wang**
Department of Industrial and
    Systems Engineering
University of Wisconsin–Madison
Madison, Wisconsin

**Xufeng Wang**
Birck Nanotechnology Center
and
School of Electrical and Computer
    Engineering
Purdue University
West Lafayette, Indiana

**Andrew M. Weiner**
School of Electrical and Computer
    Engineering
Purdue University
West Lafayette, Indiana

**Thomas D. Weldon**
The Innovation Factory
Atlanta, Georgia

**Xi-Cheng Zhang**
The Institute of Optics
University of Rochester
Rochester, New York

# 1

## *Introduction*

Shimon Y. Nof, Andrew M. Weiner, and Gary J. Cheng

**CONTENTS**

Human civilization's fascination with light is well known and understood, as illustrated in the following sample of quotes:

- We can easily forgive a child who is afraid of the dark; the real tragedy of life is when men are afraid of the light. (Plato)
- Give light, and the darkness will disappear of itself. (Desiderius Erasmus)
- I live and love in God's peculiar light. (Michelangelo)
- Edison failed 10,000 times before he made the electric light. Do not be discouraged if you fail a few times. (Napoleon Hill)
- As far as we can discern, the sole purpose of human existence is to kindle a light in the darkness of mere being. (Carl Jung)
- Knowledge is love and light and vision. (Helen Keller)
- There is a crack in everything, that's how the light gets in. (Leonard Cohen)
- The sun is gone, but I have a light. (Kurt Cobain)
- In the beginning there was nothing. God said, "Let there be light!" And there was light. There was still nothing, but you could see it a whole lot better. (Ellen DeGeneres)
- And God said, "Let there be light" and there was light, but the Electricity Board said He would have to wait until Thursday to be connected. (Spike Milligan)
- We have all the light we need, we just need to put it in practice. (Albert Pike)

Although scientific inquiry into light began in the 1600s, it was only in 1960 that photonics and lasers were discovered. With this discovery, humans have learned how to design and manipulate light for many useful applications. We are now even more fascinated by the ability to harness light energy for dealing with important problems.

This book is an inspired outcome of *The Gavriel Salvendy Second International Symposium on Frontiers in Industrial Engineering*, held May 4–5, 2012, at Purdue University (TGSIS, 2012).

The topic selected for this 2012 symposium, in proximity with the 50th anniversary of the invention of laser, was *Laser and Photonic Systems and Innovations for Better Quality of Life*. The objective was to stimulate frontier thinking, toward solutions of grand societal problems through solving engineering challenges in manufacturing, communications, security, healthcare, and medicine. Furthermore, the goal has been to stimulate and accelerate the delivery of innovations and solutions with this emerging area.

Photonics (the science, engineering, and technology of light) is an important enabling field with many diverse applications. Laser and photonics have already been recognized as game changers and potentially as significant for progress and innovations as information technology, steam power, and electronics have been. Since its invention, laser technology has been applied to overcome major obstacles in a variety of fields. Its economic impact has been growing rapidly (Savage, 2012; Tünnermann, 2012). Definition and illustrations of several terms associated with laser and photonic systems are included as introductory background in Table 1.1. Further background can be found in books by Quimby (2006), Hecht (2008), and Silfvast (2008) and in handbooks (Kasap et al., 2012; Trager, 2012). Three recent reports by the National Academy of Engineering have also addressed related topics in this emerging field (NAE, 2008, 2010, 2012).

The opportunities to solve significant problems and address challenges with creative and innovative laser and photonic systems are immense. Examples of the importance of photonics, which have emerged in just the last two decades, include the following:

Advanced manufacturing:

- High-power lasers outperform traditional processes with better quality, such as precise drilling, cutting, and welding, as well as high-tech processes, such as nanoprocessing.
- Microprocessors are fabricated today by optical lithography.
- Manufacturing of photonic products, such as optoelectronic components, visual displays, and solar equipment, could not be possible without the precision, optimal process control, and measured energy properties inherent in photonics-based processes and techniques.

**TABLE 1.1**

Laser and Photonics Basic Terminology

| Term | Description | Examples |
|------|-------------|----------|
| Light | Energy in the form of electromagnetic radiation. Visible light (to the human eye) ranges in wavelength between 390 nm (violet) and 740 nm (red). Light outside of the human visible range includes ultraviolet light, x-rays, and gamma rays (wavelengths shorter than violet) and infrared light, radio waves, and microwaves (wavelength longer than red). Light is considered to have partly wave and partly quantum properties. Light propagates in vacuum at a speed of 299,792 km/s. | Natural sources of light: sun, stars, lightning, and luminescent organisms, such as fireflies. Artificial sources of light: candles, lamps (incandescent, fluorescent, plasma, light-emitting diodes [LEDs]), lasers. |
| Optics | The science and technology of the generation, transmission, and propagation of light, mainly pertaining to imaging and vision. Scientific work in optics began in the seventeenth century. | Mirrors, infrared prisms, and imaging lenses. (Courtesy of Edmund Optics.) |
| Photonics | The science, engineering, and technology of light (its basic unit is the photon). Photonics and electronics join in optoelectronic integrated circuits. Photonics application examples include the following:<br>• Data storage (e.g., optical disks and holograms).<br>• Data transmission via fiber optics.<br>• Optical switches and light modulators for signal processing and for communication.<br>• Photonic gyroscopes in commercial aircraft that have no moving parts. | Compact discs. |

(continued)

**TABLE 1.1 (continued)**

Laser and Photonics Basic Terminology

| Term | Description | Examples |
|---|---|---|
| | Photonics covers applications of light over the entire spectrum (from ultraviolet through visible to the near, mid, and far infrared). Most applications have been in the range of visible and near-infrared light. Photonics as a field began with the invention of the laser in 1960, and the term *photonics* emerged after practical semiconductor light emitters were invented in the early 1960s, and optical fibers were developed in the 1970s. |  Several types of optical fiber: single-mode fiber (SMF), dispersion compensating fiber (DCF), multimode fiber (MMF), highly nonlinear fiber (HNLF). |
| | |  $1 \times 2$ optical switch. |

Laser

Light amplification by stimulated emission of radiation. A device generating and transmitting such visible and invisible radiation (light). It is a key technology within photonics, first demonstrated in 1960.

Laser beams

Light beams propagating mostly in one given direction in free space.

Ultrathin crystals for laser wavelength. (Courtesy of Eksma Optics Co.)

Laser marking

Methods for labeling materials with lasers. Laser marking includes permanent or temporary labeling by engraving or changing the surface of objects.

Red, green, and blue laser beams. (Wiki pedia Creative Commons.)

Laser marking on surfaces. (Courtesy of Linx Printing Technologies. Linx Printing Technologies Ltd., St. Ives, United Kingdom)

(continued)

**TABLE 1.1 (continued)**

Laser and Photonics Basic Terminology

| Term | Description | Examples |
|---|---|---|
| Laser mirrors | High-precision mirrors used in laser resonators and various optical setups. | Optical laser mirrors. (Courtesy of Exma Optics Co ) |
| Laser pointers | Appliances using laser beams for pointing at objects. | Laser pointers; alignment laser. (Courtesy of Laserglow Technologies, Toronto, Canada.) |
| Laser printers | Appliances using a laser beam to transfer digital data, text, or images onto a light-sensitive printing device. | 3D laser printer, producing models to prove a design concept or to test it. (Courtesy of Laser Imaging Inc. Houston, Texas |

Laser resonators     Optical resonators designed as basic building blocks of lasers.

Laser resonator with laser crystals, mirrors, mounts, and other accessories. (Courtesy of Kentek Corp., Pittsfield, New Hampshire)

Laser safety     Safety of those who use laser devices.

Several types of laser safety glasses.

Laser scanning     A process of shining a structured laser line over object surfaces to collect 3D data. The data are then captured by a camera sensor mounted in the laser scanner and recorded. Laser scanning is used for scientific research, stereolithography, rapid prototyping, barcode readers, laser printers, and more.

Laser scanning unit. (Courtesy of Hoya Corp., Shinjuku, Tokyo, Japan)

*(continued)*

**TABLE 1.1 (continued)**

Laser and Photonics Basic Terminology

| Term | Description | Examples |
|---|---|---|
| Optical data transmission | Transmission of information using light, for example, directly or via optical fibers. |  Short-range optical data transmission. (Courtesy of Hokuyo Automatic Co, Osaka, Japan.) |
| Optical fiber communications | Technology for information transmission via optical fibers. | Optical fiber communication devices. (Courtesy: of Mitsubishi Electric Corp, Tokyo, Japan.) |
| Optical filters | Devices designed with a wavelength-dependent transmission or reflectivity. | Filters for Nd: YAG lasers. (Courtesy of Edmund Optics Barrington, New Jersey.) |

Optical metrology — Measurements with light.

Optical metrology instrument for precision measurements. (Courtesy of Taylor Hobson Division of Ametek, Inc., Paoli, Pennsylvania.)

Optical modulators — Devices enabling manipulation of light beam properties, for example, the beam optical power or phase; useful for optical data transmission.

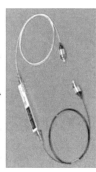

Optical modulator. (Courtesy of Fujitsu Corp, Tokyo, Japan.)

(continued)

**TABLE 1.1 (continued)**

Laser and Photonics Basic Terminology

| Term | Description | Examples |
|---|---|---|
| Optoelectronics | Interaction of electronic devices with light, also known as electro-optics. |  Photodiode (left); semiconductor optical amplifier (right). |
| Output couplers | Partially transparent laser mirrors, designed to extract output beams from laser resonators. | Partially reflecting laser mirror (right). |
| Optical resonators | Optical components designed and arranged to enable a beam of light to circulate. |  Microring resonator in silicon nitride. |

| | |
|---|---|
| Optical tweezers | Techniques for capturing, holding, and moving particles with laser beams. |
| Photons | Quanta of light energy. When atoms are excited, for example, by heating, they emit photons. |
| Photonic integrated circuits | Integrated circuits with optical functions. |
| Photonic metamaterials | Materials with nanostructure having special optical properties. |
| Power over fiber | Power delivery to electronic devices by light through an optical fiber. |
| Pulsed laser | Laser emitting light in the form of pulses (flashes of light). |

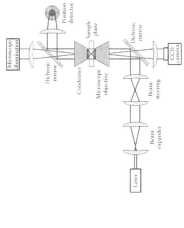

Generic optical tweezers diagram. (Wikipedia Multimedia, author: Rocky Racoon.)

Pulsed laser source (right); oscilloscope trace showing pulse train (left).

Medicine and healthcare:

- Photonics and optoelectronics enable better, higher-resolution medical imaging for early disease detection, diagnosis, and prevention.
- Brain, eye, skin, and dental surgeries rely on medical lasers to vaporize, ablate, cut, and suture.
- Laser-based DNA sequencing of a single human genome can be completed today in one day, compared to 80 years in the 1990s.

Communication:

- Photonic integrated circuits imply lower power consumption, better reconfigurability, and faster processing.
- Computer clusters in data centers communicate through high-capacity optical cables, engaging millions of lasers for signaling, to process massive computations.
- Cell phone communications, Internet work, and video chat signals are processed as optical data and transmitted through fiber-optic networks, enabling dramatic improvements in fast, reliable, and cost-effective services.

In addition to the remarkable, breakthrough innovations with laser and photonic systems and devices illustrated earlier, three unique features are enabled by laser and photonic systems and promise further great innovations. They are the following:

1. Processing at multiple scales and dimensions, from nano-, micro-, to large-scale objects, from local to remote subjects, and from 0 to 3 dimensions
2. Processing and delivery of laser fields at ultrafast speeds and frequencies and the ability to shape and reshape them
3. The ability to bring closer (almost together) the process and its process control by significantly faster sensing and communication

We describe in detail the design, development, and delivery of such innovations, processes, systems, and devices and explain the observations and their significance in advanced manufacturing (Chapters 2 and 3), biomedicine and healthcare (Chapters 4 through 6), and communication (Chapters 7 and 8). Systems, networks, and integration issues and challenges are then discussed (Chapters 9 through 15), including implication and agenda for education, research, and development.

Several presentations that were given and discussed during the symposium are summarized in the following. They highlight the scope of the field and introduce significant emerging, promising developments in terms of their design and integration challenges.

- Potential Impact of Lasers on Manufacturing in the Twenty-First Century

  Marshall G. Jones, GE Global Research

  With the invention of the laser in 1960, it is one of the youngest technologies that is being used in manufacturing today. As laser technology continues to mature, its acceptance and potential impact in manufacturing will be difficult to predict. The impact will certainly come but will greatly depend on its acceptance. In the United States, this will be paced by the education of the future manufacturing engineers and engineers in general. The rate of impact in the United States and globally will be a function of education. Europe is leading in the use of laser technology in manufacture and has set the bar high for others to reach. The benchmark of the current status globally and highlights of GE's role and others indicate significant "moving of the dial" to increase laser technology utilization and impact in manufacturing.

- Impact of Lasers on Modern Advanced Manufacturing and Materials Processing

  Yung C. Shin, Purdue University

  With the rapid advent of lasers and laser systems, there has been an explosion in the use of lasers in various manufacturing and materials processing. Lasers provide highly controlled, intense, localized heat sources, which can be used to add, subtract, and transform materials. Some of the interesting applications of lasers in modern manufacturing include surface modification, direct 3D fabrication, and precision machining. Case study results demonstrate its advantages over conventional manufacturing processes. Concurrently, physics-based modeling efforts for these processes are being developed (Shin, 2011).

- Laser Rock Drilling R&D at Argonne National Lab

  Claude B. Reed, Argonne National Laboratory

  Argonne National Laboratory (Laser Applications Laboratory) and a group of collaborators are examining the feasibility of adapting high-power laser technology to drilling for gas and oil. If drilling with lasers ultimately proves viable, it could be the most radical change in drilling technology in the last century. It was at the turn of the twentieth century when rotary drilling supplanted cable tool drilling as the petroleum industry's standard method for reaching oil and gas formations. Using lasers to bore a hole offers an entirely new approach. The novel drilling system would transfer light energy from lasers on the surface, down a borehole by a fiber-optic bundle, to a series of lenses that would direct the laser light to the rock face. Because the laser head does not contact the rock, there is no need to stop drilling to replace a mechanical bit. A laser system could contain a variety of downhole sensors—including visual imaging systems—that could communicate with the surface through the fiber-optic cabling. One of the primary objectives of current work is to obtain much more precise measurements of the energy requirements needed to transmit light from surface lasers to as much as 7000 m below the surface. Another aspect of study is to determine if sending the laser light in sharp pulses, rather than as a continuous stream, could further increase the rate of rock penetration. A third aspect of research is to determine if lasers can be used in the presence of thick fluids—called *drilling muds*. Laser-based rock drilling has already been applied in space explorations.

- Opportunity of Lasers for Scalable Manufacturing of Solar Energy Devices

Gary J. Cheng, Purdue University

Production costs, module conversion efficiencies, and adoption in commercial rooftops continue to be the most critical issues for solar energy technology. The industry continuously seeks technologies that offer scalable and high-speed solar device production to replace costly conventional processing steps. The present generation of large-scale processes has limitations related to low speed, high cost, and performance of the devices. Laser material processing technology promises to satisfy these needs for many steps along the value chain in a variety of ways. In the twenty-first century, lasers could be used as an excellent tool for solar cells, such as patterning, surface texturing, dopant diffusion, contact fusion, annealing, and crystallization (Zhang and Cheng, 2011, 2012, 2013).

- THz Wave Air Photonics: Bridging the *Gap* and Beyond

X.-C. Zhang, The Institute of Optics, University of Rochester

Historically, THz technologies were mainly used within the astronomy community for searching far-infrared radiation (cosmic background) and the laser fusion community for the diagnostics of plasmas. Since the first demonstration of THz wave time-domain spectroscopy in the late 1980s, there has been a series of significant advances (particularly in recent years) on the development of intense sources and sensitive detectors.

THz wave air photonics involves the interaction of intense femtosecond laser pulses with air. The very air that we breathe is capable of generating and detecting THz field strengths greater than 1 MV/cm and useful bandwidth from 0.1 THz to over 10 THz. THz wave generation and detection techniques using air have been developed and implemented, both as emitter and as sensor. There are still remaining challenges and future opportunities for this rapidly evolving area of research that transcends the *gap* once existing between optics and electronics. Their potential solutions include interesting national security applications (Zhang and Xu, 2009).

- Shaping Ultrafast Laser Fields for Photonic Signal Processing

A.M. Weiner, Purdue University

Lasers capable of generating picosecond and femtosecond pulses of light are now firmly established and are widely deployed for applications ranging from basic science to fiber communications to biomedical imaging. Enhanced functionalities are possible when ultrashort pulses can be programmably reshaped into user-specified waveforms. Ultrafast optics includes specifically the topic of shaping ultrafast laser fields. Because the time scales involved are much too fast for the direct application of (opto) electronic modulator technologies, all-optical approaches are employed. In particular, a Fourier synthesis waveform generation method has enjoyed considerable success. In this method, different optical frequency components contained in an ultrashort pulse are spatially separated, manipulated in parallel using a spatial light modulator, and then recombined. In addition to basic pulse shaping, some of the useful application areas include broadband photonic signal processing. This type of processing is relevant to lightwave communications and ultrabroadband radio-frequency electromagnetics (Weiner, 2009).

- Living Motion: Imaging Cellular Function in Live Tissues

David D. Nolte, Purdue University

Subcellular motions inside live tissue are sensitive indicators of cellular health and cellular response to applied drugs. Digital holography volumetrically captures these motions in tissue dynamics spectroscopy (TDS) for live-tissue drug screening. When coherent light scatters from displacing objects, the phase of the light shifts. The shift is a function of the direction and magnitude of motion. When there are many scattering objects and they are moving in random directions, the phase shifts are added randomly, leading to intensity fluctuations in the scattered light. The intensity fluctuations have statistical properties that relate to the type of motion, with different fluctuation signatures for diffusive versus directed transport and with different frequency content for faster or slower motions. TDS combines dynamic light scattering with short-coherence digital holography to capture intracellular motion inside multicellular tumor spheroid (MCT) tissue models. These spheroids are grown in bioreactors and have a proliferating shell of cells surrounding a hypoxic or necrotic core. The cellular mechanical activity becomes an endogenous imaging contrast agent for motility contrast imaging. Fluctuation spectroscopy is performed on dynamic speckle from the proliferating shell and the hypoxic core to generate drug–response spectrograms. These spectrograms are frequency versus time representations of the changes in spectral content, induced by an applied compound or by an environmental perturbation. A range of reference compounds and conditions applied to MCTs has been studied to generate drug fingerprint spectrograms with potential applications in early drug discovery. The expected benefit to drug discovery can be significant (Nolte, 2011).

- Applications of Ultrashort Laser Pulses

Marcos Dantus, Michigan State University and Biophotonic Solutions, Inc.

Ultrashort laser pulses, with durations measured in femtoseconds ($1\ fs = 10^{-15}\ s$), have highly desirable properties. First, they can be used to measure events in the time scale of atomic motion (10–1000 fs). Second, they are able to focus light in space and time and able to reach peak intensities of $10^{15}\ W/cm^2$. When such intense pulses interact with matter, they vaporize it into plasma on a time scale faster than thermal diffusion. This property allows nonthermal machining of any material. At more moderate intensities, these laser pulses are able to access nonlinear optical properties, such as two-photon absorption and intensity-induced polarization rotation, which are useful for imaging and for optical data communication and manipulation. Third, because of the uncertainty principle, ultrashort pulses have an inherent broad bandwidth that can be used for spectroscopy, metrology, and communications. These lasers have been available since the 1980s in research laboratories, where a number of proof-of-principle applications have been demonstrated, and two Nobel prizes have been awarded. Given that most applications of ultrashort pulses depend on their peak intensity elevated to a higher power (quadratic or cubic), the shorter the pulses, the more efficient they are. However, there are practical limitations preventing the use of ultrashort pulses for applications. A pulse that is 5 fs long has such a broadband width that it gets easily dispersed, stretching to 12 fs just by traveling a 1 m path in air. Work with these ultrashort pulses typically requires experts to optimize the source on a daily basis. Harnessing the unprecedented properties of ultrashort laser pulses for industrial and medical applications requires practical means for delivering the shortest possible pulses to the target without the assistance of a laser expert. Some applications, such as femto-LASIK, have already been commercialized, although with pulses that are ~700 fs long. The author of this contribution described a system that is capable of automated pulse compression at the target. The heart of this system is a pulse shaper, which is used both to implement pulse compression and to perform the ultrashort pulse waveform measurements that guide the pulse compression. This system has already been tested with pulses as short as 3 fs. Armed with such technology, recent advances have moved toward improved biomedical imaging for noninvasive biopsies, improved proteomic sequencing for medical diagnosis and drug discovery, and standoff detection of explosives (Dantus et al., 2009).

- Overview of Optics Opportunities and Challenges in Healthcare

  Peter Lorraine, GE Global Research Center

  As anyone who has attended a recent major photonics conference can attest, there is enormous research activity in biomedical optics, yet the commercial market remains fragmented and specialized compared to *traditional* diagnostic imaging modalities such as MRI, ultrasound, or x-ray CT imaging. There is a reason for optimism with several emerging applications having the potential to be widely adopted. Opportunities exist in areas including optical coherence tomography, advanced microscopy, surgical and clinical imaging, and single molecule detection. At the same time, there are technology challenges in optical systems that have to be solved. In addition, there are promising possibilities for optical subsystems in *traditional* modalities.

- Optical Telecommunication Networks

  Harshad P. Sardesai, Ciena Corporation

  Optical telecommunication networks are ubiquitous today in long-haul, metropolitan, undersea, and wireless backhaul applications. In the past decade, these networks have undergone significant structural changes in both technology and business processes to adapt to the needs of network traffic evolution and growth. This trend has opened up new engineering challenges and opportunities to ensure both network scalability and resiliency while limiting costs. In addition to hardware and system engineering, increasingly, optical networks rely on simulation and modeling tools for network design, validation, development, installation, and operations. Such network systems engineering and integration approaches are required for the successful evolution of optical networks; they present challenge areas for research and development (R&D) in component and system technologies, algorithms and optimization techniques, and software development and integration for next-generation communication networks.

- Nested Partition Optimization Method and Its Applications

  Leyuan Shi, University of Wisconsin-Madison

  Laser and photonic systems design and integration problems can often be modeled as large-scale, complex systems. Indeed, large-scale complex optimization problems arise in a vast variety of science and engineering applications. Examples range from designing telecommunication networks for multimedia transmission, planning and scheduling problems in manufacturing and military operations, and designing nanoscale devices and systems. These and other applications present large-scale, discrete optimization problems that are extremely hard to solve. To handle and solve these types of problems, both methodological and practical works have accumulated over the past decade. In an effort to address the need for an optimization methodology that would have attractive theoretical properties, such as global convergence to an optimal solution, and be powerful enough to address real-world large-scale problem, the *n*ested *p*artition (NP) method has emerged as a promising approach. Since proposing the basic idea of the NP method over 10 years ago, the methodology has been refined and applied to challenging problems from a variety of application domains, including supply chain management, logistics, healthcare delivery, and production planning and scheduling. This method is also promising for laser and photonic systems applications in these areas (Shi and Olafsson, 2008).

- Robust Coating Design in the Presence of Manufacturing Errors

  Omid Nohadani, Purdue University

  A novel robust optimization algorithm has been demonstrated, which attempts to account for expected coating errors as they can occur in the manufacturing of mirrors for ultrafast lasers. The proposed optimization method explicitly takes the possibility for layer errors into account, compromising nominal performance in order to improve robustness against layer perturbations. In theory, one could attempt to optimize a coating merit function, which involves a full Monte Carlo simulation of manufacturing statistics. However, such an approach would be computationally infeasible for all but the simplest designs. A deterministic method that explicitly considers the effects of manufacturing tolerances, while retaining the fast convergence properties inherent to a deterministic algorithm, has proven to be effective. The method is generic and can be applied to many problems that are solved through numerical simulations. To demonstrate the advancement in the presence of manufacturing errors, the performance of a proof of concept antireflection (AR) coating designed with the new robust optimization was compared with that of a conventionally optimized AR coating. It has been found that the robust algorithm produces an AR coating with a higher manufacturing yield when root mean square layer tolerances are above approximately 1 nm (Birge et al., 2011).

- Laser Systems in Precision Interactions and Collaboration

A. Bechar, J.P. Wachs, and S.Y. Nof, Purdue University

Laser technology was introduced five decades ago, and since then, it has influenced almost all aspects of our life, from health to defense, through precise analog and digital electronic systems. Laser technology has been implemented not only as an end product but also as a means of production in a variety of industries. These wide-range implementations have leveraged lasers' unique characteristics that enable multiple functionalities, such as energy transmission, high irradiance, spatial and temporal coherence, and precision. The main applications are found in medicine, information and communication technologies, quality assurance, manufacturing, defense and law enforcement, entertainment, and scientific explorations. The assimilation of laser beyond communication, processing, and discovery to interaction and collaboration systems, however, is still in its infancy. The R&D activities found in these areas are relatively limited, even though they have been recognized as vital for the increasingly complex networked engineering and service activities. Recent advances in laser technology raise the potential feasibility of harnessing certain laser solutions for new and improved integration through interaction and collaboration systems. These advances include using new wavelength bands and reducing pulse duration together with increasing precision and power efficiency, all while decreasing cost. The objective is twofold: to review and analyze the hidden opportunities in applying advanced laser techniques for precision interactions and collaboration and to evaluate the impact these techniques can have on advancing the effectiveness and quality of interaction and collaboration systems (Nof, 2009; Wachs et al., 2011).

---

## Summary

With the growth of the world population and the associated problems of sustaining and advancing civilization, furthermore, improving the quality of life for future generations, laser and photonic systems offer promising solutions. The introductory summaries presented earlier have already been further developed and discussed (Nof et al., 2013). The topics, their developments, and the challenges are explained in detail in the following chapters.

## Acknowledgment

We gratefully acknowledge the contributions to this chapter by the presenters at *The Gavriel Salvendy International Symposium on Frontiers in Industrial Engineering* May 4–5, 2012, which focused on *Laser and Photonic Systems and Innovations for Better Quality of Life* , and PhD student A.J. Metcalf for providing several photos illustrating lasers and photonics.

## References

Birge, J.R., Kärtner, F.X., and Nohadani, O. (2011). Improving thin film manufacturing yield with robust optimization. *Applied Optics*, 50(9), C36–C40.

Dantus, M., Pestov, D., and Andegeko, Y. (2009). Better results from ultrafast nonlinear microscopy, *Bio Optics World*, 2, 23–24.

Hecht, J. (2008). *Understanding Lasers: An Entry-Level Guide*, 3rd edn., Wiley-IEEE Press, Hoboken, NJ.

Kasap, S., Ruda, H., and Boucher, Y. (2012). *Cambridge Illustrated Handbook of Optoelectronics and Photonics*, Cambridge University Press, New York.

NAE (2008). *Nanophotonics: Accessibility and Applicability*, National Academies Press, Washington, DC.

NAE (2010). *Seeing Photons: Progress and Limits of Visible and Infrared Sensor Arrays,* National Academies Press, Washington, DC.

NAE (2012). *Optics and Photonics: Essential Technologies for Our Nation*, National Academies Press, Washington, DC.

Nof, S.Y. (2009). *Springer Handbook of Automation*, Springer Publishers, New York.

Nof, S.Y., Cheng, G.J., Weiner, A.M., Chen, X.W., Bechar, A., Jones, M.G., Reed, C.B. et al. (2013). Laser and photonic systems integration: Emerging innovations and framework for research and education. *Human Factors and Ergonomics in Manufacturing & Service Industries* 23(6), 483–516.

Nolte, D.D. (2011). *Optical Interferometry for Biology and Medicine*, Springer, New York.

Quimby, R.S. (2006). *Photonics and Lasers: An Introduction*, Wiley-Interscience, Hoboken, NJ.

Savage, N. (2012). Business news: Laser market sees record sales. *Nature Photonics*, 6(4), 215–216.

Shi, L. and Olafsson, S. (2008). *Nested Partitions Method: Theory and Applications*, Springer, Dordrecht, the Netherlands.

Shin, Y.C. (2011). Laser assisted machining—LAM benefits a wide range of difficult-to-machine materials. *Industrial Laser Solutions*, 26(1), 18–22 .

Silfvast, W.T. (2008). *Laser Fundamentals*, 2nd edn., Cambridge University Press, New York.

TGSIS, (2012). *The Gavriel Salvendy International Symposium on Frontiers in Industrial Engineering*, https://engnierring.purdue.edu/ie/AboutUs/TGSIS.

Trager, F. (2012). *Springer Handbook of Lasers and Optics*, 2nd edn., Springer, New York.

Tünnermann, A. (2012). The engine of growth for the entire industry. *Laser Technik Journal*, 9(1), 1.

Wachs, J.P., Kölsch, M., Stern, H., and Edan, Y. (2011).Vision-based hand gesture applications: Challenges and innovations. *Communications of the ACM*, 54(2), 60–71.

Weiner, A.M. (2009). *Ultrafast Optics*, John Wiley & Sons, Hoboken, NJ.

Zhang, M.Y., Nian, Q., Cheng, G.J. (2012), "Room Temperature Deposition of Alumina-doped zinc oxide (AZO) on flexible substrates by Direct Pulsed Laser recrystallization", *Applied Physics Letter*, 100, 151902.

Zhang, X.-C. and Xu, J.-Z. (2009). *Introduction to THz Wave Photonics*, Springer, New York.

Zhang, Y. and Cheng, G.J. (2011). Highly conductive and transparent alumina-doped ZnO films processed by direct pulsed laser recrystallization at room temperature. *Applied Physics Letter*, 99, 051904.

Zhang, Y. and Cheng, G.J. (2013). Direct pulsed laser crystallization of nanocrystals for absorbent layers in photovoltaics: Multiphysics simulation and experiment. *Journal of Applied Physics*, 113 (19), 193506.

# 2

## *Laser-Based Manufacturing Systems for Nanomaterials and Nanostructures*

Prashant Kumar and Gary J. Cheng

### CONTENTS

### Introduction

Materials can be manipulated by processing at extreme temperature and pressure that alters their microstructure. Nanoscience and technology has gained its significance as many innovative technologies are capable of delivering materials with desired properties. Nano-initiatives taken by several leading economies have given a huge dream to people that all major problems facing the current manufacturing and design would possibly be solved one day. To exploit the nanosize effect, one would have to fabricate nanocomponents and integrate them to achieve various functional devices. Material manipulation, design, and their integration have to be carried out reproducibly with precision. Laser is a special class of electromagnetic waves that are monochromatic and coherent in nature. Lasers are available with various photon energy, beam fluence, pulse width, and pulse rate. Apart from these characteristics, one can even vary its polarizations. Such degree of control of laser makes it very special as an energy delivery source. Laser–matter interaction can be exploited for various fruitful pursuits for micro-nanoscale manipulation of materials and structures, which can find many applications in micro- and nanosystems. In this chapter, we would elaborate on how various laser-based nanomanufacturing systems fabricate nanomaterials and nanostructures.

We will discuss the basic principles and experimental findings in literature in several key laser systems for the manufacturing of nanomaterials or nanostructures, namely, (1) laser surface nanostructuring, (2) laser crystallization, (3) laser sintering of nanomaterials, (4) nanomaterial-integrated laser shock peening (LSP), and (5) laser nanolithography.

## Laser Surface Nanostructuring

Due to the varieties of applications emerging for nanostructured surfaces, there is a great demand for developing new technologies for nanostructuring. For example, nanostructured glass can be used for dirt repellenting, toughening, tissue culturing, better adhesion to metals, etc. Laser-based techniques for nanostructuring of surfaces offer a great deal of flexibility and are gradually becoming lucrative.[1] For example, nanostructured polymer surfaces are in demand due to their enhanced biocompatibility. As polymers are being used in flexible electronics, wettability of metals to polymer surface has been a concern that could be resolved via surface nanostructuring. Femtosecond laser and excimer laser have extensively exploited for nanostructuring of polymer surfaces.[2,3] Krajnovich et al.[4] used KrF laser photoablation to achieve conical nanostructures in polyimide as shown in Figure 2.1. It was observed that the cone becomes sharper when the inclination of laser beam is raised. At 45° inclination, sharp cones have been formed as shown in Figure 2.1.

Rebollar et al.[5] have carried out laser nanostructuring of Polystyrol. After 6000 laser pulses, ripples were observed on the surface due to the constructive interference of laser being incident and that scattered from the surface. At the

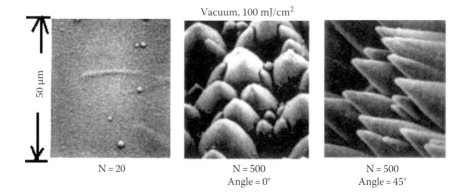

Vacuum, 100 mJ/cm$^2$

| 50 μm | N = 20 | N = 500<br>Angle = 0° | N = 500<br>Angle = 45° |

**FIGURE 2.1**
Excimer laser nanostructuring of polyimide at different inclination. (From Krajnovich, D.J. and Vázquez, J.E., *J. Appl. Phys.*, 73, 3001, 1993.)

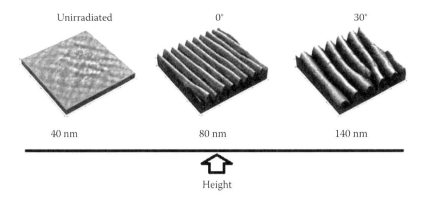

| Unirradiated | 0° | 30° |
| --- | --- | --- |
| 40 nm | 80 nm | 140 nm |

Height

**FIGURE 2.2**
Laser nanostructuring of polystyrol at different laser inclination. (From Rebollar, E. et al., *Biomaterials*, 29, 1796, 2008.)

sites of constructive interference, polymer chains were found to break into smaller pieces and bulge. Ripple height that is 80 nm for 0° inclination of laser increases to 140 nm when inclination is raised to 30°, as shown in Figure 2.2.

Semiconductors having tremendous applications in electronic and in optoelectronic device provide a great platform of opportunity for nanostructuring. For example, silicon, which is a non-emitting material, can become an emitting material if features on the surface can be constructed in its quantum confinement regimes. The physical behavior of surface, such as adhesion of metals, electrical conductivity, and optical absorption, depends on the extent of nanostructuring and feature dimensions therein. In addition, such nanostructured surfaces can find applications in biomedical industry. Silicon has been nanostructured by excimer laser below and above its ablation threshold. Figure 2.3a shows a schematic diagram for nanostructuring of surfaces by direct laser irradiation. One point to be noted is that laser nanostructuring also depends on the medium being used for the purpose. Vacuum is usually used. In case one intends to oxidize surface, one can do laser irradiation in oxygen too. Below ablation threshold itself, melting occurs and one can create nanopores as shown in Figure 2.3b or nanoparticles on the surface as shown in Figure 2.3c. Pore's density and dimensions are dependent on laser wavelength, laser fluence, number of laser pulses used for the purpose, and pulse rate.[6]

In case of single laser pulse irradiation, laser fluence would determine the peak temperature and laser fluence and pulse rate would determine the cooling rate. For multiple laser pulse irradiation of surface, heat would accumulate with the increased number of laser pulses. When the peak temperature is lower than the vapor point but above the melting point, melting—partial solidification—may take place and so on and so forth. As a result, surface morphology, phase structure, and crystalline structure changes may change

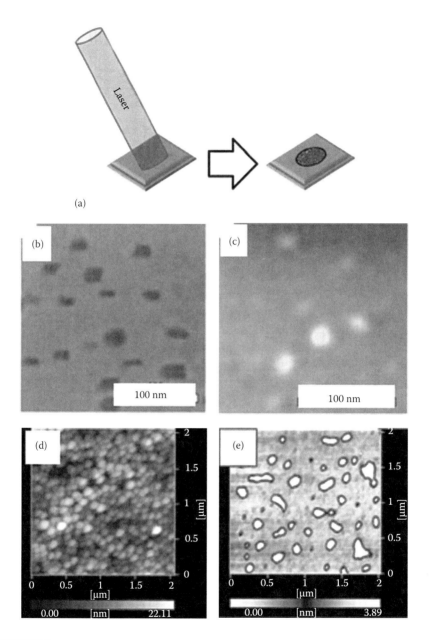

**FIGURE 2.3**
**(See color insert.)** (a) Schematic diagram for laser nanostructuring of silicon surface; laser nanostructuring at laser fluence below ablation threshold at (b) 0.10 J/cm$^2$ and (c) 0.25 J/cm$^2$ and above ablation threshold at 2 J/cm$^2$ in (d) air medium and (e) in water medium. (From Kumar, P., *Appl. Phys. A*, 99, 245, 2010; Kumar, P. et al., *J. Nanosci. Nanotechnol.*, 9, 3224, 2009; Kumar, P., *Adv. Sci. Lett.*, 3, 67, 2010.)

significantly. When the peak temperature is higher than the vapor point, the nanostructuring process is relatively complicated and can be considered to be a combination of several processes such as fragmentation of materials, material melting, material evaporation, and recondensation of evaporated vapor back on the surface due to air back pressure. When laser intensity is high enough to trigger ablation directly, materials could have direct phase transformation from solid to vapor or ionized gas, which does not involve melting and solidification and the surface is much cleaner as a result.

Silicon when nanostructured in air medium at the laser fluence of 2 J/cm[2] yields nanoparticulated surface[7] as shown in Figure 2.3d. When laser irradiation is carried out in a water medium,[8] one achieves different surface morphology as shown in Figure 2.3e. One can conclude that such morphology in a water medium results due to laser absorption by water and also due to cleaning of material surface by laser-pressure-induced water flow while laser irradiation is going on. Laser irradiation can also be employed for nanostructuring of metallic surfaces. One can use different pulse width of laser apart from the variation in laser fluence and pulse rate. Ultrashort lasers such as femtosecond lasers have been used for laser nanostructuring of metal surface.[9] Laser nanostructuring can be used to create photonic crystals in materials like sapphire. Korte et al.[10] have employed femtosecond laser nanostructuring for the purpose. Periodically, nanostructured surface with subwavelength surface nanofeatures is shown in Figure 2.4.

Laser nanostructuring provides a novel tool for achieving rough surface even for very high-melting-point materials, which otherwise is difficult to achieve by other contemporary techniques. For example, ceramic materials such as alumina has melting point above 2000°C and graphite has an exceedingly high melting point of 3656°C–3697°C. Excimer laser nanostructuring of

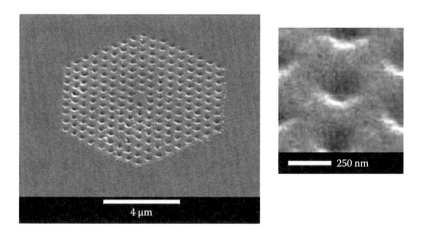

**FIGURE 2.4**
Femtosecond laser-induced subwavelength nanostructured sapphire surface. Right panel shows zoomed image. (From Korte, F. et al., *Appl. Phys. A*, 77, 229, 2003.)

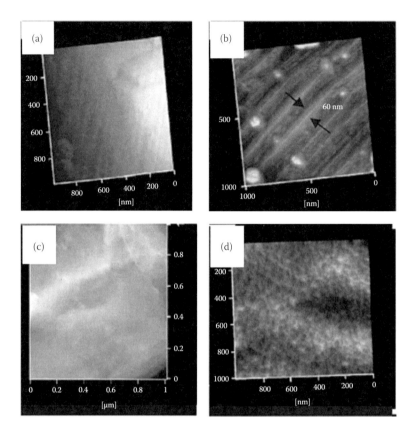

**FIGURE 2.5**
Excimer laser nanostructuring of alumina surface (a) before and (b) after laser irradiation and for graphite surface (c) before and (d) after laser irradiation. (From Krishna, M.G. and Kumar, P., Non-lithographic techniques for nanostructuring thin films and surfaces, in *Emerging Nanotechnology for Manufacturing*, Eds. W. Ahmed and M.J. Jackson, Elsevier Inc., Oxford, U.K. pp. 93–110, 2009.)

alumina and graphite surfaces has been demonstrated.[11] Figure 2.5a and b shows alumina surface before and after laser treatment. Similarly, Figure 2.5c and d shows graphite surface before and after laser treatment.

## Laser Crystallization

With the thrust of miniaturization in nanoscale to exploit various applications such as in nanophotonics, optoelectronics, and display applications, one would love to have thin film thickness of the functional materials as minimum as possible. However, the irony is that when thickness is being sliced, functionality is marginalized. Defects are generated in chemical vapor deposition and physical vapor deposition (PVD). In addition, crystalline thin films with low defects are desired in many devices that need to be made on flexible substrates. Annealing is needed after thin film deposition to reduce crystal defects. Several annealing processes, such as furnace annealing, thermal rapid annealing, and induction annealing, have been developed. However, these annealing techniques have serious limitations, such as low speed, low throughput, nonselectivity, and nonfeasibility for low-temperature substrates. In addition, these annealing processes may not be feasible for *on-device* semiconducting material coating such as silicon or GaAs thin films or transparent conducting oxide (TCO) thin films. Laser-induced crystallization provides an alternative that could address these issues simultaneously. Figure 2.6 shows the schematic of laser crystallization process that is being investigated for many functional materials including semiconducting, magnetic, plasmonic, optoelectronic, superconducting, and ferroelectric materials. Efficiency to achieve large crystals, fast crystallization, scalability, and versatile nature are some of the key attributes of laser crystallization. In addition, one can use small beam diameter to write crystalline lines on amorphous background. The flexibility offered by laser crystallization makes it an ideal solution for many industrial applications.

Laser crystallization of silicon thin film coated on glass[12] was demonstrated long back. Kim et al.[13] demonstrated that such laser crystallization can amend silicon thin films for thin film transistor applications. Dassow et al.[14] also carried out research along the same line. They achieved electron field mobility of 410 $cm^2/V$ s. Dassow et al. reported crystallization rate as high as 35 $cm^2/s$ under laser fluence of 1.25 $J/cm^2$ and pulse rate of 100 kHz. Adikaari et al.[15] reported on the thickness-dependent laser crystallization for amorphous thin silicon films. As one can visualize from the research reported by Peng et al.[16], there is a huge grain growth taking place as one gradually increase laser fluence, as shown in Figure 2.7.

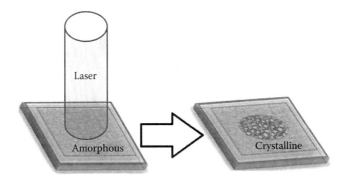

**FIGURE 2.6**
Schematic diagram for laser crystallization of amorphous thin films.

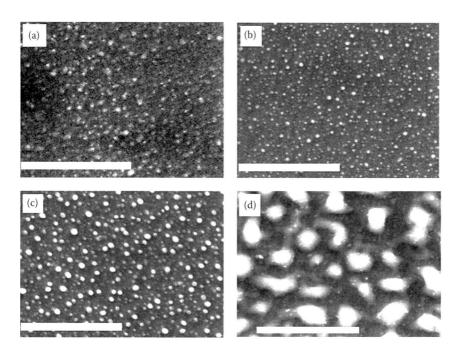

**FIGURE 2.7**
SEM images of α-Si thin films irradiated at laser energy densities of (a) 204 mJ/cm², (b) 240 mJ/cm², (c) 303 mJ/cm², and (d) 480 mJ/cm². Scale bar in each panel is 600 nm. (From Peng, Y.C. et al., *Semicond. Sci. Technol.*, 19, 759, 2004.)

Jin et al.[17] demonstrated the effect of laser fluence on crystallization kinetics as shown in Figure 2.8. They carried out a detailed Raman spectroscopy of laser-crystallized silicon thin film for this purpose. Amorphous silicon thin film shows a broad Raman peak without any sharp feature in it, which upon laser crystallization starts exhibiting a sharp feature close to Raman shift of 518 cm⁻¹. At this point of discussion, it has to be noted that the extent of oxidation, hydrogenation, and defects present in the system also contribute to the final crystallinity exhibited by laser-crystallized amorphous silicon thin films.

Apart from the huge grain growth and crystallization for silicon thin film case,[18] thin films of plasmonic materials such as gold and indium thin films have also been crystallized employing excimer laser crystallization.[19] Plasmonic thin films of gold and indium are of significance to their applications in advanced plasmonic waveguides at micro- or nanoscales. There are some other creative applications of plasmonic materials such as in lithography or light amplification. Figure 2.9 shows grain growth during laser crystallization process for gold thin films (Figure 2.9a–c), indium thin films (Figure 2.9d–f), and silicon thin films (Figure 2.9g–i), respectively. Apart from laser parameters, the overall average crystallite size in laser-crystallized thin films is dependent on several factors including the thin film material

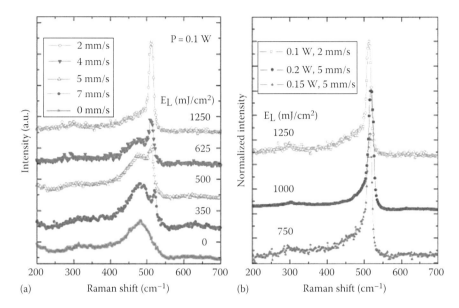

**FIGURE 2.8**

**(See color insert.)** Raman spectra of P-doped *a*-Si:H films with two-layer structures (P-doped films (II)) on glass substrates (a) before and after laser annealing at laser power of 0.1 W with different scan speed and (b) after laser annealing at laser fluence from 0.75 to 1.25 J/cm². (From Jin, J., *Appl. Surf. Sci.*, 256, 3453, 2010.)

(its density, melting point, thermal conductivity), thin film thickness, and substrate material (thermally insulating material helps in retaining heat in thin film itself). Laser energy delivered to the thin film system provides the material the necessary trigger to overcome the energy barrier for crystallization. Porosity and defects too influence the laser crystallization process. Ultrafast laser has been used for crystallization of amorphous thin silicon films by Choi et al.[20] In situ TEM study of laser-induced crystallization was carried out by Taheri et al.[21]

GaAs, a direct band gap semiconductor, has many applications in optical and optoelectronic devices. However, PVD GaAs thin films coated on glass at room temperature (RT) for lower thickness usually suffer from crystallinity degradation, which can potentially be solved by laser crystallization. Pirzada et al.[22] have reported laser fluence and thickness dependence of laser crystallization for GaAs thin films. Extensive study of microstructure and texture development during multiple-pulse excimer laser crystallization has been carried out.[23] Figure 2.10 shows secondary grain growth while laser crystallization of amorphous GaAs thin film takes place.

Transparent conducting thin films are of great attention in photonic and optoelectronic industry. Light transmits through them and can act or excite the material beneath them and at the same time, the top TCOs can work as electrodes. Such physical properties make them very special material for

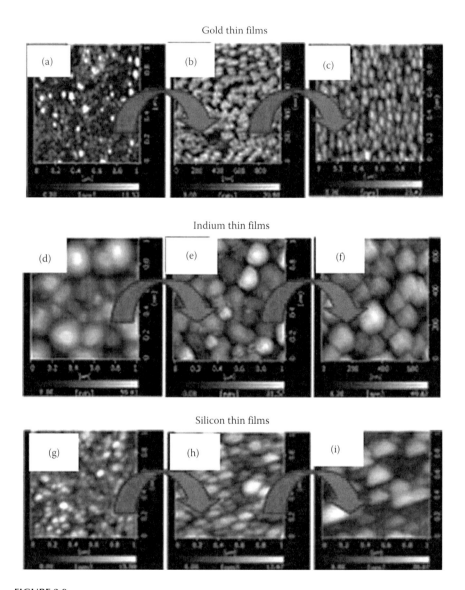

**FIGURE 2.9**
Excimer laser crystallization of (a)–(c) gold thin film, (d)–(f) indium thin film, and (g)–(i) silicon thin film. (From Kumar, P., *Adv. Sci. Lett.*, 3, 62, 2010; Kumar, P. and Krishna, M.G., *Phys. Status Solidi A*, 207, 947, 2010.)

applications such as in solar cells, for example. Indium tin oxide (ITO) and alumina-doped zinc oxide (AZO) are some of this class of materials. Recently, Zhang et al.[24] have used pulsed laser crystallization of AZO thin films at RT. Zhang et al.[25] used laser for direct pulsed laser recrystallization (DPLR) of RT AZO thin films coated by PLD as shown in Figure 2.11.

 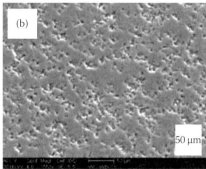

**FIGURE 2.10**
SEM micrographs showing secondary grain growth in 0.1 μm thick GaAs thin film as a result of multiple-pulse irradiation, (a) initial grain structure "single pulse," and (b) after 40 pulses. (From Pirzada, D. and Cheng, G.J., *J. Appl. Phys.*, 105, 093114, 2009.)

Thus, laser crystallization seems to solve many problems related to low crystallinity and high defects density in thin films, especially at ambient deposition conditions and on substrate where there is a large lattice mismatch between the layers under epitaxial deposition processes. The flexibility, selectivity, controllability, and scalability make the laser crystallization technique an ideal technique for various industrial scale applications.

## Laser Sintering of Nanomaterials

Laser sintering employs additive manufacturing and hence facilitates 3D printing of desired design. Such additive manufacturing simplifies the process, as one does not really require breaking the material for manufacturing. Sequential addition of components by laser sintering process makes the production of complex structures feasible. In the context of nanomaterials processing, laser sintering provides a huge opportunity as it offers a great extent of controlled fabrication and design of nanomaterials. The interesting aspect of laser sintering is that one can intuitively create a new class of hybrid nanomaterials retaining good qualities of the constituents. The advantage of the laser sintering process of nanomaterials would be a (1) great degree of control as the process involves many laser parameters, (2) local processing without actually affecting other components, (3) fast process within a few minutes, and (4) optical process and hence clean fabrication. Nanomaterials have a different melting points due to the difference in density and enhanced surface area. Since thermal properties of nanoparticles are different from their elemental form, size dependence has to be considered in laser sintering of nanomaterials and their composite forms. Figure 2.12 shows the schematic

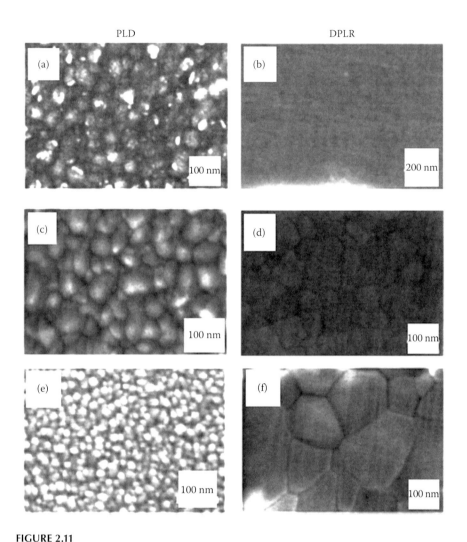

**FIGURE 2.11**
AZO on Kapton deposited by PLD (a) and processed by 25 mJ/cm$^2$ with 15 laser pulses (b); AZO on SLG deposited by PLD (c) and processed by 30 mJ/cm$^2$ with 30 laser pulses (d); AZO on Al foil deposited by PLD (e) and processed by 60 mJ/cm$^2$ with 100 laser pulses (f). (From Zhang, M.Y. et al., *Appl. Phys. Lett.*, 100, 151902, 2012.)

diagram of the laser sintering process for nanoparticle ink where one uses laser power of own choice for sintering.

Bioceramics implants coated with functional gradient nanomaterials by laser sintering possess many attributes, such as stronger interfacial bonding and prolonged lifetime, due to their superior control of biocompatibility and mechanical properties. Size effect would be pronounced on properties such as thermal conductivity, electrical conductivity, heat capacity, and melting point. Zhang et al.[26] have demonstrated based on their experiments and

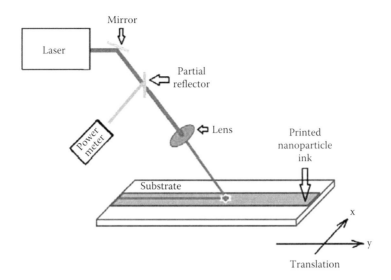

**FIGURE 2.12**
Schematic diagram depicting laser sintering process.

multiphysics theoretical models that nanoparticle size affects the heat distribution and conduction rate. Smaller nanoparticles are demonstrated to be better heat distributers as shown in Figure 2.13a. In addition, normalized resistive heating is orders of magnitude higher for 80 nm diameter as compared to that for the 600 nm diameter, as shown in Figure 2.13b. One should keep in mind that apart from laser–nanoparticle interaction, the important factor here is that the melting point of smaller nanoparticle is reduced considerably.

Titanium (Ti) nanoparticles heat up comparatively more than hydroxyapatite (HAp) nanoparticles as shown in Figure 2.14 and heat conduction in the combined system gives rise to equilibrium when temperatures of nano-Ti and nano-HAp are both the same. Such thermal equilibrium is reached due to heat transfer at the nanoscale interface where they tightly bind under laser irradiation. Melting point, thermal conductivity, and heat capacity for bulk Ti and Ti nanoparticles were used for the simulations carried out by Zhang et al.[26] Laser energy intensity needed to achieve the melting temperature was shown to increase as a function of nanoparticle size.

Laser sintering has been investigated as a viable approach for biocompatible ceramic materials. Laser-based synthesis of biphasic calcium phosphate (BCP)/Ti nanocomposite starting from HAp and Ti nanoparticles has been demonstrated.[27] Continuous wave Nd–YAG laser was used for the purpose. Due to excellent laser–nanoparticle interaction and enhanced sinterability of the HAp/Ti nanocomposite system, low-temperature coating has been realized. Coating–substrate interfacial bonding was observed

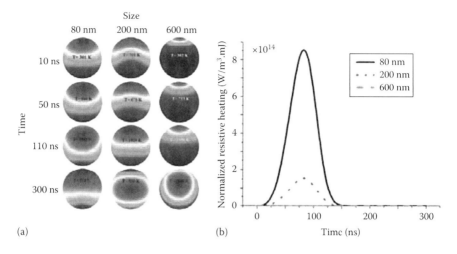

(a)            (b)

**FIGURE 2.13**
**(See color insert.)** (a) Temperature distribution for nanoparticles of varying diameter with time and (b) comparison of normalized resistive heating for nanoparticles with varying diameter. (From Zhang, M.Y. and Cheng, G.J., *J. Appl. Phys.*, 108, 113112, 2010.)

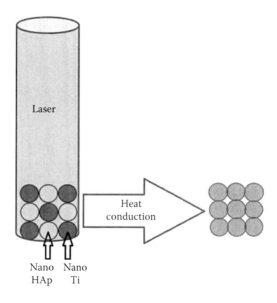

**FIGURE 2.14**
Schematic diagram for the mechanism of heat flow while laser sintering of nanoparticles of two different materials with different thermal conductivity takes place.

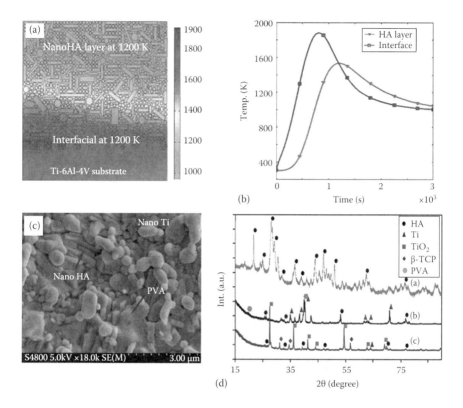

**FIGURE 2.15**
**(See color insert.)** (a) A model showing heating of HAp and Ti nanoparticles, (b) rise and fall of temperature, (c) SEM image of the nanocomposite of HAp and Ti nanoparticles, and (d) x-ray diffraction pattern showing various materials components in the nanocomposite. (From Zhang, M.Y. et al., *ACS Appl. Mater. Interfaces*, 3, 339, 2011.)

to be twice stronger than the commercially available plasma-sprayed coatings. Figure 2.15a shows a model of the nanocomposite–substrate interface. Figure 2.15b shows the rise and fall of temperature for the system. Figure 2.15c depicts SEM image of the final product and Figure 2.15d shows the XRD pattern for the product.

To achieve printed electronic circuitry at low cost and in a scalable manner, new and innovative technologies are being sought after. The laser sintering process provides a huge opportunity in this regard, which will make it possible to write metals in a targeted manner. Control of local temperature makes the laser sintering process a powerful tool in delivering targeted deposition of metal lines or dots with a good finish. Such laser-based printing would eliminate the need of vacuum deposition of thin films and the photolithographic steps one has to follow otherwise. Laser sintering is an environment-friendly technique and gives rise to RT manufacturing. Ko et al.[28] employed laser sintering for inkjet-printed nanoparticles. As one

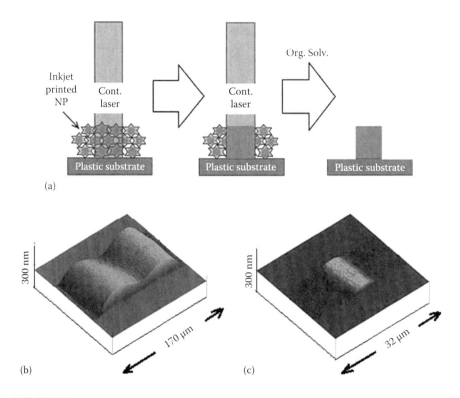

**FIGURE 2.16**

(a) Schematic diagram for laser sintering of inkjet-printed nanoparticle ink using a continuous laser. AFM topographic images of (b) an original inkjet-printed nanoparticle line and (c) a selectively laser-sintered line. (From Ko, S.H. et al., *Nanotechnology*, 18, 345202, 2007.)

can see in Figure 2.16a, one can write metallic lines or dots precisely. Figure 2.16b and c shows original inkjet-printed line and developed laser-sintered pattern, respectively.

To achieve good inkjet printing, dispersing agents and carrier fluids are being used, which has to be taken out of the system. Post-deposition photolithography has to be carried out to achieve a device of one's choice. Also, different device components are written with different ink and some of them degrade at relatively lower temperature. Laser sintering simultaneously solves all these concerns and therefore is deemed to be one of the viable options for targeted printing of electronic circuit on flexible substrates. Kumpulainen et al.[29] have used continuous laser as well as pulsed laser for this purpose. It was demonstrated by Kumpulainen et al. that laser sintering is suitable for selected printing without hampering the prospect of the use of fragile components produced by other technologies. Figure 2.17 shows laser-printed lines by a continuous laser and a pulsed laser.

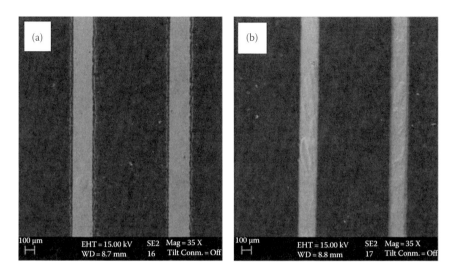

**FIGURE 2.17**
Laser-sintered lines by (a) continuous wave laser and (b) pulsed laser. (From Kumpulainen, T., *Opt. Laser Technol.*, 43, 570, 2011.)

## Nanostructure-Integrated Laser Shock Peening

LSP process is a surface-strengthening process by introducing work-hardening and compressive residual stress. As a result, the mechanical properties, such as strength, hardness, fatigue life, and stress corrosion resistance, could be enhanced. LSP employs lasers with high laser fluence and small pulse width to be focused on an ablative coating, which absorbs the heat energy from the laser beam. A confining media either water or a solid glass with adequate refractive indices direct the shock wave energy into the surface of the material. When the laser is fired periodically, the laser beam passes through the confining media, explodes the ablative material, and then creates a shock wave that deforms the material. Figure 2.18 shows the schematic diagram for LSP where plasma is formed in ablative coating and this coating absorbs all heat and passes on the pressure shock wave to the material underneath.

One of the variants of LSP is nanostructure-integrated LSP, such as warm LSP (WLSP). The nanostructures interact with laser shock wave and generate enhanced mechanical properties, such as fatigue resistance and stress corrosion resistance. WLSP induces nanoscale precipitation and high-density dislocation arrangement, which gives rise to pronounced surface strength than the LSP.[30] Due to dislocation pinning effect, there is minimal relaxation of residual stress and surface dislocation. WLSP results in improved fatigue life due mainly to enhanced residual stress stability and surface strength apart

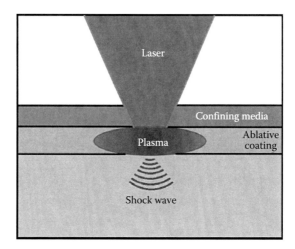

**FIGURE 2.18**
Schematic diagram showing laser peening process. (From Liao, Y., *J. Appl. Phys.*, 108, 063518, 2010.)

from compressive residual stress itself and high-density dislocations. Highly dense nucleation precipitation with dense dislocation structure at circumference has been observed by Liao et al.[31] when WLSP was performed on aluminum alloy 6061. The effect of processing parameters such as processing temperature and strain rate on the nucleation process was investigated. The rod and spherical precipitate is shown in Figure 2.19a and b.

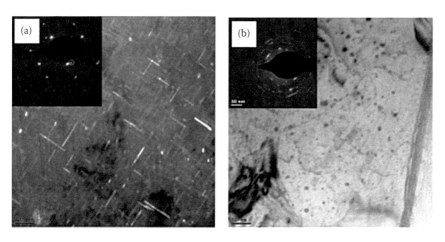

**FIGURE 2.19**
TEM images showing precipitate structures in AA6061 processed by WLSP at 160°C and corresponding diffraction patterns: (a) both rod and spherical precipitates in the T6 sample and (b) spherical precipitates in the solutionized sample.

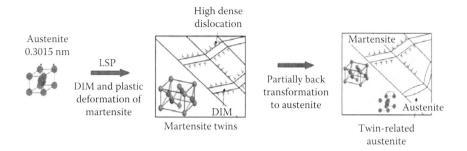

**FIGURE 2.20**
Schematic showing martensite transformation in NiTi shape memory alloy while laser peening takes place. (From Liao, Y., *J. Appl. Phys.*, 112, 033515, 2012.)

WLSP of steel (AISI 4140) has also been carried out by Ye et al.[32] Steel undergoes fatigue life improvement after WLSP. LSP followed by post-deformation annealing of NiTi shape memory alloys have been carried out by Ye et al.[33] They observed bimodal nanocrystallization due to such treatments. The mechanism of thermal-engineered LSP was studied in detail by Liao et al.[34] due to its advantage in achieving enhanced fatigue performance. Liao et al.[35] have carried out a detailed study on the LSP-driven localized-deformation-induced martensite in NiTi shape memory alloy as shown in Figure 2.20. Higher laser fluence and low temperature are favorable to yield higher DIM volume fraction. Higher laser fluence seems to enhance the chemical driving force for the transition. Microscale shape memory effect was observed after post-LSP treatment.

Ultrafast femtosecond laser was used for LSP processing of steel by Lee et al.[36] Zinc was used as ablative coating and water was used as confining media. A high numerical aperture aspheric lens was used to avoid optical breakdown of water. Mechanical hardness was investigated before and after the LSP processing. Figure 2.21 summarizes the results achieved by Lee et al.

## Laser Nanolithography

Nanoarchitectures such as nanodots, nanowires, nanogrids, or any other patterns generated on the substrate have a variety of applications in photonics, electronics, optoelectronics, magnetic memory, catalysis, field emission, chemical and biosensing, etc. However, to precisely pinpoint and quantify the effect exploited due to nanofeatures on one hand and to achieve the same signal every time, one needs to achieve periodic patterns. Such spatially defined nanofeatures are in great demand for various nano-applications

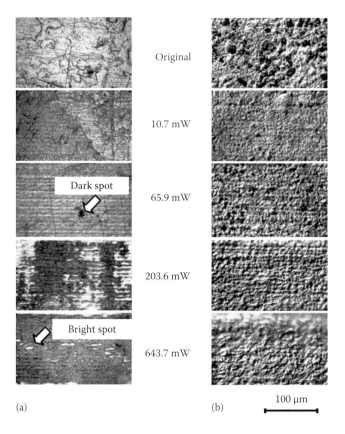

**FIGURE 2.21**
**(See color insert.)** Top surface of shock-peened specimens for select fluencies for (a) galva-nized and (b) galvannealed steel. The term "Original" refers to the top surfaces before LSP process. (From Lee, D. and Asibu, E.K., Jr., *J. Laser Appl.*, 23, 022004-1, 2011.)

especially for their device applications. Laser, being a high-power energy source with option of precise delivery of a particular energy reproducibly, gives us a great opportunity for exploiting it for lithography purposes. One can use laser with mask and achieve patterns, which would be governed by mask dimensions apart from laser parameters. Maskless lithography is also possible with laser exploiting the optical nature of laser light such as interference and diffraction. Maskless-laser-based lithography can pri-marily be divided in far-field and near-field techniques. Laser-interference lithography (LIL) is one of the maskless approaches to achieve large-scale nanofabrication that has the capability to yield periodic nanofeatures such as nanodots or nanowires or other 2D nanofeatures. Interference of two laser beams gives rise to standing wave patterns that is usually exploited for laser interference lithography. Laser wavelength and the relative angle between the two lasers decide the minimum spacing between the periodic nanofeatures:

$$\Delta = \frac{\lambda}{2n}\sin\left(\frac{\alpha}{2}\right) \tag{2.1}$$

Photoresists are photosensitive polymers used for recording the interference pattern. Micro- or nanoscale features can thus be achieved for a large area using LIL technique. The variation of angles between the lasers can yield various shapes of nanofeatures such as circular, hexagonal, and square. Figure 2.22a shows a schematic diagram of the simplest form of LIL (where two lasers are incident on prism faces and interfere and form pattern at the base of the prism). Figure 2.22b shows a nanocone structure fabricated by LIL.[37]

LIL often suffers from resolution deficit, which has been addressed by many ways including the use of extreme ultraviolet (EUV) source. Such reduced wavelength lasers when exposed to photoresist give rise to greater ablation depth in the order of 100 nm on silicon-based photoresist. When one combines EUV laser and Lloyd's mirror interferometer as shown in schematic diagram Figure 2.23a, fine nanostructures of average size of 60 nm were achieved[38] on PMMA in reproducible manner as shown in Figure 2.23b–d.

Boor et al.[39] combined LIL to the metal-assisted etching and achieved vertically standing silicon nanowires in a periodic manner that can be used for many applications such as in solar cell and photoemiting diodes. Nanowire diameter was observed to be in the range of 60–70 nm as shown in Figure 2.24. Such accuracy obtained by combining two technologies is unprecedented.

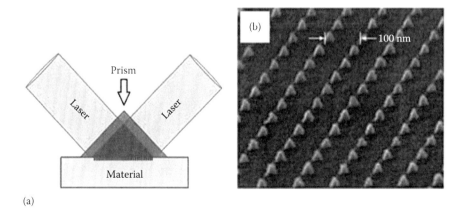

(a)

**FIGURE 2.22**
(a) Schematic diagram for laser interference lithography, (b) a nanocone structure fabricated by laser interference lithography (height 40 nm and width 30 nm). (From Savas, T.A. et al., *J. Appl. Phys.*, 85, 6160, 1999.)

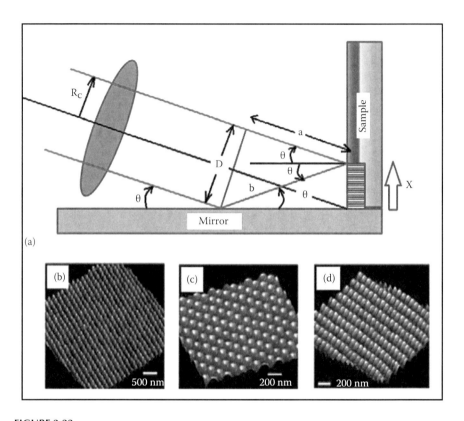

**FIGURE 2.23**

**(See color insert.)** 2D nanopatterns on PMMA produced by EUV laser interference lithography using Lloyd's mirror interferometer (schematic shown in (a)) with two exposures at different angles, (b) dots with 60 nm FWHM feature size and a period of 150 nm, (c) regular-shaped dots, and (d) elongated dots. (From Marconi, M.C. and Wachulak, P.C., *Prog. Quant. Elect.*, 34, 173, 2010.)

Among many possible exploitations of the near field of laser, nonpropagating evanescent wave interference lithography (EIL) is one of them. Diffraction limit is beaten by this technique. In this technique, total internal reflection (TIR) is exploited for laser-based evanescent wave near-field interference lithography. The schematic diagram of such lithography is shown in Figure 2.25a and b shows the AFM image of 2D nanostructures so obtained by EIL while TIR of four p-polarized incident laser beams occur.[40]

Surface plasmon interference lithography (SPIL) is yet another lithographic technique that exploits the near field of laser. Collective electron oscillations are called surface plasmons (SPs). Such waves set up when plasmonic metals form interface with dielectrics. Such SP waves and evanescent waves couple and give rise to the field enhancement. One can exploit such interactions of laser light with the SP to achieve SPIL. SP waves have wavelength that

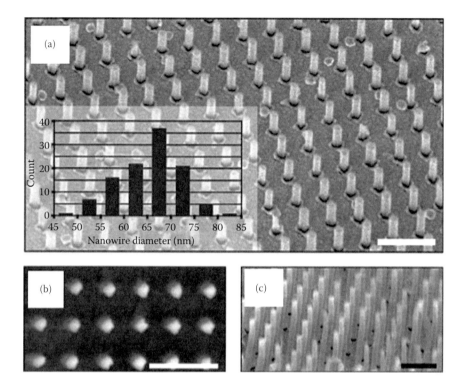

**FIGURE 2.24**
(a) SEM micrograph of silicon nanowires produced using laser interference lithography combined with metal-assisted etching, with a periodicity of p = 244 nm and a diameter d = 66 nm. The overlay presents the size distribution. (b) and (c) show the same sample after longer etching time in top view and tilted view, respectively. Note that the wires in (c) have a length of about 1 mm, corresponding to an aspect ratio of 1:15. The scale bars are 500 nm. (From Boor, J.D. et al., *Nanotechnol.*, 21, 095302, 2010.)

is shorter than that for the exciting laser itself. SPIL technique beats EIL in energy transmission and also in fabrication depth. Figure 2.26 shows a periodic nanopattern generated on silicon by SPIL technique employing Al mask and UV Ar ion laser.[41]

Laser when incident on small and sharp features gives rise to local electric field enhancement. This effect can thus be used to indirectly use laser for lithography purposes. Laser incident on scanning force microscopic tips have been used to write lines at laser fluence values lower than 0.1 J/cm². Such field enhancement between the SPM tip and the sample gives rise to high local temperature that melts or evaporates materials underneath the SPM tip depending on the laser fluence being used. Thus, a precise lithography is achieved using laser in conjunction with SPM tip. Chimmalgi et al.[42] have used this technology to write sharp lines on 25 nm gold thin films as shown in Figure 2.27.

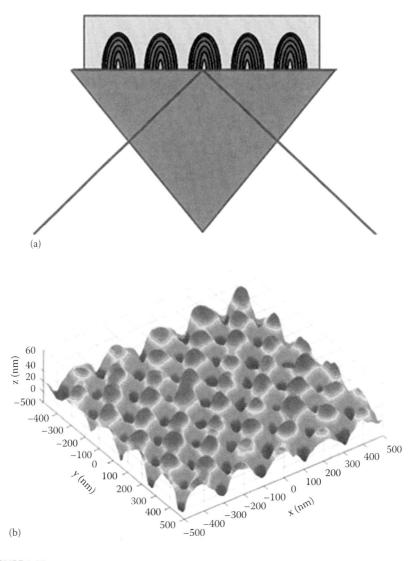

(a)

(b)

**FIGURE 2.25**
**(See color insert.)** (a) Schematic diagram for EIL for TIR of four p-polarized incident beams and (b) AFM image of the nanostructures so obtained. (From Chua, J.K. and Murukeshan, J.K., *Micro Nano Lett.*, 4, 210, 2009.)

## Summary and Outlook

Laser, a special class of electromagnetic radiation, interacts with nanomaterials drastically different as compared to that for its bulk counterpart. Photothermal and photochemical transformations could take place while

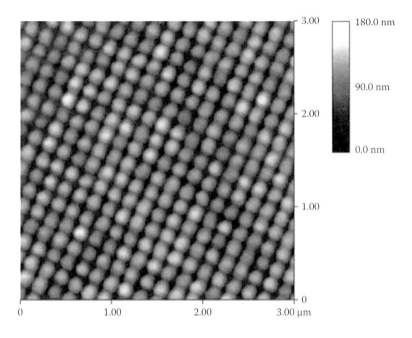

**FIGURE 2.26**
An AFM image of periodic nanodot array produced on Si by SPIL technique, using an Al mask and an UV argon ion laser. (From Sreekanth, K.V., *Appl. Opt.*, 49, 6710, 2010.)

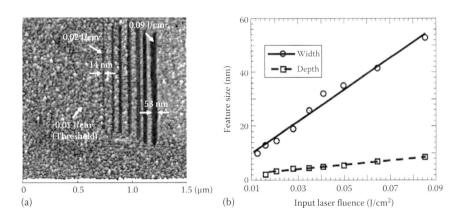

**FIGURE 2.27**
Surface nanostructuring of 25 nm thick Au film for laser fluence between 0.012 and 0.086 J/cm² and the dependence of lateral feature size on laser fluence. (From Chimmalgi, A. et al., *J. Appl. Phys.*, 97, 104319, 2005.)

laser interacts with nanomaterials. We have attempted to brief on some of the key laser-based nanomanufacturing systems.

Laser nanostructuring is being developed as a standard scalable technique to achieve nanostructured surfaces with added functionality. Such nanostructured semiconducting surfaces find applications in better adhesion of thin films in electronic industries and as light-emitting devices. Nanostructured insulators such as glass and polymers are useful for bioapplications such as tissue culturing or modulating optical behavior of materials. Laser crystallization of amorphous thin films of functional materials at marginal thickness opens the door for miniaturization of devices where these materials are being sought for. Functional materials such as plasmonic, magnetic, transparent conducting, shape memory alloys, and ferroelectric or superconducting materials can instantly be laser crystallized in a scalable manner in a single-step processing. Apart from scalability and high throughput, laser crystallization provides high selectivity without affecting other device components. Laser sintering of materials is one of the techniques being used in making printed electronics with precise patterns. We have discussed how nanoparticle size reduction affects heat diffusion and hence the sintering process with relevant examples. LSP, which exploits laser ablation induced pressure, has been applied to many engineering materials. It was observed that nanostructure-integrated LSP gives rise to enhanced mechanical properties due to the unique interaction between the nanomaterials and shock wave. Nanolithography is an important patterning process to integrate nanofeatures to devices. We have described with a few examples that laser-based nanolithography provides a precise, scalable, efficient, and accurate approach as compared to other contemporary lithographic techniques.

Laser-based manufacturing systems have been demonstrated to be capable of manipulating materials at nanoscale, which in turn changes their physical behavior. Laser's contribution toward recent developments of lithographic techniques, which is bound to solve device integration issues, is noteworthy. With the advancements in emergent laser-based technologies, a great promise has been shown toward bringing nanotechnology solutions to a meaningful level. The stage is now set for such techniques to be a primary driving force for new innovation and novel applications.

## References

1. I. N. Zavestovskaya, Laser nanostructuring of materials surfaces, *Quantum Electronics*, 40, 942 (2010).
2. J. Heitz, E. Arenholz, D. Biuerle, R. Sauerbrey, and H. M. Phillips, Femtosecond excimer-laser-induced structure formation on polymers, *Applied Physics A*, 59, 289–293 (1994).

3. J. Heitz, E. Arenholz, D. Bäuerle, and K. Schilcher, Growth of excimer-laser-induced dendritic structures on PET, *Applied Surface Science*, 81, 103–106 (1994).

4. D. J. Krajnovich and J. E. Vázquez, Formation of intrinsic surface defects during 248 nm photoablation of polyimide, *Journal of Applied Physics*, 73, 3001 (1993).

5. E. Rebollar, I. Frischauf, M. Olbrich, T. Peterbauer, S. Hering, J. Preiner, P. Hinterdorfer, C. Romanin, and J. Heitz, Proliferation of aligned mammalian cells on laser-nanostructured polystyrene, *Biomaterials*, 29, 1796–1806 (2008).

6. P. Kumar, Surface modulation of silicon surface by excimer laser at laser fluence below ablation threshold, *Applied Physics A*, 99, 245–250 (2010).

7. P. Kumar, M. G. Krishna, and A. Bhattacharya, Excimer laser induced nanostructuring of silicon surfaces, *Journal of Nanoscience and Nanotechnology*, 9, 3224–3232 (2009).

8. P. Kumar, Microscopic study of the effect of media on laser nanostructuring of silicon surface, *Advanced Science Letters*, 3, 67–70 (2010).

9. A. Y. Vorobyev and C. Guo, Femtosecond laser nanostructuring of metals, *Optics Express*, 14, 2164 (2006).

10. F. Korte, J. Serbin, J. Koch, A. Egbert, C. Fallnich, A. Ostendorf, and B. N. Chichkov, Towards nanostructuring with femtosecond laser pulses, *Applied Physics A*, 77, 229–235 (2003).

11. M. G. Krishna and P. Kumar, Non-lithographic techniques for nanostructuring thin films and surfaces, in *Emerging Nanotechnology for Manufacturing*, W. Ahmed and M. J. Jackson (Eds.), pp. 93–110, Elsevier Inc., Oxford, U.K. (2009).

12. R. A. Lemons, M. A. Bosch, A. H. Dayem, J. K. Grogan, and P. M. Mankiewich, Laser crystallization of Si films on glass, *Applied Physics Letters*, 40, 469 (1982).

13. H. J. Kim and J. S. Im, New excimer-laser-crystallization method for producing large-grained and grain boundary-location-controlled Si films for thin film transistors, *Applied Physics Letters*, 68, 1513 (1996).

14. R. Dassow, J. R. Kohler, Y. Helen, K. Mourgues, O. Bonnaud, T. M. Brahim, and J. H. Werner, Laser crystallization of silicon for high-performance thin-film transistors, *Semiconductor Science Technology*, 15, L31–L34 (2000).

15. A. A. D. T. Adikaari and S. R. P. Silva, Thickness dependence of properties of excimer laser crystallized nano-polycrystalline silicon, *Journal of Applied Physics*, 97, 114305 (2005).

16. Y. C. Peng, G. S. Fu, W. Yu, S. Q. Li, and Y. L. Wang, Crystallization of amorphous Si films by pulsed laser annealing and their structural characteristics, *Semiconductor Science Technology*, 19, 759–763 (2004).

17. J. Jin, Z. Yuan, L. Huang, S. Chen, W. Shi, Z. Cao, and Q. Lou, Laser crystallization of amorphous silicon films investigated by Raman spectroscopy and atomic force microscopy, *Applied Surface Science*, 256, 3453 (2010).

18. P. Kumar, Electric field and excimer laser nanostructuring of Ni and Si thin films, *Advanced Science Letters*, 3, 62–66 (2010).

19. P. Kumar and M. G. Krishna, A comparative study of laser- and electric-field-induced effects on the crystallinity, surface morphology and plasmon resonance of indium and gold thin films, *Physica Status Solidi A*, 207, 947–954 (2010).

20. T. Y. Choi, D. J. Hwang, and C. P. Grigoropoulos, Ultrafast laser-induced crystallization of amorphous silicon films, *Optical Engineering*, 42, 3383–3388 (2003).

21. M. L. Taheri, S. McGowan, L. Nikolova, J. E. Evans, N. Teslich, J. P. Lu, T. LaGrange, F. Rosei, B. J. Siwick, and N. D. Browning, In situ laser crystallization of amorphous silicon: Controlled nanosecond studies in the dynamic transmission electron microscope, *Applied Physics Letters*, 97, 032102 (2010).

22. D. Pirzada, P. Trivedi, D. Field, and G. J. Cheng, Effect of film thickness and laser energy density on the microstructure of a-GaAs films after excimer laser crystallization, *Journal of Applied Physics*, 102, 013519 (2007).

23. D. Pirzada and G. J. Cheng, Microstructure and texture developments in multiple pulses excimer laser crystallization of GaAs thin films, *Journal of Applied Physics*, 105, 093114 (2009).

24. M. Y. Zhang and G. J. Cheng, Highly conductive and transparent alumina-doped ZnO films processed by direct pulsed laser recrystallization at room temperature, *Applied Physics Letters*, 99, 051904 (2011).

25. M. Y. Zhang, Q. Nian, and G. J. Cheng, Room temperature deposition of alumina-doped zinc oxide on flexible substrates by direct pulsed laser recrystallization, *Applied Physics Letters*, 100, 151902 (2012).

26. M. Y. Zhang and G. J. Cheng, Nanoscale size dependence on pulsed laser sintering of hydroxyapatite/titanium particles on metal implants, *Journal of Applied Physics*, 108, 113112 (2010).

27. M. Y. Zhang, C. Ye, U. J. Erasquin, T. Huynh, C. Cai, and G. J. Cheng, Laser engineered multilayer coating of biphasic calcium phosphate/titanium nanocomposite on metal substrates, *ACS Applied Materials & Interfaces*, 3, 339–350 (2011).

28. S. H. Ko, H. Pan, C. P. Grigoropoulos, C. K. Luscombe, J. M. J. Frechet, and D. Poulikakos, All-inkjet-printed flexible electronics fabrication on a polymer substrate by low-temperature high-resolution selective laser sintering of metal nanoparticles, *Nanotechnology*, 18, 345202 (2007).

29. T. Kumpulainen, J. Pekkanen, J. Valkama, J. Laakso, R. Tuokko, and M. Mantysalo, Low temperature nanoparticle sintering with continuous wave and pulse lasers, *Optics & Laser Technology*, 43, 570–576 (2011).

30. C. Ye, Y. Liao, and G. J. Cheng, Warm laser shock peening driven nanostructures and their effects on fatigue performance in aluminum alloy 6160, *Advanced Engineering Materials*, 12, 291 (2010).

31. Y. Liao, C. Ye, B.-J. Kim, S. Suslov, E. A. Stach, and G. J. Cheng, Nucleation of highly dense nanoscale precipitates based on warm laser shock peening, *Journal of Applied Physics*, 108, 063518 (2010).

32. C. Ye, S. Suslov, B. J. Kim, E. Stach, and G. J. Cheng, Fatigue performance improvement in AISI 4140 steel by dynamic strain aging and dynamic precipitation during warm laser shock peening, *Acta Materialia*, 59, 1014–1025 (2010).

33. C. Ye, S. Suslov, X. Fei, and G. J. Cheng, Bimodal nanocrystallization of NiTi shape memory alloy by laser shock peening and post deformation annealing, *Acta Materialia*, 59, 7219–7227 (2011).

34. Y. L. Liao, C. Ye, S. Suslov, and G. J. Cheng, The mechanisms of thermal engineered laser shock peening for enhanced fatigue performance, *Acta Materialia*, 60, 4997–5009 (2012).

35. Y. Liao, C. Ye, D. Lin, S. Suslov, and G. J. Cheng, Deformation induced martensite in NiTi and its shape memory effects generated by low temperature laser shock peening, *Journal of Applied Physics*, 112, 033515 (2012).

36. D. Lee and E. K. Asibu, Jr., Experimental investigation of laser shock peening using femtosecond laser pulses, *Journal of Laser Applications*, 23, 022004-1 (2011).

37. T. A. Savas, M. Farhoud, H. I. Smith, M. Hwang, and C. A. Ross, Properties of large area nanomagnetic arrays with 100 nm period made by interferometric lithography, *Journal of Applied Physics*, 85, 6160 (1999).
38. M. C. Marconi and P. C. Wachulak, Extreme ultraviolet lithography with table top lasers, *Progress in Quantum Electronics*, 34, 173 (2010).
39. J. D. Boor, N. Geyer, J. V. Wittemann, U. Gorsele, and V. Schmidt, Sub-100 nm silicon nanowires by laser interference lithography and metal-assisted etching, *Nanotechnology*, 21, 095302 (2010).
40. J. K. Chua and J. K. Murukeshan, UV laser assisted multiple evanescent waves lithography for near field nano patterning, *Micro and Nano Letters*, 4, 210–214 (2009).
41. K. V. Sreekanth, J. K. Chua, and V. M. Murukeshan, Interferometric lithography of nanoscale feature patterning—A comparative analysis between laser interference, evanescent wave interference and surface plasmon interference, *Applied Optics*, 49, 6710–6717 (2010).
42. A. Chimmalgi, C. P. Grigoropoulos, and K. Komvopoulos, Surface nanostructuring by nano-/femtosecond laser-assisted scanning force microscopy, *Journal of Applied Physics*, 97, 104319 (2005).

# 3

## Photonic Systems for Crystalline Silicon and Thin-Film Photovoltaic Manufacturing

Peter Bermel, Gary J. Cheng, and Xufeng Wang

### CONTENTS

### Introduction

In this chapter, we will discuss the impact that photonics have on photovoltaic (PV) manufacturing. The broad field of photonics is now commonly thought to encompass classical optics (Hecht, 2001), nanophotonics (Joannopoulos et al., 2007), and metamaterials (Cai and Shalaev, 2009). Over the last few decades, photonic technology has become an increasingly integral part of PV manufacturing, both in the crystalline silicon and thin-film arenas.

PV manufacturing has attracted increasing interest recently, thanks in part to its tremendous growth over the past decade. Innovative German feed-in tariffs have greatly facilitated this market growth, from 2.2 GW globally (373 MW EU) in 2002 to an estimated 37–38 GW installed globally in 2013. The last few years have also marked the arrival of low-cost manufacturing from China, with multicrystalline silicon average selling prices dropping from \$3.55/Wp in 2006 to \$0.66/Wp in 2013. The impact has been to make PVs competitive in price per kilowatt-hour with a wide variety of incumbent technologies in certain markets—an important concept known as *grid parity*.

Here, we will discuss the contributions of photonic technologies toward these dazzling achievements, while also identifying some key areas for future growth. For example, while most photonic tools employed in PV manufacturing at present are derived from classical optics, newer energy conversion technologies being developed in nanophotonic and metamaterial platforms may also become an integral part of PV manufacturing going forward.

This chapter will proceed along the following structure. First, we will discuss photonic technologies employed in crystalline silicon manufacturing, which include wafer dicing for dividing raw ingots into substrates for further processing; doping of the resultant wafers to suppress recombination; laser-fired and grooved contacts, which allow the creation of a back surface field that suppresses surface recombination and improves efficiency; and wrap-through technologies. The discussion will then continue onto thin-film manufacturing, including laser scribing and laser edge isolation and deletion. Finally, characterization techniques needed for postmanufacturing quality assurance are discussed. These include photoluminescence for material identification, electroluminescence for shunt detection, quantum efficiency (QE) measurements, and solar simulation.

## Crystalline Silicon Manufacturing

Ever since the first silicon-based PV cell was demonstrated by Bell Labs in 1954, synergy with the semiconductor industry has led crystalline silicon-based PV to play a dominant role over the last 59 years. The role of photonics in crystalline silicon, however, is somewhat more recent, in that it is mainly driven by laser-based technologies. The laser was invented in 1960 by Theodore Maiman, drawing on theoretical and experimental work performed by Schawlow and Townes. Since then, the key metrics for laser performance, including operating wavelength, optical power, coherence, and affordability, have all improved by orders of magnitude compared to the earliest experiments. This technology has now become ubiquitous enough that it can now replace incumbent manufacturing technologies while also adding new efficiency-boosting features to crystalline silicon cells that were not previously possible to include in manufacturing without photonics.

## Wafer Dicing

All crystalline silicon solar cells are built on the foundation of the silicon wafer. A properly prepared silicon wafer offers the following combination of features, necessary for high-performance PV cells: ultrahigh purity, high carrier mobility, and low nonradiative recombination. It is also helpful but not absolutely mandatory to have a monotonically varying band structure with a back surface field to enhance carrier extraction near the terminals.

Underlying all these features is the raw material requirement of ultrahigh-purity crystalline silicon. Generally, electronics-grade polysilicon with eight nines of purity is melted and then extracted as a large mass through the float-zone or Czochralski process (Streetman and Banerjee, 2000). Afterwards the resulting silicon boule must then be diced into pieces with a thickness that ensures sufficient optical absorption without excessive carrier recombination. In the world record–setting 25.0% efficient crystalline silicon cell, known as the PERL cell from Martin Green's group at the University of New South Wales, a thickness of 450 µm is employed.

Typically, wafers are created from the boule via mechanical sawing. In this process, a thin diamond-tipper blade saw or wire saw is used to physically separate pieces with the target thickness from the boule, one at a time, in rapid succession (Streetman and Banerjee, 2000). Wafers emerging from this process require further postprocessing to be rendered flat enough for photolithographic processing and chamfering along the edges to reduce the probability of chipping during processing. Furthermore, in many cases, a chemical–mechanical polish with a slurry of fine particles (e.g., silica or ceria) is employed to achieve a flat surface on at least one side. In many cases, this is followed by etching with an alkaline (high pH) solution, in order to roughen the surface to assist with light trapping. However, this process also has certain drawbacks. One is that approximately 50% of the material is wasted as saw dust, which is technically known as kerf loss. The second is that it requires multiple mechanical and chemical processing steps, which take time and even require rare earths (such as cerium) for some versions of the process.

A new alternative to the die saw method is laser-based dicing. In this approach, the energy of the laser beam replaces the mechanical sawing action. This can be thought of metaphorically as an *optical knife*. There are two basic strategies for laser-based dicing: conventional and stealth dicing. In conventional laser-based dicing, one utilizes a wavelength that is strongly absorbed by the target, which locally heats it enough to cause vaporization and removal from the target area, which then clears the way for further cutting. Since silicon has an indirect bandgap wavelength of 1107 nm, such a laser would generally be in the visible or near-infrared.

In stealth laser-based dicing, one instead utilizes a wavelength that is typically not absorbed directly by the substrate but requires a nonlinear absorption process (generally two-photon absorption). In this case, multiple beams from different directions can be focused to create localized point-like or linear high-intensity regions inside the material. The nonlinear effect then ensures that the cutting action will occur only at the high-intensity regions.

## Wafer Doping

In order to introduce the correct band structure for crystalline silicon wafers, necessary for high performance, dopant diffusion is required. Most commonly, the silicon wafers are grown as p-type and then doped with an n-type front region through phosphorous gas (POCl) diffusion. In such cases, it is well known that the diffusing particles will observe Brownian motion. This process is quantified by Fick's law and allows the doping profile to be tuned by adjusting both the doping time and temperature. The latter is generally a particularly sensitive parameter, since such processes generally follow the Arrhenius equation, with an exponential dependence on temperature. At the same time, most substrates have a limited *thermal budget*, that is, a maximum time–temperature product that they can bear during processing. As a result, most doping is performed using a rapid-thermal annealing machine (Streetman and Banerjee, 2000).

An alternative method that offers greater control over doping profile is known as ion implantation. In this process, dopants are accelerated by a high voltage and separated by a magnetic field en route to the target. Adjusting the type, energy, and angle of the ions can fine-tune the implanted doping profile. However, the cost of this method is very high, due to the need for a beamline and relatively low throughput. As such, it is not generally deployed in semiconductor manufacturing.

In laser doping, one combines the best of both worlds: the relatively modest costs of dopant diffusion with the more precise control and lack of thermal stress associated with ion implantation. In this approach, a doping precursor is first deposited on the target, to serve as an ion source. Then the process is engaged by scanning the entire surface where doping is required with a pulsed laser beam. It is currently believed that the laser pulse will rapidly and locally heat the doping precursor as well as the top surface of the silicon. Some of the doping precursors will then rapidly diffuse into the target, while the rest evaporates. At the end, one is left with a doped silicon wafer and an unchanged substrate.

The most popular tool for laser-based doping is a green Nd:YAG laser, operating at 530 nm. It combines a high repetition rate of 10 kHz and a small spot diameter of 0.1 mm with sufficient beam fluence. Alternatives include ruby lasers (694 nm, 5 Hz repetition, and 9 mm beam spot) as well as excimer lasers (ArF, 193 nm, 10 Hz repetition, 20 mm$^2$ spot).

## Laser-Fired and Grooved Contacts

Discussion will then proceed to laser-fired and grooved contacts and the results of the induced back surface field on cell performance.

Laser-fired contacts offer a unique approach to improving the performance of solar cells, based on redesigning the design of the back contacts. The common, straightforward method of placing a eutectic metal backing directly behind a silicon wafer has several disadvantages. First, the reflection is generally suboptimal, because the group velocity of light in silicon is well below that of air. Second, the work function of the metal can create a Schottky barrier, which impedes the smooth flow of current through the structure to the electrical contacts. Third, the surface recombination velocity associated with the silicon–metal interface can be relatively high, which degrades the PV cell power conversion efficiency (particularly the open-circuit voltage). With laser-fired contacts, one first creates a low-refractive-index layer behind the silicon wafer, either through native oxide growth or through deposition. Then holes are drilled in the low-index material with laser cutting, for example, with an ultraviolet wavelength excimer laser. Then the eutectic metal is then used to fill in the holes and coat the back. This strategy has several advantages, which include increasing the reflectivity of the base metal via the higher group velocity in the low-index material, decreasing or eliminating any Schottky barriers previously present, and creating a back surface field to facilitate the flow of electrons and decrease surface recombination. The combined effect is a quantifiable improvement in the short-circuit current.

Laser-grooved contacts are of interest for similar reasons. In this approach, invented by Martin Green and Stuart Wenham of the University of New South Wales, one uses lasers to groove the front of the wafer. This provides an opportunity for secondary dopant diffusion near the surface, followed by electroplating. Several advantages can be achieved from this approach, including improved collection of long-wavelength photons; decreased shadowing, due to the narrower width of the busbars; and higher current-carrying capacity/lower series resistance, thanks to the greater cross-sectional area of the contacts.

## Wrap-Through Technologies

Wrap-through technologies (metal wrap-through [MWT] and emitter wrap-through [EWT]) will then be shown to offer significant advantages in terms of reduced shadowing and improved module-level integration.

The MWT cell design concept was devised by van Kerschaver and collaborators (1998). MWT functions by shifting half of the front contacts of a solar cell toward the back, in order to reduce shadowing losses by a proportional fraction. In this process, holes are periodically drilled in the PV active material using a high-throughput laser. Next, a metal paste (known as a eutectic) is filled in from behind, in order to connect the front and back regions of the cell. The net effect is to increase power production more than the additional cost of this processing step. While the idea has been well known for some time, the first prototype production came online in 2007 at Fraunhofer ISE in Germany. The multicrystalline silicon cells produced offered a 16% average efficiency, about a 0.5% absolute improvement over the standard design without MWT (Clement et al., 2010).

EWT was proposed by James M. Gee and collaborators (1993). It consists of interdigitated p- and n-type metallic back contact grids, with highly conductive vias drilled to the n+-doped front layers. This, in many ways, resembles laser backfired contacts but with deeper drilling for the n-type grid. This design offers several features that contribute to high performance, including low shadowing and low resistivity due to the geometry of the contacting approach, natural alignment between the drilled vias and metallized n-type grid, and separate emitter and contact diffusion steps for maximum fabrication control (Gee et al., 1993). For adequate industrial throughput, it has been estimated that to achieve a via spacing of 1 mm in each direction, each individual via must be drilled in less than 6 ms.

## Thin-Film Manufacturing

Thin-film photovoltaics (PVs) represent a paradigm shift in PV manufacturing. In the last decade, quality has greatly improved in most market segments, including cadmium telluride (CdTe), as promising new technologies demonstrate enhanced commercial viability, including gallium arsenide (GaAs), copper indium gallium diselenide (CIGS), and organic heterojunctions.

In this part, a different set of laser-based techniques for thin-film manufacturing (micromorph, CdTe, CIGS, and CZTS) will be discussed. The primary focus will be on cell definition and module integration approaches.

### Laser Scribing

In particular, laser scribing will be shown to play a critical role in enabling monolithic integration over very large sheets of glass, necessary for economic competitiveness.

For monolithic integration of PV modules, it is important to have an approach to connect the front and back of a large array of cells. This allows for maximum power output with much lower currents (for greater efficiency and safety), plus significantly higher voltages (since the maximum power point voltage of a single silicon cell is approximately 0.6 V, whereas home outlets have a 110 V rms output). This goal can be achieved by scribing three cuts at each line of interconnection. The first cut is an isolation step performed before PV material deposition. It divides the front transparent conductive oxide (TCO) into electrically disconnected stripes. After the PV material is deposited, a second cut is performed through it. This creates a via from the back of the cell to the front TCO. Subsequently, a metal backing is conformally deposited all along the back surface. Finally, a third cut is performed in the metal backing, closely aligned with the first two cuts in order to force electricity to flow through the PV active material. A slight drawback to this method is small *dead* regions caused by the surface area associated with the scribe features, which limits the density of scribe lines to approximately 1 cm spacing.

There are three serious candidates for large-scale production of scribe lines: photolithography, mechanical scribing, and laser scribing. In the first method, photolithography, a photoresist is deposited and then spun to uniformity. Then an optical mask and light source is used to create the desired pattern in the photoresist. A selective etching process (such as a chemical wet etch or reactive ion etch) is then employed to remove the underlying material. Finally, the remaining photoresist is removed. However, this approach is fairly time-consuming, complex, low throughput, and expensive—as a result, it is not widely used. The second method, mechanical scribing, is much simpler, in that it uses a sharp blade embedded on a stylus to physically separate the deposited layers. In principle, this approach is highly appealing in terms of simplicity and cost. While this approach works well for soft materials, including metallic pastes and certain active PV materials (particularly CdTe), it is not as suitable for etching TCOs with high haze factors, due to their mechanical roughness, as required for the first scribing step. In those cases, the channels formed are jagged and not accurate enough to meet the specifications of high-end manufacturing. As a result, a third strategy, based on laser patterning, has come to be much more widely used for one or more of the scribing steps. In this strategy, a high-intensity pulsed laser with picosecond durations is employed for cold ablation of the targeted surface—that is, the laser is delivered so quickly that the thin film does not suffer any heat damage. The net result is to create clean, precisely defined channels on the surface of each layer at high speed (on orders of meters per second) and to a depth controlled to an accuracy of 20 nm (Shin et al., 2011).

In terms of the lasers suitable for scribing, it has been found that a wide variety can be used, including Nd:YAG at both 1064 and 532 nm (frequency-doubled), copper vapor at both 510.6 and 578.2 nm, and excimer lasers using xenon chloride (308 nm) and krypton fluoride (248 nm) (Compaan et al., 2002). Generally, scribing can be achieved with pulse energies approximately ten times the threshold for surface damage at the wavelength used. The damage threshold for opaque materials is typically 0.5 J/cm$^2$ over a broad range of wavelengths. However, the thresholds are significantly higher for optically transparent materials such as TCOs, due to a lack of absorption, unless one reaches ultraviolet wavelengths with excimer lasers. For a scribe width of 40 μm, 200 mJ delivered energy is required for each meter of scribing. Since Nd:YAG lasers typically deliver 1 mJ, a rapid succession of pulses is required to scribe appreciable distances on the solar cell. At a repetition rate of 1 kHz, each line will take 0.2 s to scribe. Since excimer lasers generally deliver much more pulse energy (e.g., 400 mJ) than Nd:YAG lasers, it generally suffices to print even the largest form factor modules in a single pulse. At a repetition rate of 50 Hz, each line will take 0.02 s to scribe.

## Laser Edge Isolation/Deletion

Laser edge isolation/deletion has been shown to be helpful in maximizing performance and also have some cross industry relevance in crystalline silicon manufacturing. Emerging techniques such as laser-induced *in situ* crystallization, interference lithography, and laser-LIGA are also discussed.

Laser edge isolation can be achieved by using a laser scribe (from Section 3.3.1) to define a groove around the outermost edge of the solar cell. It is important for the groove to extend deep enough through the n+ front-doped region into the bulk p-type region, in order to minimize shunting.

## Laser Crystallization

Laser-induced *in situ* crystallization is a technique under recent development, in which a high-power pulsed laser is utilized to increase the grain size and decrease the defect density of a deposited thin film of PV material (Cheng et al., 2012 and Cheng and Zhang, 2013). The current thin-film PV has the following problems: (1) Low formation temperature leads to recombination of excess minority carriers in the active absorber, and (2) low reaction kinetics of near vacuum vapor processes limits maximum achievable throughputs in current thin-film PV manufacturing. Here, we will develop direct laser crystallization (DLC) of nanoparticle inks, a high-speed crystallization technique in which nanoparticles are rapidly heated and crystallized at ambient conditions. The novelties are as follows: (1) higher performance due to much lower defects, (2) high speed and scalability, (3) lower cost—due to the much better energy and materials usage, nanoparticles have much higher sinterability than bulk materials—and (4) selective heating, leading to scalable process on various substrates and different layers without affecting other layers.

Targeted materials under current investigation include aluminum-doped zinc oxide (AZO) and CIGS. Recent studies of laser crystallization in AZO have shown that the DLC process can increase the mobility significantly, thanks to a reduction in defect density and strain (Zhang and Cheng, 2011, 2012). The schematic of the process, simulation, and experimental results are shown in Figure 3.1.

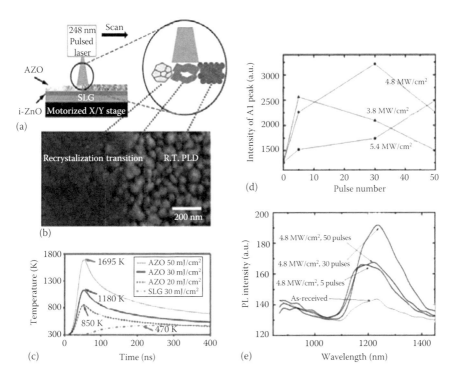

**FIGURE 3.1**

DLC of AZO nanoparticles: (a) schematic of DLC; (b) crystalline structures after DLC; (c) temperature profile at various layers; (d) crystallinity by Raman spectrum; and (e) photoluminescence. (From Zhang, M.Y. and Cheng, G.J., *Appl. Phys. Lett.*, 99, 051904, 2011; 2012.)

## Laser Nanolithography

Fabricating high-performance 2D and 3D nanophotonic structures requires advanced lithography tools. While steppers are an industry standard, they are generally high in cost and suited primarily for large-scale manufacturing of 2D structures. Other lithography tools, such as laser writers and e-beam lithography, can be much lower in cost but also are impeded by throughput that is many orders of magnitude lower as well.

As an alternative, we are building the high-accuracy and high-throughput laser interference lithography approach depicted in Figure 3.2c. The principle behind this method is depicted in Figure 3.2a: two coherent laser beams of wavelength λ, when intersecting at a particular angle θ, will form a standing wave with a periodicity λ/(2 sin θ). This standing wave can be recorded in a layer of photosensitive material, known as the photoresist. Upon immersion in a special chemical, the so-called developer, the areas of photoresist that experienced large optical intensities will be washed away, while areas that did not will remain unaffected. This will result in a grating pattern

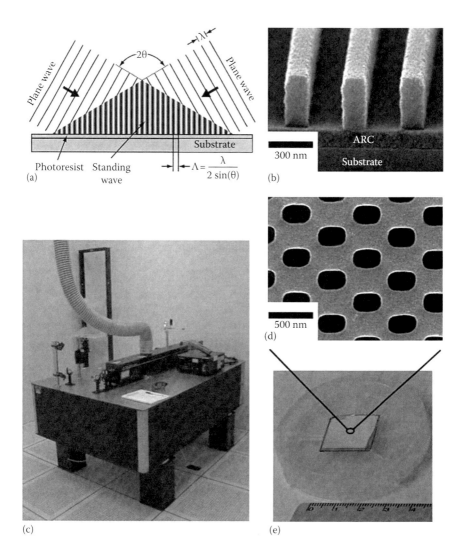

**FIGURE 3.2**
Interference lithography setup for 3D-integrated nanophotonic fabrication: (a) The basic concept is to create a standing wave that will be recorded in a photosensitive film (i.e., the photoresist). (b) SEM of a grating pattern formed by developing away exposed resist. (c) Snapshot of an interference lithography system at Purdue's Birck Nanotechnology Center. (d) A 2D grid pattern formed by interference lithography after two consecutive exposures. (e) Snapshot illustrating the extensive, uniform coverage of this technique over 4 cm$^2$.

(Figure 3.2b). If two consecutive exposures are conducted before the photoresist is developed, a grid pattern, such as the one in Figure 3.2d, will appear. This pattern will be the starting point for our angled etching. An exposure setup with a 248 nm wavelength KrF excimer laser inside a class-10 clean room, run jointly by Prof. Bermel and Prof. Qi, is depicted in Figure 3.2e. The iridescent green color evident in the photograph demonstrates the presence of uniform wavelength-scale 2D periodicity.

## Laser-LIGA (Lithography, Electroplating, and Molding)

Laser-LIGA is a technique for forming high-aspect-ratio nanostructures, such as encapsulant materials for protecting PV modules from environmental degradation. It involves the following three steps: lithography, electroforming, and molding (Kathuria, 2004). In the original version of LIGA, the lithography step was performed using synchrotron x-rays for high spatial resolution at the nanoscale and utilized a specially designed x-ray mask, using high-atomic-weight scattering materials, and an x-ray-sensitive resist, such as PMMA. In laser-LIGA, an analogous process is pursued, in which lasers serve an analogous role as x-rays. However, the laser-based process is substantially more widely accessible and much lower cost, due to the relative ubiquity of laser sources such as KrF excimer lasers, as well as the relatively commonality of masks and resists (e.g., SU-8) that are suitable for the UV wavelengths produced by such sources. In either case, the next step consists of electroplating the lithographically formed structure with a metal of interest—common choices include nickel, copper, and gold. The resulting metallized structures can then be used as a mold for mass-producing inverted structures made from common injection molding materials, such as epoxy or polyethylene.

---

## Characterization Techniques

In this section, new and emerging techniques for PV cell and module characterization are discussed: photoluminescence for material identification, infrared shunt detection, QE measurements, and broadband solar simulation.

## Photoluminescence

Photoluminescence (PL) is a useful procedure for rapid in-line materials characterization, particularly for relatively novel PV materials. Key information that can be extracted from PL includes electronic bandgap, minority carrier lifetime, and quasi-Fermi level splitting. In this approach, one injects current into the material of interest and measures the energy spectrum of the resulting photons emitted. The rationale for this approach is that in a reasonably high-quality semiconductor material, the radiative lifetime will at least be within a few orders of magnitude of the nonradiative lifetime. This means that if both electrons and holes are present in the conduction and valence bands, respectively, there will be a nonnegligible probability of radiative recombination. The current injection serves to increase this probability substantially, improving the signal-to-noise and detection rate. The resulting emission spectrum will begin very nearly at the bandgap (neglecting excitons) but extend at least one or two thermal voltages above that energy as well. The line shape of the resulting spectrum will reflect the joint density of states of the system, which can be used to work backward to determine the relative flatness of the conduction and valence bands. If periodic photonic excitations are applied and the resulting re-emitted photons are resolved carefully in time, this can give rise to the related time-resolved PL (TRPL) technique, which directly measures minority carrier lifetime under both high-injection and low-injection conditions.

On a related note, electroluminescence techniques can be used to measure the external fluorescence efficiency (also known as the external radiation efficiency [ERE]). This quantity is significant because it sets an upper bound to the device performance (specifically, the open-circuit voltage).

## Shunt Detection

Shunts in PV cells allow current to bypass its desired path through an extra electrical connection appearing through errors in the PV manufacturing process. Shunts can cause substantial degradation in performance at the cell level. At the module level, the effects can be even worse, since standard series interconnections mean that individual cells with significant shunting tend to degrade the performance of their neighbors as well.

In order to address this issue, one can employ infrared photography in order to both detect and repair these shunts. Effectively, the cells are contacted and reverse biased, in order to force current through the cell. They are then imaged with an appropriate infrared camera to detect localized heating. For operation near room temperature, sensitivity out to 9 μm, near the peak of the thermal emission spectrum, is ideal, although sensitivity to 5 μm may be sufficient. This wavelength sensitivity can be achieved through several routes, including platinum silicide and mercury CdTe-based detectors. In many cases, it is important to record the images immediately after the current is injected (within a few seconds), or the heating can spread throughout the cell, obscuring the source of the problem.

In order to address the problems detected through this technique, several strategies are possible. Especially with thin films, one can potentially use a point-like laser scribe to etch away or remove the shunt region. An alternative is to electrically isolate it, although the exact procedure depends on the level at which the shunt is detected. For shunts detected at the cell level, one can add another electrical contact to reduce the impact of the shunt or at the module level by adding a bypass diode to connect the neighbors of the shunted cell.

## Quantum Efficiency Measurements

A pair of important measurements of the spectral response of a solar cell can be grouped together as QE measurements. The first QE measurement is internal quantum efficiency (IQE), which captures the probability that an absorbed photon is collected as a corresponding unit of electrical current. The second QE measurement, external quantum efficiency (EQE), is the probability that an incident photon from outside will be converted into a corresponding unit of current. It is also equal to the product of the IQE and the absorption probability $A(\lambda)$. Both QE measurements can range from 0% to 100%, but EQE will always be bounded from above by the IQE.

Measuring QE accurately is a nontrivial task. While several strategies have been implemented experimentally, the basic premise of most techniques is as follows. Generally, since IQE cannot be measured directly, one generally measures EQE, transmission, and reflection at each wavelength separately. Then the IQE is obtained as the quotient of the EQE and the inferred absorption. Finding the EQE requires a setup with a few key elements: a calibrated source lamp with significant emission over the range of interest, a monochromator allowing one to select a single wavelength at a time, a receiver calibrated over that same range, and the ability to contact and measure the targeted solar cell. It is also highly recommended to have a computer-based controller to automate this measurement procedure over the full spectrum of interest. Assuming the calibration for both source and receiver is accurate, the electrical current measured can be divided by the current from the reference system and multiplied by the reference EQE, in order to find the EQE for the sample of interest.

Detecting absorption as a means to subsequently determine IQE can take on different levels of complexity. For flat structures, it is generally sufficient to measure specular reflection and transmission only, which for a point-like source can be detected at specific locations easily predicted by geometric optics. Furthermore, most solar cells have negligible transmission due to a metallic back contact and can be assumed to vanish. For textured structures, however, diffuse reflection becomes a key component of loss. For greatest accuracy, one must then integrate the reflection over all angles to obtain all reflected rays. This function is generally performed by an integrating sphere, in which a source and detector are placed in specific small locations within a large sphere and the rest of the interior is coated with a bright white diffuse scatterer. Given proper calibration, this method can accurately determine the full reflection (combining both diffuse and specular components). In any case, once all reflection components are accounted for, the absorption is taken to be unity less the total reflection. The EQE is then divided by the absorption to obtain the IQE measurement.

**Solar Simulation**

While the EQE measured in the previous section can be used to predict the short-circuit current with reasonably good accuracy, the overall power conversion efficiency often cannot be, due to uncertainties in the open-circuit voltage and fill factor. This gives rise to the need for solar simulation testing. In this procedure, one creates a standardized spectrum of light in which to operate a PV cell and then measures the output current as a function of the bias voltage applied to the cell (known as the light I–V curve). The current at zero applied voltage is the short-circuit current $I_{sc}$, and the voltage when the current vanishes is the open-circuit voltage $V_{oc}$. Since output power is the product of the current and voltage, the maximum power output can also be easily obtained from this test. The quotient of the maximum power output with respect to the product of $I_{sc}$ and $V_{oc}$ is the fill factor.

Clearly, the standard spectrum of light must be well defined in terms of relative wavelength intensities, as well as overall intensity, for this procedure to make sense. Great variability with location, weather, time of year, and time of day means that careful studies are needed to establish standards consistent with the real operating conditions of solar cells. ASTM-G173 established the two most well-known and accepted standards: AM0 (air mass zero) for outer space, and AM1.5 (air mass one and a half) for terrestrial applications. The corresponding powers for each case are 1.366 and 1 kW/m². The AM1.5 spectrum has less light, particularly in the blue portion of the spectrum, due to Rayleigh scattering, as well as some atmospheric absorption in certain wavelength bands (e.g., from water vapor). In order to reproduce the AM1.5 spectrum in a laboratory, one generally needs a broadband high-thermal-temperature source (to mimic the sun), such as a xenon arc lamp. The output light must then be spectrally filtered with a series of filters to closely match the target spectrum.

# Conclusion

In conclusion, we demonstrated a broad range of applications for photonic and laser technologies toward PV manufacturing. For crystalline silicon manufacturing, it was shown that laser-based wafer dicing offers improved materials usage, laser-based doping provides an alternative to high-temperature diffusion suitable for reducing thermal budgets, laser-fired and grooved contacts suppress surface recombination, and EWT and MWT technologies offer lowered shadowing and series resistance. It was also shown that thin-film manufacturing can benefit from laser scribing for series interconnection, laser edge isolation and deletion for reduced surface recombination, and DLC for improved minority carrier lifetimes and mobilities, as well as two related lithography techniques, interference lithography and laser-LIGA. Finally, four optical characterization techniques for quality assurance in PV manufacturing were discussed: photoluminescence for material identification, infrared shunt detection, QE measurements, and solar simulation.

## Acknowledgments

The authors thank Ashraf Alam, Ryyan Khan, Mark Lundstrom, Minghao Qi, and Ali Shakouri for useful discussions. Support was provided by the Department of Energy, under DOE Cooperative Agreement No. DE-EE0004946 ("PVMI Bay Area PV Consortium"), as well as the Semiconductor Research Corporation, under Research Task No. 2110.006 ("Network for Photovoltaic Technologies.")

## References

Bartlome, R., Strahm, B., Sinquin, Y., Feltrin, A., and C. Ballif, Laser applications in thin-film photovoltaics, *Appl. Phys. B* 100, 427–436 (2010).

Cai, W.-S. and V. Shalaev, Optical Metamaterials: *Fundamentals and Applications.* Springer, New York, (2009).

Cheng, G.J., M.Y. Yang, and Y. Zhang, High speed laser crystallization of nanoparticle inks for thin film solar cells, U.S. Patent application number US2012-0021559.

Cheng, G.J. and M.Y. Zhang, Laser crystallization of thin film on various substrates, U.S. Patent application number US 2013-0075377.

Clement, F. M. Menkoe, D. Erath, T. Kubera, R. Hoenig, W. Kwapil, W. Wolke, D. Biro, R. Preu, High throughput via-metallization technique for multi-crystalline metal wrap through (MWT) silicon solar cells exceeding 16% efficiency, *Sol. Energy Mater. Sol. Cells* 94, 51–56 (2010).

Compaan, A.D., I. Matulionis, and S. Nakade, Optimization of laser scribing for thin-film PV modules, Final Technical Progress Report, NREL/SR-520-24842, University of Toledo, Toledo, OH (available from NTIS, 1998).

Gee, J.M., W.K. Schubert, and P.A. Basore, Emitter wrap-through solar cell, *23rd IEEE Photovoltaic Specialists Conference*, Louisville, KY, pp. 265–270 (1993).

Hecht, E., *Optics*, 4th edn. (Addison-Wesley, Reading, MA, 2001).

Hermann, J.M. Benfarah, S. Bruneau, E. Axente, G. Coustillier, T. Itina, J.-F. Guillemoles, and P. Alloncle, Comparative investigation of solar cell thin film processing using nanosecond and femtosecond lasers, *J. Phys. D: Appl. Phys.* 39, 453–460 (2006).

Huber, H.P., F. Herrnberger, S. Kery, and S. Zoppel, Selective structuring of thin-film solar cells by ultrafast laser ablation, *SPIE* 6881, Commercial and Biomedical Applications of Ultrafast Lasers VIII, 17 (2008).

Joannopoulos, J.D., J.N. Winn, R.B. Meade, and S.G. Johnson, *Photonic Crystals: Molding the Flow of Light* (Princeton University Press, Princeton, NJ, 2007).

Kathuria, Y.P., L3: Laser, LIGA, and lithography in microstructuring, *J. Indian Inst. Sci.* 84, 77–87 (2004).

Shin, Y.C., G.J. Cheng, W. Hu, M.Y. Zhang, and S. Lee, High Precision Scribing of Thin Film Solar Cells by a Picosecond Laser, *Proceedings of NSF Engineering Research and Innovation Conference*, Atlanta, GA (2011).

Streetman, B.G. and S. Banerjee, *Solid State Electronic Devices* (Prentice Hall, Englewood Cliffs, NJ, 2000).

van Kerschaver, E., R. Einhaus, J. Szlufcik, I. Nijs, and R. Mertens, A. A novel silicon solar cell structure with both external polarity contacts on the back surface, *Proceedings of the Second World Conference on Photovoltaic Energy Conversion*, Vienna, Austria, Vol. 2, pp. 1479–1482 (1998).

van Kerschaver, E., S. DeWolf, and J. Szlufcik, Toward back contact silicon solar cells with screen-printed metallization, *28th IEEE Photovoltaic Specialists Conference*, Anchorage, AK, pp. 209–212 (2000).

Wiedeman, S., R.G. Wendt, and J.S. Britt, Module interconnects on flexible substrates *AIP Conference Proceedings*, Vol. 462, pp. 17–22 (1999).

Wolden, C.A., J. Kurtin, J.B. Baxter, I. Repins, S.E. Shaheen, J.T. Torvik, A.A. Rockett, V.M. Fthenakis, and E.S. Aydil. Photovoltaic manufacturing: Present status, future prospects, and research needs, *J. Vac. Sci. Technol. A* 29, 030801 (2011).

Zhang, M.Y. and G.J. Cheng, Highly conductive and transparent alumina-doped ZnO films processed by direct pulsed laser recrystallization at room temperature, *Appl. Phys. Lett.* 99, 051904 (2011).

Zhang, M.Y., Q. Nian, and G.J. Cheng, Room temperature deposition of alumina-doped zinc oxide (AZO) on flexible substrates by direct pulsed laser recrystallization, *Appl. Phys. Lett.* 100, 151902 (2012).

# 4

## Optics in Healthcare: A Systems Perspective

Peter Lorraine

### CONTENTS

### Introduction

Healthcare optics sits at the nexus of multiple disciplines including biology, chemistry, medicine, and engineering. Although a long partner with medicine and biology, the field is undergoing rapid transformations today as interdisciplinary communities come together to solve problems and establish new platforms that others can build on.

### Healthcare Optics

Optics has a long and valued role in biology and healthcare—the invention of the microscope-transformed human understanding of the nature of life and disease. Any attendee at a recent national meeting will attest that there is lots of amazing research going on. The ability to differentiate tissues from their spectral response, the importance of direct observation as the core of medicine, and the ability of light to interrogate and penetrate tissue underlie this activity.

Healthcare optics is entering a new era where interdisciplinary technology and a systems view are becoming critical and enabling new applications and markets.

From a current business perspective, laser and photonic systems in healthcare look different than the primary markets in diagnostic imaging. Diagnostic imaging is dominated today by four major modalities: x-ray, ultrasound, computed tomography, magnetic resonance imaging—each with an annual market over $3.5B and nuclear imaging at $1.5B.[1] These large modalities have largely matured and have a combined annual growth rate of roughly 5%. The biomedical optics space, however, aggregates to over $2B but consists of many smaller markets and unique applications of technology. Some of these new opportunities are growing at over 30% and may emerge as significant markets in time. The breadth of healthcare optics applications is large and changing rapidly.

One of the larger biomedical optics markets that exists today is optical microscopy at roughly $750 M/year. The primary medical application is in histopathology where modern technology for acquisition and analysis of images is just beginning to displace a well-established workflow. The application of advanced system automation, molecular pathology detection of proteomic and genomic signatures, and digital analysis is ready to transform the conventional pathology lab. In addition, research microscopy is benefitting from new imaging approaches including computational, light-field, and adaptive optics. Superresolution approaches make possible imaging of structures well below the diffraction limit of light, and the growth of new approaches for spectroscopy is enabling new applications.

Biomedical optics is justifiably an area of intense research interest today and markets are evolving as new technology is introduced. In this chapter, we will look at specific examples of the systems involved in advanced microscopy and patient imaging and the new applications emerging. Specifically, we will discuss the impact of automation, new biochemical detection approaches, and image and statistical analysis. We will conclude with a look at several emerging techniques.

## Healthcare Optics Systems Examples

### Digital Pathology Platform

Histopathology—the study of the cellular organization of tissue samples—is a critical part of hospital workflow and is used to evaluate biopsy samples for diseases like cancer, to evaluate unknown samples during surgery, and to examine excised masses. Until recently, the overall workflow has been largely unchanged over the last 100 years albeit with more sophisticated microscopes and handling equipment. The typical workflow has been to take a tissue sample and stabilize it through a process called *fixation* by immersing for 24 h in formalin that prevents postmortem decay and typically increases the mechanical stability. The samples are then sliced into 3 mm tissue blocks and slowly perfused with paraffin. Slices are obtained using a microtome, mounted to slides, and stained to create image contrast. Hematoxylin and eosin (H&E) are commonly used to stain nuclei dark purple and cytoplasm pink. A pathologist then examines the glass slides under a microscope. The slides are retained.

Recently, digital pathology systems[2] have entered the market—the GE Omnyx system is shown in Figure 4.1. A digital pathology system creates a high-quality digital image of the slide that is examined on a pathology workstation that can display the image, handle case management, and provide image analysis tools. The use of digital images has obvious advantages including the ability to share images, viewing and analysis at remote locations,

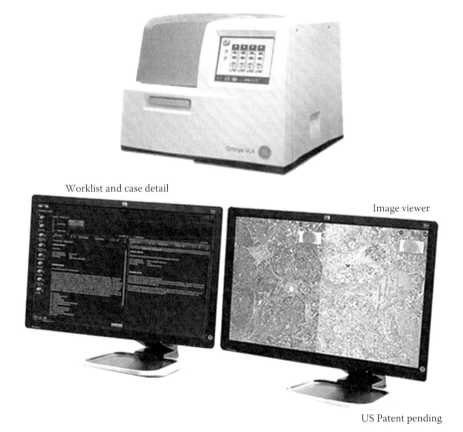

Worklist and case detail

Image viewer

US Patent pending

**FIGURE 4.1**
GE Omnyx—a modern digital pathology platform. The system consists of an automated slide scanner with image stitching, autofocus, and 30 s/slide throughput and a pathology workstation for case management, image viewing, and analysis tools.

workflow streamlining, and the ability to apply quantitative tools for image analysis. Conventional pathology imaging produces exceptional images with consistent focus, color, and ease of use. The combination of fine focus adjustments and the inherent accommodation of the eye let the pathologist stay in focus anywhere and pan and zoom smoothly.

In this section, we will look at the technical challenges required to provide accurate focusing for whole slide imaging that arise from the limited depth of field of high numerical aperture objectives and the inherent tissue flatness variability. The requirement to achieve high slide throughput—critical to economic justification—and high image quality required new technology.

The 15 mm × 15 mm tissue area on a slide is much larger that the roughly 1 mm diameter field of view of a 20× objective. Imaging the whole slide requires acquisition of multiple images that are aligned and combined to create a complete view of the sample that can be panned and zoomed digitally in a way similar to that experienced with a direct optical microscope. Beyond macroscopic deviations such as stage or slide tilt, tissue variations are of the order of few microns over a single 1 mm diameter view. The autofocus problem is to automatically compensate for this as a pathologist would by adjusting the fine focus and doing so rapidly. Overall image quality is a complex function of sample preparation, illumination uniformity and level, camera/sensor quality and specifications, and the quality of the focus. Achieving a high-quality focus quickly is the most difficult of these challenges.

The image quality and resolving power needed by pathologists require high numerical aperture objectives. The high numerical aperture creates a short depth of focus—the range over which an acceptable image is formed. The tight depth of focus makes compensation of tissue topography variations important. The use of lower numerical objectives is not generally acceptable. In practice, focal positioning in the $z$-direction with 0.5 μm of the ideal is needed to avoid noticeable artifacts.

Autofocus can be performed by either direct measurement of the distance between the sample and the objective, by calculation of the distance from an image or other sensor signal, or direct measure of focal performance from an image. Reflection-based autofocus typically reflects a laser or other beacon signal from the supporting slide. This is turned into a measurement of standoff by a method such as triangulation. Typically, this does not work well for tissue samples that are not at a constant distance from the reference glass surface but can compensate for macroscopic errors such as stage tilt.

Image-based autofocusing derives a control signal from the tissue image directly rather than from the surface of the glass slide. Typically, a stack of images is acquired at several $z$-positions, and a calculation is performed to determine the optimum position. A figure of merit is derived from the image that indicates the quality of focus. Image-based autofocus can produce very-high-quality focused images of the surface of the tissue, but the time to acquire the image stack and develop the control signal has been too long in the past.

The selection of a figure of merit used in the autofocus algorithm is very important. Yazdanfar[3] has evaluated several figures of merit and the performance in pathology imaging. The simplest figure of merit is the contrast defined as

$$C = \frac{s_{max} - s_{min}}{s_{max} + s_{min}} \tag{4.1}$$

where $s_{max}$ and $s_{min}$ are the grayscale maximum and minimum pixel values. Image variance is defined in the following equation:

$$V = \frac{1}{\mu} \sum_{i=1}^{N} \sum_{j=1}^{M} \left[ s(i,j) - \mu \right]^2 \tag{4.2}$$

where
s(i, j) is the grayscale value at coordinates (i, j)
N and M are the image dimensions in the i and j directions

A third figure of merit is image entropy given by the following equation:

$$E = -\sum_{k}^{L} p_k \log(p_k) \tag{4.3}$$

where $p_k$ is the probability of a grayscale value falling in the $k$th histogram bin. The fourth figure of merit we will look at is the Brenner gradient in the following equation:

$$B = \sum_{i=1}^{N} \sum_{j=1}^{M} \left[ s(i,j) - s(i+m,j) \right]^2 \tag{4.4}$$

where $m$ is an offset distance. The Brenner gradient is a simple edge detector that when $m=2$ measures the difference between a pixel and its next-nearest neighbor. Figure 4.2 shows a comparison of these four figures of merit for a set of images through the focal plane. The Brenner gradient is the most sensitive to focus with a narrow peak and good shape. Yazdanfar[3] has experimentally observed that the Brenner gradient can be approximated well by a Lorentzian

$$f(z) = \frac{\alpha}{\beta + (z - z_0)^2} \tag{4.5}$$

and that as the reciprocal of this is a quadratic, three images are adequate to determine the focal plane. Figure 4.3 shows the experimental verification of this fit and shows the accuracy of the focal position estimation. Figure 4.4

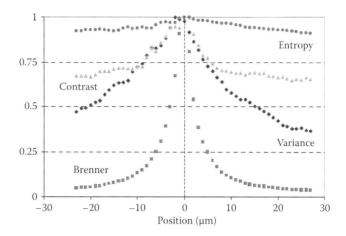

**FIGURE 4.2**
Comparison of focal figures of merit[3]—the Brenner gradient is the most sensitive to focus and is symmetric about the peak.

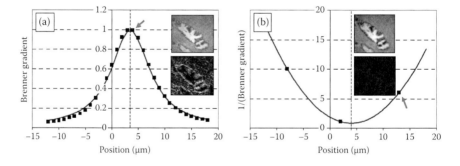

**FIGURE 4.3**
Performance of autofocus using Brenner gradient.[3] (a) Lorentzian fit to Brenner gradient shows sample is 3.53 µm from optical system-designed focus. (b) Reciprocal of the Brenner gradient used to calculate focal position using a parabolic fit and three images. Inset grayscale images are obtained at position indicated by arrow—Brenner gradient is sum of image data in black and white images.

shows the application of this algorithm to a real tissue sample that is 10 µm out of focus. The algorithm develops an accurate estimation of focal position that is stable to 0.02 µm when applied multiple times.

Consider the ideal case where image-based autofocus is performed at each imaged field of view or tile in the slide. This involves moving to the correct location, settling out vibration, and acquiring a stack of five images, for example. For a high-quality sensor with a frame acquisition time of 100 ms, this will require more than 0.5 s/location for a surveying focus at each tile

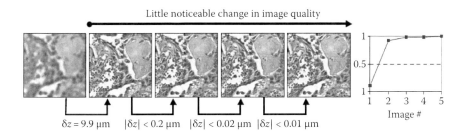

**FIGURE 4.4**
Rapid convergence and stability of autofocus algorithm.[3] Initial image is ~10 µm out of focus—algorithm is applied four times with little residual motion of $z$-axis.

location. Acquisition time varies with camera sensor area, illumination levels, and frame readout times for high pixel count sensors. As it takes 500 tiles to cover the 15 mm × 15 mm area, this translates to as much as 300 s when motion overhead is included. While high-quality images are possible, getting acquisition times below 60 s is challenging.

To reduce the time burden, instead of focusing on every tile, approaches have been adopted to create a sparse map of focal positions prior to scanning. Survey points are chosen, or a spacing of multiple tiles is chosen to approximate the focal solution for the slide to be scanned. A continuous focal map can be generated by interpolation from these sparse samples, but the underlying topography may deviate. Speed can also be gained by skipping 3–4 tiles before computing a new focus. The underlying assumption is that neighboring tiles have approximately the same focal offset. Studies have shown that variations of more than 1 µm are common from tile to tile—a value that produces noticeable image degradation. The operator may be allowed to select locations of where the topography is changing or image quality is desired to generate another type of focal map. This step is not acceptable for high-volume automated scanning.

An image sensor typically has time associated with readout of the data where no image is being acquired. In a conventional, single camera system, the same sensor is used for both autofocus images at multiple $z$-positions, as well as acquisition of the focused image with significant impact on the overall imaging speed. GE has developed a dual sensor system that takes advantage of the readout time of the camera. In this system, one sensor is used to acquire the high-resolution image. The second sensor creates a focus control using the time when the high-resolution image is being read out. To make this work, an optimal approach is required to minimize the number of focal $z$-positions required to estimate the focus correction. The Brenner gradient figure of merit performs well with only three $z$-images. The translation stages can be kept moving continuously by using a pulsed light source to eliminate motion blur. The acquisition sequence consists of acquisition of a high-resolution image in focus and during readout of that image acquisition of three $z$-position images with the focus sensor and generation of a focus correction. As the slide translation stage is in motion at all times, the three autofocus $z$-position images are acquired at slightly different locations. The overlap region of all three images is used to develop the focal control signal. This approach is called *independent dual sensor* (IDS).

The advantages of IDS are clear. Each image tile is acquired in focus, and the overall image quality is higher than for focal mapping and approximation approaches described earlier. This results in fewer focus faults. A study compared the reference approach of stopping and generating optimal focus settings at each scan position using 50 $z$-position estimation images with the IDS approach. Both approaches found the same optimal focal position and produced comparable images. IDS has a major advantage in speed of acquisition as it eliminates focus image acquisition overhead and motion overhead associated with stopping and starting and allowing vibration to settle. The overhead reduction translates into significant savings when scanning a 15 mm × 15 mm image that has 600 separate image tiles at 20× and 2400 image tiles at 40×. The system is fully automatic and does not require pathologist intervention to select focal positions or map tissue areas in the slide.

An unofficial industry standard for *scan speed* is the time required to scan a 15 mm × 15 mm area at 20× magnification. The time between loading a slide and producing an image for pathologist viewing is the time-to-first-image and includes postprocessing to adjust image contrast and color. Time-to-first-image is generally not critical except in the case of fresh frozen pathology sections where a surgeon is waiting on the outcome. Scan speed is the average time per slide for a set of slides to be imaged—a scan speed of 60 s would suggest that the time to process 10 slides would be around 10 min. The overall throughput of the system is a key economic metric.

The economic advantages of digital pathology are driven by productivity arguments offset by the capital acquisition costs of the pathology scanner. This consideration makes the difference between 90 and 60 s/slide significant. A laboratory planning a throughput of 100 slides/day (~500,000 per year) requires two scanners if the throughput is 60 s/slide. At 90 s/slide, an additional scanner for a total of three is required with four required at 120 s/slide. A scanner that produces exceptional images without the necessary throughput will not be justifiable.

We have seen that accurate and rapid focusing is critical to high-quality whole slide imaging because of the dual factors of tissue surface topography and the limited field of view of high numerical aperture objectives. Conventional line scanning and tiled scan systems are challenged to keep the sample within the depth of focus of the system at a high throughput. We have developed a new approach that uses two sensors—one used for continuous focusing and the other for image acquisition—that enables consistent high-quality focusing at a high throughput.

The balancing of diverse systems is important for digital pathology. The end user is less interested in specific technical aspects of the hardware than in the overall application performance where knowledge of the pathology application requirements, the pathologist work environment, and economic arguments is necessary to displace existing processes. As we will see in later sections, once meeting the requirements to replace the traditional process, new applications can be enabled.

## Application of Molecular Biology and Digital Imaging to Colon Cancer

A key technology that has been added to pathology imaging in recent years is molecular imaging—the ability to see key proteins both on and inside cells that can give insights into disease. The advent of digital imaging technology enables the combination of molecular sensitivity with quantitation and more sophisticated analysis. In this section, we will describe an approach that reveals additional information about prostate cancer and identifies various patient groups with more information than conventional histopathology.

Conventional pathology imaging uses hematoxylin to stain nuclei purple and eosin to stain cytoplasm pink. A pathologist will examine these or similar images and do a semiquantitative analysis of a tumor section and assign a *severity* class to the cancer (0, +1, +2, +3). As the treatment course and prognosis for different classes may differ, this assignment is important.

The advantages of a quantitative molecular analysis can be seen in the study by Ginty.[4] Hepatocyte growth factor is a molecule that regulates important cell activities including cell growth, motility (movement), and morphogenesis. It is known to play an important role in several tumor processes and progressions including growth of new blood vessels (angiogenesis) and matrix invasion of surrounding tissues.[5] The receptor for this growth factor is called MET and is a proto-oncogene—a gene that through mutation can help start cancers. An unusual expression of MET is known in many cancers, and overexpression is known to be associated with reduced survival time for a variety of cancers including ovarian, cervical, and esophageal.[6–11] In colon cancer, studies have not shown an association of overexpression with outcome that might be expected. Ginty reports on a new study that looks at the specific spatial distribution of MET within colon cancer cells reasoning that while the total amount might not be predictive, the distribution might be significant.

The study used a variety of stains to help identify different regions in the cell as shown in Figure 4.5. Pan-cadherin stains membrane red, while 4′,6-diamidino-2-phenylindole (DAPI) shows nuclei as blue. MET in tissue is detected with an antibody to MET that is attached to a green fluorescent dye. The images are processed and segmented to identify membrane and cytoplasm regions, nuclei, and unconnected membrane regions. A statistical analysis was performed on a variety of colon cancer samples to identify

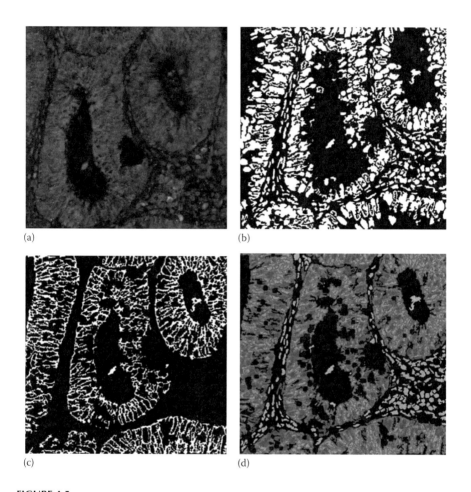

(a)

(b)

(c)

(d)

**FIGURE 4.5**
**(See color insert.)** Colon cancer tissue images stained to show distribution of MET protein and segmentation.[4] (a) Combined image with red (membrane), blue (nuclei), and green (MET protein); (b) probability map for nuclei; (c) probability map for membrane; and (d) compartments with red (membrane), blue (epithelial nuclei), gray (stromal nuclei), and green (cytoplasm). Pink regions are excluded from quantification and black is background and extracellular matrix.

MET levels in various parts of the cell and to identify any associations between the levels and distribution and cancer stage and severity.

The results are shown in Figure 4.6 and are revealing. The graphs are grouped by cancer stage scores (1, 2, 3) and show survival probability versus time. Stage 3, as expected, has a worse survival probability with time. Ginty[4] identifies a parameter based on the amount of MET in cytoplasm to membrane levels (MET membrane to cytoplasm parameter) as statistically significant for patient survival. In Figure 4.6b, the solid curve represents overall survival

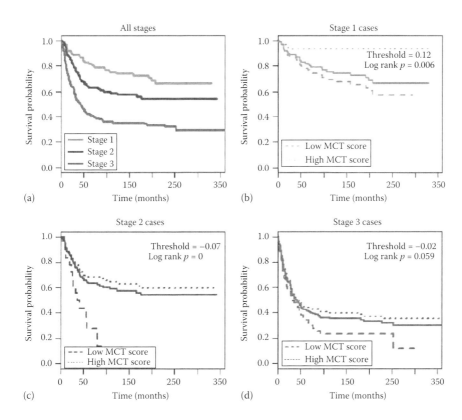

**FIGURE 4.6**
Survival curves with death due to disease[4] for (a) Stage I, II, and III patients and (b) Stage I, (c) Stage II, and (d) Stage III patients subdivided into high- and low-risk populations based on MET MCT scores.

probability. When separated by MET MCT scores, two distinct populations are visible. The effect is even more pronounced with Stage 2 cases.

This molecular test, combined with automated image processing and segmentation, can identify different populations within each cancer stage. A patient with a low MET MCT score is expected to have a lower survival probability—information that might suggest different treatment plans. Further, understanding of the distribution of MET might suggest mechanisms for tumor progression and therapy development.

This is a single example illustrative of the important role that molecular detection can have that is made possible by digital acquisition systems and data analysis packages. In this example, the quantitative analysis of the spatial distribution of this growth factor was key. Ultimately, this is leading toward improved diagnosis and treatment not possible in the past.

## Digital Stain-Level Adjustments

The ability to work with digital images enables new types of processing impossible with the traditional direct optical microscope. In conventional histopathology, tissue samples are prepared, stained to increase contrast of structures of interest, and directly imaged. In the conventional workflow, all decisions are final—too much stain or the wrong balance of several stains once selected cannot be undone. In this section, we will look at the work by Bilgin[12] that demonstrates a new freedom gained once images are digital and contain the right information. This work demonstrates computationally modifying the levels of signal derived from specific stains without reprocessing tissue samples.

H&E staining is relatively common and familiar to pathologists. The relative ease in preparation and low cost and the widespread use in both training and practice make this a good example. Hematoxylin has a blue color that is enhanced by mixing with metal salts—the specific salt and concentration affect the color. Hematoxylin stains chromatin blue purple and reveals ribosomes and nuclei. Eosin stains cytoplasm, red blood cells, collagen, and muscle fibers pink and is used in combination with hematoxylin. Other interactions give rise to areas with subtle yellow and brown coloration that is significant.

There is a lot of variability in how H&E staining is implemented and the results obtained. Different processes for staining, specific manufacturer formulation, and sample preparation all impact the relative colors and contrast. Part of this is due to pathologist preference—some prefer images with sharp nuclear contrast and a light background, while others are more comfortable with a dark background. Some of the variability is due to undesired process or chemistry deviations and results in undesired image characteristics.

Adjusting separately the red green blue (RGB) channels from the camera sensor can change image appearance, but what is sought here is the ability to adjust the image as if the amount of hematoxylin or eosin in the original slide had been changed while maintaining the appropriate color significance of the image. Bilgin shows an example of how this can be accomplished.

The RGB image is assumed to be given by the following equation:

$$\begin{bmatrix} R \\ G \\ B \end{bmatrix} = \begin{bmatrix} M_R \\ M_G \\ M_B \end{bmatrix} - \begin{bmatrix} w_{11} & w_{12} \\ w_{21} & w_{22} \\ w_{31} & w_{32} \end{bmatrix} \begin{bmatrix} E \\ H \end{bmatrix} + \begin{bmatrix} d_R \\ d_G \\ d_B \end{bmatrix} \tag{4.6}$$

where

$[R, G, B]^T$ represents the RGB channel intensities
$[M_R, M_G, M_B]^T$ is the absorption-free intensity (in the absence of stain)
$[d_R, d_G, d_B]^T$ is the residual due to dark current

The amount of eosin ($E$) and ($H$) is solved for. This is rewritten as

$$\begin{bmatrix} R' \\ G' \\ B' \end{bmatrix} = \begin{bmatrix} w_{11} & w_{12} \\ w_{21} & w_{22} \\ w_{31} & w_{32} \end{bmatrix} \begin{bmatrix} E \\ H \end{bmatrix} - \begin{bmatrix} d_R \\ d_G \\ d_B \end{bmatrix} \tag{4.7}$$

where

$$\begin{bmatrix} R' \\ G' \\ B' \end{bmatrix} = \begin{bmatrix} M_R & -R \\ M_G & -G \\ M_B & -B \end{bmatrix}$$

Equation 4.7 can be written as

$$A = WS - D \tag{4.8}$$

$S$ is the $[E, H]^{\mathsf{T}}$ levels that are being solved for. This is solved for iteratively in a process where an initial estimate for $S$ is made, and $W$ and $D$ are solved for. Then, using these values for $W$ and $D$, $S$ is solved for. This is called a non-negative matrix factoring because of the constraint that none of the values in Equation 4.8 are negative—image channel intensities are always positive and eosin and hematoxylin reduce signals. The overall process is shown in Figure 4.7.

Figure 4.8 shows the output of the factoring where an initial H&E image acquired on an RGB sensor (shown in gray scale) is on the top row, row 2 is the eosin component image, and row 3 is the hematoxylin image component. Note how the nuclei in the tissue images are visible in row 3 but not in row 2. Once separated, scaling each in intensity and then applying the forward transformation given in Equation 4.6 can recombine the two images. Figure 4.9 shows the impact of digitally adjusting the relative eosin and hematoxylin levels in a typical tissue slice containing a fold of tissue. A pathologist can tune the system according to their preferences for stain levels without distorting the information color encodes.

There are many similar applications that are possible once data are digital. Fluorescent microscopy uses stains that attach to specific proteins or cell components and fluoresce rather than simply attenuate like H&E do. Most tissue has a weak background fluorescence called *autofluorescence* that degrades the image and makes quantitative measurements at low levels very difficult. Pang[13] has shown that the autofluorescent background can be solved for and removed robustly with two images acquired before and after staining.

## Image-Guided Surgery

The previous examples have shown the benefits to pathology tissue imaging of using digital microscopy, biochemistry to stain tissues of interest, and

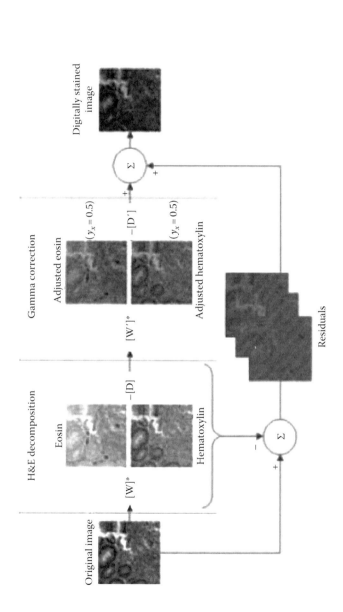

**FIGURE 4.7**
RGB H&E-stained image unmixed into components using nonnegative matrix factorization.[12] Once separated, components can be balanced and recombined.

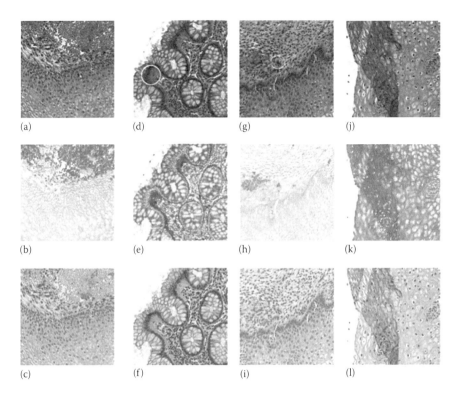

**FIGURE 4.8**

H&E decomposition of tissues.[12] Columns 1, 3, and 4 are benign cervix tissues, and column 2 is colon polyp with high dysplasia. Row 1 is original image, row 2 is eosin component, and row 3 is hematoxylin component showing nuclei.

computer processing to extract and enhance information. We will now look at a different way of combining these three technologies to provide a direct benefit in surgery.

During surgery, contrast across tissue types can be poor. The goal of fluorescence image-guided surgery (FIGS) is to employ a fluorescent agent that attaches to tissues of interest and to visualize those tissues during surgery. Applications have mapping sentinel lymph nodes with tumors, coronary artery graft assessment, and tumor margin visualization.[14–16]

Wang[17] describes a compact instrument developed for surgical use. The type of device that is employed for this work is shown in Figure 4.10. The system has a fiber for delivery of both the light to excite the fluorescent agent and the white light used for general illumination. Dichroic elements are used to separate the fluorescent return light from the background image, and optical elements are selected and aligned to provide image registration. For this system with methylene blue as the agent, a 500 mW 660 nm laser diode was used for excitation.

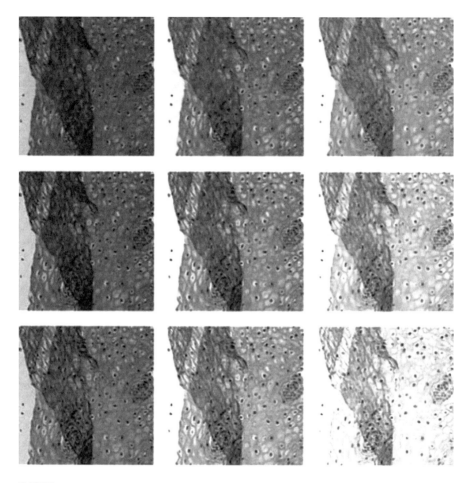

**FIGURE 4.9**
**(See color insert.)** Adjusted H&E images of cervix tissue with folded region.[12] Left to right corresponds to increasing and decreasing hematoxylin levels and top to bottom to decreasing eosin levels. A pathologist may digitally adjust stain levels to better understand a tissue without processing a new sample.

Wang used this equipment to visualize the condition of rat heart tissue intraoperatively. Methylene blue circulates in the vasculature and is taken up by health tissue but not by scar tissue. Dynamic visualization of the methylene blue fluorescence following injection can be used to directly visualize the scarred area. Figure 4.11 shows the images as presented to the surgeon where the top images correspond to immediately after injection and the bottom ones to 60 s later. The circle indicates ischemic (damaged) regions of the heart. Note that the actual fluorescence of the tissue is in the near-infrared and it is visualized as green in the merged view.

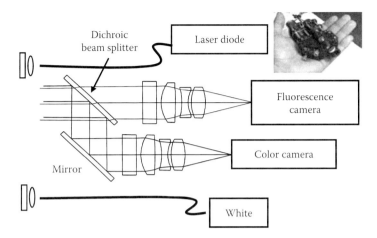

**FIGURE 4.10**

A compact FIGS system[17] showing fluorescent and white light imaging paths. Inset shows the experimental prototype.

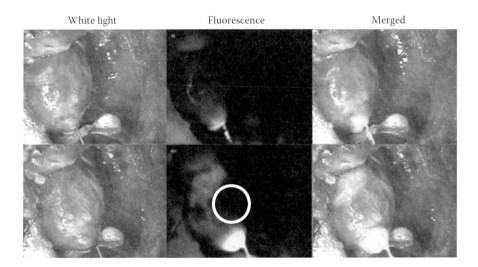

**FIGURE 4.11**

**(See color insert.)** Preclinical FIGS[17] for vasculature mapping in rat heart at time of injection (top) and 60 s after injection (bottom). Fluorescent regions show circulating methylene blue and vasculature, while ischemic regions (circle) remain dark.

A future application of this technology might be for cardiovascular stem cell therapy where stem cells or cardiomyocyte progenitor cells might be implanted to regenerate heart function. Visualization of the damaged tissue regions is critical for this type of therapy.

## Conclusions

The examples we have looked at show the interplay between systems and technology. A solution requires technology from optics, biology, chemistry, electrical engineering, and computer science. End users in this space tend to be focused on applications—solving a specific problem—and rarely demand a specific new technology. Biomedical optics is a complex field where engineers and physical scientists need to work hand in hand with biologists or healthcare professionals to understand their problems. A common microscopy problem in drug screening is identifying and counting the cells in clusters in a growth medium. The problem involves compensating for optical aberrations across an extended depth of field and developing sophisticated analysis algorithms—yet the high-level problem is what the customer demands a solution for without bias as to how it is achieved. The performance of the complete system is what matters.

In the example of digital pathology imaging, the simple concept of creating a digital image that matches the quality of a histopathology microscope direct image has long-reaching consequences. Once introduced as a labor or cost-saving tool or as a means of examining samples from a distance, opportunities for training emerge. Once the data are in a digital format, more sophisticated analysis impossible for direct human evaluation is possible—counting the number of cells and performing segmentation across an entire sample or processing to correct stain levels become possible. Digital pathology becomes a platform that whole new applications can be built on and that in turn demands new technology.

Like many areas of modern technology, the application of high-power computation is transforming healthcare optics and will continue to do so in the coming years. Direct spectroscopy can tell us much about biology but cannot match the ability of molecular imaging to see specific proteins—both areas will continue to grow in importance.

It is not possible to say whether truly transformational technologies will emerge at some point. Scatter remains a barrier for deeper examination at high resolution. The ultimate desire to image the workings of living cells deep inside functioning tissue inside patients or animals and to see molecular mechanisms at work at high resolution remains distant. New technology that surmounts this problem would be very important, but new approaches that are tolerant of scatter yet produce valuable information may be equally transformational and yield large markets.

## Acknowledgments

The author would like to acknowledge the contributions of numerous colleagues including Siavash Yazdanfar, Robert Filkins, Mike Montalto, Fiona Ginty, Jens Richter, and Xinghua Wang.

## References

1. Diagnostic Imaging Market (CT, MRI, X-Ray & Ultrasound—Competitive Landscape & Global Forecasts 2010–2016, markets@markets.com, http://www.marketsandmarkets.com/Market-Reports/diagnostic-imaging-market-411.html (July 2011).
2. M. C. Montalto, R. R. McKay, and R. J. Filkins, Autofocus methods of whole slide imaging systems and the introduction of a second-generation independent dual sensor scanning method, *J. Pathol. Inform.* **2**, 44 (2011).
3. S. Yazdanfar, K. B. Kenny, K. Tasimi, A. D. Corwin, E. L. Dixon, and R. J. Filkins, Simple and robust image-based autofocusing for digital microscopy, *Opt. Express* **16**, 8670–8677 (2008).
4. F. Ginty, S. Adak, A. Can, M. Gerdes, M. Larsen, H. Cline, R. Filkins, Z. Pang, Q. Li, and M. C. Montalto, The relative distribution of membranous and cytoplasmic met is a prognostic indicator in stage I and II colon cancer, *Clin. Cancer Res.* **14**, 3814–3822 (2008).
5. C. F. Gao and G. F.VandeWoude, HGF/SF-Met signaling in tumor progression, *Cell Res.* **15**, 49–51 (2005).
6. R. L. Camp, E. B. Rimm, and D. L. Rimm, Met expression is associated with poor outcome in patients with axillary lymph node negative breast carcinoma, *Cancer* **86**, 2259–2265 (1999).
7. C. N. Qian, X. Guo, B. Cao et al., Met protein expression level correlates with survival in patients with late stage nasopharyngeal carcinoma, *Cancer Res.* **62**, 589–596 (2002).
8. M. R. Anderson, R. Harrison, P. A. Atherfold et al., Met receptor signaling: A key effector in esophageal adenocarcinoma. *Clin. Cancer Res.* **12**, 5936–5943 (2006).
9. Y. Miyata, H. Kanetake, and S. Kanda, Presence of phosphorylated hepatocyte growth factor receptor/c-Met is associated with tumor progression and survival in patients with conventional renal cell carcinoma, *Clin. Cancer Res.* **12**, 4876–4881 (2006).
10. J. Y. Kang, M. Dolled-Filhart, I. T. Ocal et al., Tissue microarray analysis of hepatocyte growth factor/Met pathway components reveals a role for Met, matriptase, and hepatocyte growth factor activator inhibitor 1 in the progression of node-negative breast cancer, *Cancer Res.* **63**, 1101–1105 (2003).
11. D. L. Rimm, Tissue microarray-based studies of patients with lymph node negative breast carcinoma show that met expression is associated with worse outcome but is not correlated with epidermal growth factor family receptors, *Cancer* **97**, 1841–1848 (2003).

12. C. C. Bilgin, J. Rittscher, R. Filkins, and A. Can, Digitally adjusting chromogenic dye proportions in brightfield microscopy images, *J. Microsc.* **245**, 319–330 (2012).
13. Z. Pang, N. E. Laplante, and R. J. Filkins, Dark pixel intensity determination and its applications in normalizing different exposure time and autofluorescence removal, *J. Microsc.* **246**, 1–10 (2012).
14. W. Stummer, U. Pichlmeier, T. Meinel, O. D. Wiestler, F. Zanella, and H.-J. Reulen, Fluorescence-guided surgery with 5-aminolevulinic acid for resection of malignant glioma: A randomised controlled multicentre phase III trial, *Lancet Oncol.* **7**, 392–401 (2006).
15. E. M. Sevick-Muraca, R. Sharma, J. C. Rasmussen et al., Imaging of lymph flow in breast cancer patients with microdose administration of a near-infrared fluorophore: Feasibility study, *Radiology* **246**, 734–741 (2008).
16. S. L. Troyan, V. Kianzad, S. L. Gibbs-Strauss, S. Gioux, A. Matsui, R. Oketokoun, L. Ngo, A. Khamene, F. Azar, and J. V. Frangioni, The FLARE™ intraoperative near-infrared fluorescence imaging system: A first-in-human clinical trial in breast cancer sentinel lymph node mapping, *Ann. Surg. Oncol.* **16**, 2943–2952 (2009).
17. X. Wang, S. Bhaumik, Q. Li, V. Staudinger, and S. Yazdanfar, Compact instrument for fluorescence image-guided surgery, *J. Biomed. Opt.* **15(2)**, 020509 (2010).

# 5

## Biomedical Applications of Coherent Light Scattering

David D. Nolte

### CONTENTS

### Biological Relevance in Drug Screening

The cost of drug-discovery R&D is rising exponentially, while the number of FDA approved drugs is growing only linearly [1]. Total industry R&D expenses have risen to around $50 billion per year, doubling every 10 years, while the number of new approved drugs per year has risen from 10 in 1950 to around 20 today (with a spike above 50 in the mid-1990s). The efficiency of the drug development industry, defined as the number of new drugs approved per total yearly cost, has been whimsically called *Eroom's law* (Figure 5.1), as the inverse of the well-known *Moore's law* of the electronics industry [2]. Whereas the efficiency of chip manufacturing doubles every 18 months, the efficiency of the drug development industry halves every 9 years. Although it is perhaps unfair to compare the drug-discovery market to the electronics industry because of the greater biological complexity, this business trend is unsustainable, and there is lively debate over the causes and the cures [3]. Most likely, there are many interacting causes that will require more than one cure. However, one factor that has emerged consistently is the need for improved biological relevance in early screening [4].

Cellular systems are highly complex, with high redundancy and dense cross talk among signaling pathways [5]. Biochemical target-based high-content screening can isolate single mechanisms in important pathways but often fails to capture integrated system-wide responses. Phenotypic

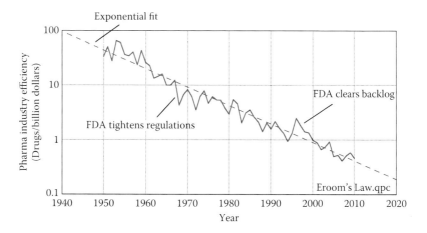

**FIGURE 5.1**

The so-called *Eroom's law* of pharmaceuticals from Ref. [2]. High investments in drug discovery R&D along with diminishing rate of new drug approvals are the main causes of decreasing efficiency in the pharmaceutical industry. (From Scannell, J.W., *Nat. Rev. Drug Discov.*, 11, 191, March 2012.)

profiling, on the other hand, presents a systems biology approach that has more biological relevance by capturing multimodal influence of therapeutics [6]. Ironically, phenotypic profiling is anachronistic, harking back to the days before genomics provided isolated targets. Nonetheless, it remains today one of the most successful approaches for the discovery of new drugs [7].

Although phenotypic profiling cannot be a panacea for the ills of the pharmaceutical industry, it does provide a path forward to improve efficiency. But before phenotypic profiling can fulfill this promise, its own efficiency needs to be improved. For instance, most phenotypic profiling continues to be performed on 2D culture, even though 2D monolayer culture on flat hard surfaces does not respond to applied drugs in the same way as cells in their natural 3D environment (Figure 5.2). This is in part because genomic profiles are not preserved in primary monolayer cultures [8–10]. There have been several comparative transcriptomic studies that have tracked the expression of genes associated with cell survival, proliferation, differentiation, and resistance to therapy that are expressed differently in 2D cultures relative to 3D culture. For example, 3D culture from cell lines of epithelial ovarian cancer [11,12], hepatocellular carcinoma [13–15], or colon cancer [16] displays expression profiles more like those from tumor tissues than when grown in 2D. In addition, the 3D environment of 3D culture presents different pharmacokinetics than 2D monolayer culture and produces differences in cancer drug sensitivities [17–20].

The challenge for 3D culture is how to extract high-content phenotypic responses from inside heterogeneous tissue. Conventional microscopies, such as confocal [21], or nonlinear microscopies, such as two-photon [22] or

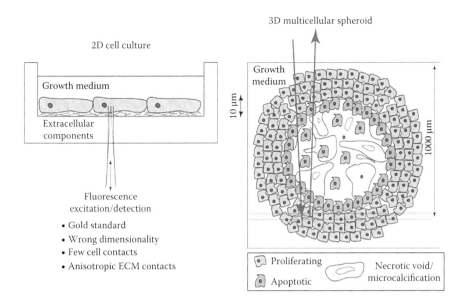

**FIGURE 5.2**
The 2D cell culture versus 3D multicellular spheroids. The 2D screening fails to provide precise information about intercellular contacts and cell signaling. For example, *missed drugs* occur because promising compounds fail to respond appropriately in 2D screening.

CARS microscopy [23], penetrate only a few hundred microns into tissue and lose imaging resolution with increasing penetration. Optical coherence tomography (OCT) can image more than 1 mm deep into tissue but also loses imaging resolution with increasing penetration [24]. Fluorescence lifetime imaging microscopy (FLIM) provides signals from some depth inside tissue, but retaining spatial resolution with increasing depth remains a challenge [25]. Holographic imaging uses holographic recording media, for example, photorefractive (PR) multiple quantum wells [84], for optical coherence imaging (OCI). Unlike OCT, where speckles are undesirable side effects of the coherent imaging, the holographic imaging detection methods are highly susceptible to speckles as signals and sources of information. Figure 5.3 depicts the evolution of holographic imaging of tissue.

An approach alternative to imaging *form* is to image *function* and in particular functional motions. Motion is ubiquitous in all living things and occurs across broad spatial and temporal scales. At one extreme, motions of molecules during Brownian diffusion occur across nanometers at microsecond scales, while at the other extreme, motions of metastatic crawling cells occur across millimeters, taking many hours. As one spans these scales, many different functional processes are taking place: molecular diffusion; molecular polymerization or depolymerization of the cytoskeleton; segregation of enzymes into vesicles, exocytosis, and endocytosis; shepherding of vesicles by molecular motors; active transport of mitochondria; cytoskeletal forces

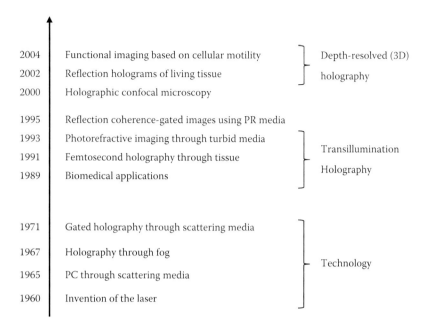

**FIGURE 5.3**
Time line of holographic imaging of tissue. From the invention of laser in 1960 to the application of lasers in scattering media in 1965, transillumination holography in the 1990s (which
were based on shadow-casting but not 3D), the use of PR materials for looking in the tissue in
1995, and development of the first reflection holograms of the living tissues in 2002 and beyond
to PR and digital holography and functional imaging based on cellular motility.

pushing and pulling on the nucleus; undulations of the cell membrane; formation or alteration of cell-to-cell adhesions; deformation of the cell, cell
division, and ultimately the movement of individual cells through tissue.
All of these very different types of cellular dynamics are active and useful
indicators of the functioning behavior of cells.

Imaging motion in 3D tissue is simpler than imaging structure or performing molecular imaging, because motion modulates coherent light through
the Doppler effect. When light scatters from an object that is displacing, the
phase of the light is modified. If the light has coherence, then the motion-
induced phase shifts of one light path interfere with the phase shifts of other
light paths in constructive and destructive interference. Even light that is
multiply scattered in tissue carries a record of the different types of motions
that the light encountered. By measuring the fluctuating phase of light scattered from living tissue, the different types of motion across the different
space and time scales can be measured. The trade-off for greater depth of
penetration into 3D tissue is reduced spatial imaging resolution. But volumetric imaging of intracellular motions in tissue is still possible, within limits, by using low-coherence interferometry that can select light from specified depths by using coherence-gating approaches [26] (e.g., see Figure 5.4).

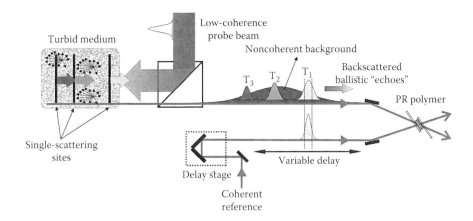

**FIGURE 5.4**
General principle of coherence-gated holographic imaging using a PR polymer device as coherence filter. (From Salvador, M. et al., *Opt. Express*, 17, 11834, July 6, 2009.)

## Dynamic Light Scattering from Cells and Tissue

Dynamic light scattering (DLS) is a characterization tool for remotely extracting dynamic information from inside media, such as internal displacements, diffusivities, and velocities [27,28]. In the dilute limit, in which light is scattered only once, DLS is called quasielastic light scattering (QELS) [29]. In biological applications, QELS is useful for optically thin systems, such as cell monolayer cultures, in which it can provide information on diffusion and directed motion in the cytosol [30,31] and in the nucleus [32] as well as undulations of the plasma membrane [33–37]. However, as the optical thickness of a sample increases beyond one optical transport length, multiple scattering becomes significant, and the diffusive character of light begins to dominate [38–40]. In this multiple-scattering limit diffusing-wave spectroscopy (DWS) [41] and diffuse correlation spectroscopy (DCS) [42–44], continue to carry motional signatures from inside tissue in the field and intensity fluctuations of the scattered light.

DLS is a fundamental process by which light scattered from displacing scatterers experiences phase shifts related to the magnitude and direction of displacement [27]. Living tissue is a dense collection of scatterers that are executing many different types of motion, leading to a rich collection of phase shifts. For the slow motions inside a cell (microns per second), the phase drift associated with displacements is equivalent to an ultralow-frequency (ULF) Doppler shift in the light frequency. The Doppler frequency offsets are too small to measure directly, but by using interferometry (e.g., digital holography), the phase modulation is measured as a heterodyne intensity

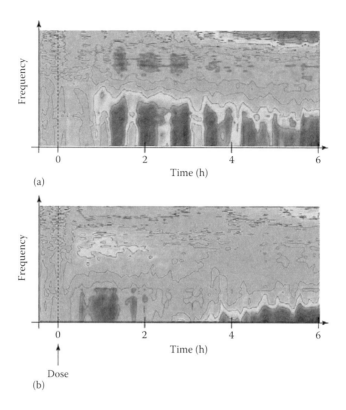

**FIGURE 5.5**
**(See color insert.)** Drug-response spectrograms for phenotypic profiling of (a) iodoacetate and (b) cytochalasin D; Time–frequency analysis is performed on the time-dependent signals received from fluctuating speckles. The difference between the drug responses is highlighted according to their *relatively unique* voice prints and signatures in the spectrograms.

fluctuation signal. The fluctuating intensities of the backscattered speckle contain the motional information of the scatterers. Figure 5.5 shows different tissue dynamic spectrograms (frequency vs. time) corresponding to osteogenic sarcoma tissue responding to two different drugs: iodoacetate and cytochalasin D.

Diffusion of very small organelles, as well as molecular diffusion, occurs with backscatter frequencies up to 100 Hz, and frequencies from molecular diffusion extend into the kilohertz range. Organelle and vesicle motion, in which molecular motors move organelles along the cytoskeleton, occurs at speeds of microns per second [32,45–48] that produce characteristic frequencies generally above 1 Hz and extending to tens of Hz. Larger mitochondria and organelles have slightly lower backscattering frequencies. Membrane undulations are a common feature of cellular motions, leading to the phenomenon of flicker [49–53]. The characteristic frequency for membrane undulations tends to be in the range around 0.01–0.1 Hz [47,50,54].

Force relaxation and cellular rheology occur at even lower frequencies. This spatial–temporal trend is only semiquantitative, but it provides a general principle that may help disentangle the mixtures of frequencies that arise from multiple dynamic light-scattering mechanisms (Figure 5.6).

Even in the presence of multiple scattering, dynamic light-scattering information still can be extracted [55,56] using DWS [57,58] and an equivalent formulation of multiscattering DLS called DCS [42,44,59,60]. There are two principles underpinning DWS: (1) each scattering event along a path contributes stochastically to the phase shift of the photon following that path, and (2) all the possible paths that a photon can take must be considered in the total correlation function. In this way, the stochastic phase modulations accumulate in the phase of the photon and hence can be expressed as a product of stochastic exponentials. Also, the density of these possible paths is found as a solution to the photon diffusion equation.

In biological applications, DCS has been used to assess tissue response to burns [61] and brain activity [62], to monitor blood flow [43,44], and to monitor tissue structure [63]. DCS uses high-coherence laser sources and measures the temporal diffusing field autocorrelation function. The primary uses have been for macroscopic measurements of blood flow, which has been validated through comparisons with Doppler flowmetry and ultrasound [64–66]. A technique related to DCS, but that more directly uses speckle imaging, is speckle contrast imaging (SCI) [43]. This technique is used primarily for imaging of vasculature in vivo because the fast motions of blood cells tend to blur the speckle contrast in images acquired with a long exposure time [67]. This approach is particularly well suited for imaging of blood flow.

OCT is an imaging technique that forms images by scanning and rastering. Speckle in OCT has long been considered an unwanted side effect of the coherent imaging, and many approaches have been explored to reduce speckle [68–70]. However, speckle decorrelation also can be studied in OCT data to provide similar information as provided by DCS. This has been used to measure intracellular rheology [31] and to find dynamic signatures of apoptosis [71]. OCT also could be sensitive to speckle contrast caused by motions that occur faster than the scan rate, but it is more commonly used in a Doppler mode for blood flow detection [72,73].

## Motility Contrast Imaging of Live Tissue

Motility contrast imaging (MCI) is a hybrid of OCT and DWS. It performs tissue-scale functional imaging by using subcellular motion as an endogenous form of imaging contrast [74,75]. MCI of live tissue is based on two principles: DLS from the motions of the constituents of living cells and coherence-gated depth resolution [76] that isolates the motional signals from

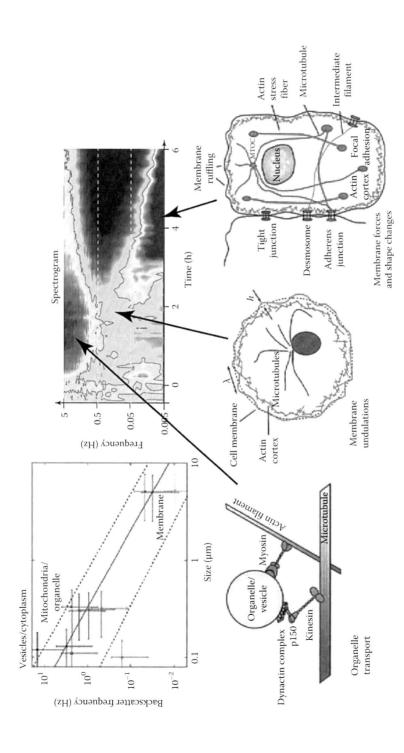

**FIGURE 5.6**

**(See color insert.)** Light-scattering mechanisms showing backscatter frequencies as a function of size for vesicle transport, organelle motions, membrane undulations, and membrane forces and shape changes.

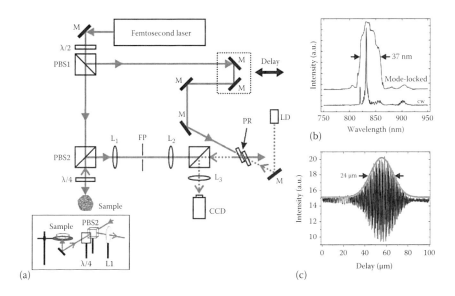

**FIGURE 5.7**

(a) Experimental layout for holographic OCI. M, mirror. λ/2, half-wave plate. λ/4, quarter-wave plate. PBS, polarization beam splitter. $L_1$–$L_3$, lenses. PR, photorefractive polymer device. LD, continuous wave laser diode. FP, Fourier plane. The inset shows the positioning of the sample from a side view. (b) Output spectra of the Clark-MXR, NJA-5 femtosecond Ti–sapphire laser in cw and mode-locked operation. (c) Corresponding fringe pattern (interferogram) of the Ti–sapphire laser in mode-locked operation using a mirror as target. (From Salvador, M. et al., *Opt. Express*, 17, 11834, July 6, 2009.)

specific depths (Figure 5.7). The frequency shifts measured by dynamic imaging are ULF Doppler signals caused by light scattering off the movements of intracellular constituents.

The holographic capture of depth-resolved images from optically thick live tissues has evolved through several stages, from OCI to MCI and tissue dynamics spectroscopy (TDS).

OCI uses coherence-gated holography to optically section tissue up to 1 mm deep [77,78]. It is a full-frame imaging approach, closely related to en face OCT [79,80], but relies on high-contrast speckle to provide high sensitivity to motion [81]. The first implementations of OCI used holographic recording media [82] such as PR quantum wells [83] to capture the coherent backscatter and separate it from the diffuse background. Digital holography [84–87] replaced the recording media and has become the mainstay of current implementations of OCI [88]. Highly dynamic speckle was observed in OCI of living tissues caused by intracellular motions [74] and was used directly as an endogenous imaging contrast in MCI that could track the effects of antimitotic drugs on tissue health [75].

In OCI, the coherence gate is formed by a low-coherence light source detected by digital holography on a charge-coupled device (CCD) camera chip [89]. Only light scattered from a chosen depth is recorded in the digital hologram. When DLS and coherence gating are combined, intracellular motion is measured volumetrically up to 1 mm deep inside tissue with a spatial resolution of 30 µm. Low-coherence light is split at a beam splitter into a signal arm that illuminates the sample and a reference arm that has a variable-delay retroreflector. Light scattered from the sample intersects with the reference beam at the plane of the CCD electronic chip [88]. Low-coherence light has the property that only signal paths with the same path length as the reference arm produce interference fringes [90]. Therefore, the selected depth is defined by the path length from the beam splitter to the CCD plane. To extract information from other depths simply requires a shift in the retroreflector mirror in the reference arm. The low-coherence light source is either a 100 fs Ti–sapphire laser or a Superlum superluminescent diode, both operating at a center wavelength of 840 nm. The hologram is recorded on the CCD chip, *readout* by a digital Fourier transform algorithm, and stored as a layer in a volumetric data set. Figure 5.7 shows the optical setup with a PR polymer device.

MCI is a form of DLS in tissues. DLS is performed as QELS when light is predominantly singly scattered and as DWS [55,57] or diffusing correlation spectroscopy (DCS) [61] when light is multiply scattered. QELS has been applied mainly to single cells or monolayer cultures to study motion in the nucleus [32], the cytosol [31], cell motion [91], and membrane fluctuations [37]. DWS and DCS probe deeply into tissue and have been used to study actin filament networks [92], imaging dynamic heterogeneities [93], and brain activity [62]. The transition from single scattering to multiple scattering is important for backscatter applications [38,94]. An example of MCI is shown in Figure 5.8 for a 0.8 mm diameter multicellular tumor spheroid of osteogenic sarcoma. The figure shows motility maps at a selection of depths through the tumor spheroid.

MCI is a new form of functional imaging that shares some features in common with functional magnetic resonance imaging (f-MRI). MRI is based on the precession of proton spins in a magnetic field, and the raw signal is composed of radio-frequency currents in a pickup coil. The functional aspects of f-MRI are extracted by using statistical methods to measure locally varying changes in precession decay rates based on local oxygen activity. Functional MCI (f-MCI) is also statistical in nature that measures local dynamic activity through intensity fluctuations. In this way, f-MCI is a statistically based functional imaging in the same way as f-MRI. The notable difference, as shown in Figure 5.9, is the penetration depth of the two modes of imaging. f-MRI penetrates through tens of centimeters, while f-MCI penetrates through only 1 mm of tissue. However, the functional information that is retrieved in both methods are very different, with MRI best suited to brain imaging and MCI best suited to in vitro assays of living tissue.

While MCI captures the overall motion inside tissue, significant information is contained in the fluctuating intensity that relates to different types of

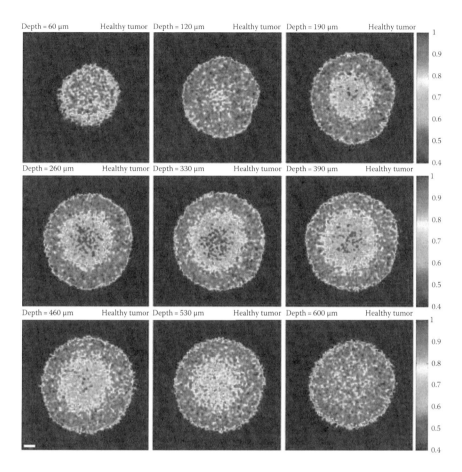

**FIGURE 5.8**
**(See color insert.)** MCI of a 0.8 mm diameter tumor; cross sections of tissue activity are color-coded red, as active, and blue, as quiet.

intracellular motion. Therefore, we extended motility contrast into TDS by analyzing fluctuating speckle intensity time series into individual frequency components. Different frequencies relate to different types of motion, and TDS can time-resolve changes in these motions as tissues evolve under environmental or pharmacological perturbations [95,96]. TDS produces unique fingerprints of the action of specific drugs on specific cell lines [97]. These drug fingerprints give insight into drug mechanisms of action (MoA) and provide the training set for phenotypic profiling [98] and the classification and discovery of potential new drugs. Figure 5.10 depicts a spectrogram collage, as libraries or knowledge bases of drugs MoA. Figure 5.11 shows hierarchical clustering and classification of different types of compounds and drugs. The block-diagonal similarity matrix displays their similarities and differences according to feature vectors derived from the drug-response

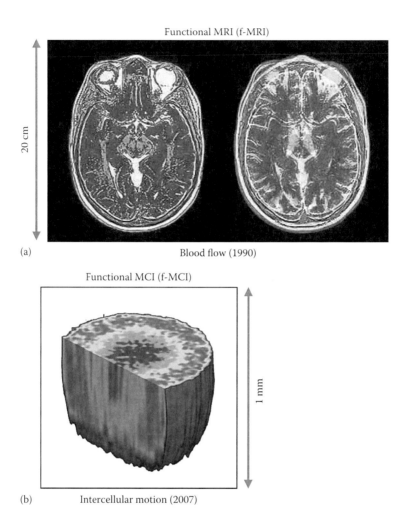

(a)                                    Blood flow (1990)

(b)            Intercellular motion (2007)

**FIGURE 5.9**
**(See color insert.)** Functional imaging: (a) f-MRI the blood flow in the brain and (b) f-MCI volumetric imaging of cellular motion in a tumor spheroid; reds and oranges represent high motility in the outer shell surrounding the low-motility region of the necrotic core.

spectrograms. This clustering and classification is known as *phenotypic profiling*, that is, capturing the physiological responses of cells and tissues to the drugs. Phenotypic profiling is an alternative approach for targeted drug discovery.

For the past decade, the major pharmaceutical companies have invested heavily in target-based drug discovery, but with disappointing returns and fewer-than-expected discoveries (with some notable exceptions like Gleevec [99,100]). Target-based drug discovery is a bottom-up approach that starts with specific molecular targets in signaling pathways and develops drugs

| Compound or condition | Action |
| --- | --- |
| Temperature 24°C to 37°C | Increased motility |
| pH 6 | Weak acidic |
| KCN 20 ug/mL | Inhibits electron transport |
| Iodoacetate 1 ug/mL. | Inhibits glycolysis |
| Osm 508 | Hypertonic cell dessication |
| Osm 428 | Hypertonic cell dessication |
| Iodoacetate 10 ug/mL | Inhibits glycolysis |
| KCN 200 ug/mL | Inhibit electron transport |
| TNF 5 ug/mL. | Cytokine apoptosis induction |
| Cytochalasin 10 ug/mL | Antiactin, antimitotic |
| Cytochalasin 1 ug/mL | Antiactin, antimitotic |
| Iodoacetate 40 ug/mL | Inhibits glycolysis |
| pH 9 | Strong basic |
| Nocodazole 0.01 ug/mL. | Antitubulin, antimitotic |
| Osm 154 | Strong hypotonic cell swelling |
| Osm 77 | Strong hypotonic cell swelling |
| Cytochalasin 50 ug/mL | Antiactin, antimitotic, apoptosis |
| pH 5 | Strong acidic |
| Colchicine 10 ug/mL | Antitubulin, antimitotic |
| Iodoacetate 20 ug/mL | Inhibit glycolysis |
| Nocodazole 1 ug/mL. | Antitubulin, antimitotic |
| Taxol 10 ug/mL | Tubulin stabilization, antimitotic |
| Cycloheximide | Apoptosis induction |
| Nocodazole 0.1 ug/mL | Antitubulin, antimitotic |
| Colchicine 1 ug/mL. | Antitubulin, antimitotic |
| Nocodazole 10 ug/mL. | Antitubulin, antimitotic |
| Taxol 1 ug/mL. | Tubulin stabilization, antimitotic |
| pH 8 | Weak basic |

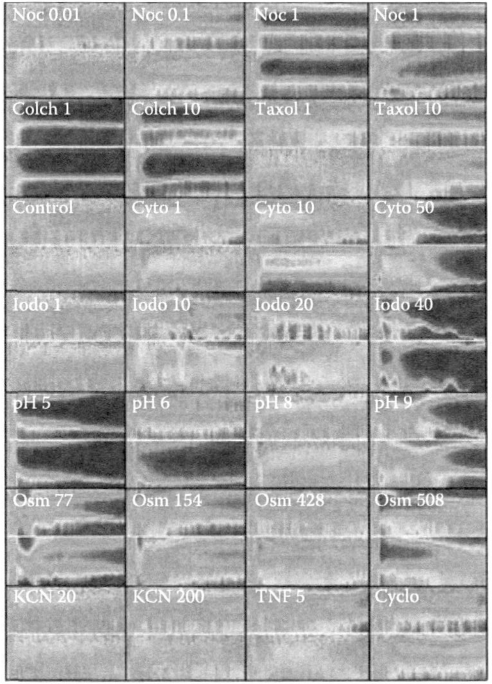

**FIGURE 5.10**

Spectrogram collage of different drugs and compounds, with a table showing their conditions and actions.

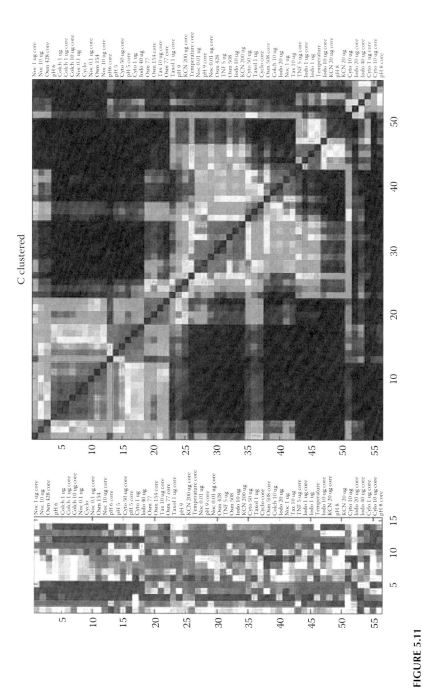

**FIGURE 5.11**

Hierarchical clustering of drugs and compounds after unsupervised clustering of drugs and responses. Phenotypic profiling is based on feature vectors and the clustered similarity matrix.

that enhance or inhibit that target to produce desired downstream effects. The greatest problem with this approach is the probability of off-target effects of the drug that ultimately prevent its clinical use.

The opposite of target-based drug discovery is *phenotypic profiling* that measures the broad-spectrum cellular response to drugs. In a recent study of all drugs approved by the FDA since 1999, it was found that 2/3 of those approved were developed through phenotypic profiling and only 1/3 by target-based approaches [7]. The 2D monolayer cultures are the current industry gold standard for phenotypic profiling, but the deficiencies are well known, with the wrong dimensionality, the wrong cell shapes, and the subsequent modified biochemistries that are not representative of natural tissues and lead to nonrepresentative drug responses [101,102]. While there are tentative moves to 3D cultures, extracting information from dense 3D tissue has been a bottleneck. Tissue dynamic imaging and spectroscopy are poised to remove that bottleneck by extracting multiparameter physiological responses inside live tissue. TDS has the hallmarks of a successful phenotypic screening method: high sensitivity and specificity, label-free operation, three dimensionality, noninvasive, high content, and compatibility with high-throughput automation.

## Emerging Trends and Challenges

Biodynamic imaging, as a new form of biomedical imaging contrast, is an emerging approach to measurement of intracellular motions through propagation of light through biological tissues. Potential applications are in the field of cancer therapeutics, developing new approaches to cancer care and therapy through combining the physics of coherence with the biophysics of cancer. More than 60% of chemotherapy patients gain no benefit from first-line cancer drugs, since drugs developed for single genetic or protein usually fail. Biodynamic imaging can select better drugs that are personalized to the patient using phenotypic profiling, increasing positive outcomes for cancer patients, thus improving their quality of life. In this approach, the details of all affected protein signaling pathways may not be discoverable, but the overall efficiency of the drug may be monitored, despite the heterogeneity and within all of the diverse microenvironments associated with specific cancers.

The recent developments in TDS have uncovered new opportunities to study the functional influence of new drug compounds on living tissue, as an alternative to target-based drug discovery. TDS performs label-free noninvasive measurements of the intracellular dynamics inside living tissue and evaluates their responses to applied drugs. The unique feature of TDS lies in its ability to bridge the gap in the drug-discovery pipeline between high-throughput 2D monolayer cultures and preclinical animal models, where drug failures are much more expensive. The 3D tissue-based drug screens are required to bridge this gap, since 3D tissues are more representative of cellular and tissue responses to applied drugs than those in 2D monolayer culture. TDS is an emerging screening technology, entering the marketplace for early drug discovery. The field of use of TDS is *hit confirmation and expansion*, where screens are required to test hits for therapeutic efficacy (does it modify a selected target?) and specificity (does it *only* affect a selected target?). The target is *high-content phenotypic profiling* of physiological response to drug candidates.

Application of TDI and MCI is envisioned to go beyond the presented examples in this chapter. All living tissue is characterized by internal motions that are highly specific to the functions driving that motion. In the future, emergence of a new type of microscopy, called *biodynamic microscopy*, is anticipated, as a multipurpose tool in life science laboratories. A recent example is in the field of *in vitro fertilization* for assessing the variability of embryos prior to implantation. This could improve the success rate for pregnancies, minimizing the associated costs and complications. Such applications suggest a promising utility for biodynamic imaging in the future.

## Acknowledgment

The author acknowledges support from NSF CBET-0756005 and wishes to thank doctoral student and PRISM Center researcher Mohsen Moghaddam for help in preparing this chapter manuscript.

## References

1. B. Munos, Lessons from 60 years of pharmaceutical innovation, *Nature Reviews Drug Discovery*, 8, 959–968, December 2009.
2. J. W. Scannell, A. Blanckley, H. Boldon, and B. Warrington, Diagnosing the decline in pharmaceutical R&D efficiency, *Nature Reviews Drug Discovery*, 11, 191–200, March 2012.
3. S. M. Paul, D. S. Mytelka, C. T. Dunwiddie, C. C. Persinger, B. H. Munos, S. R. Lindborg, and A. L. Schacht, How to improve R&D productivity: The pharmaceutical industry's grand challenge, *Nature Reviews Drug Discovery*, 9, 203–214, March 2010.
4. E. C. Butcher, Can cell systems biology rescue drug discovery?, *Nature Reviews Drug Discovery*, 4, 461–467, June 2005.
5. C. Ainsworth, Networking for new drugs, *Nature Medicine*, 17, 1166–1168, October 2011.

6. C. T. Keith, A. A. Borisy, and B. R. Stockwell, Multicomponent therapeutics for networked systems, *Nature Reviews Drug Discovery*, 4, U71–U10, January 2005.

7. D. C. Swinney and J. Anthony, How were new medicines discovered?, *Nature Reviews Drug Discovery*, 10, 507–519, July 2011.

8. P. Hamer, S. Leenstra, C. J. F. Van Noorden, and A. H. Zwinderman, Organotypic glioma spheroids for screening of experimental therapies: How many spheroids and sections are required?, *Cytometry Part A*, 75A, 528–534, June 2009.

9. J. Poland, P. Sinha, A. Siegert, M. Schnolzer, U. Korf, and S. Hauptmann, Comparison of protein expression profiles between monolayer and spheroid cell culture of HT-29 cells revealed fragmentation of CK18 in three-dimensional cell culture, *Electrophoresis*, 23, 1174–1184, April 2002.

10. K. Dardousis, C. Voolstra, M. Roengvoraphoj, A. Sekandarzad, S. Mesghenna, J. Winkler, Y. Ko, J. Hescheler, and A. Sachinidis, Identification of differentially expressed genes involved in the formation of multicellular tumor spheroids by HT-29 colon carcinoma cells, *Molecular Therapy*, 15, 94–102, January 2007.

11. N. A. L. Cody, M. Zietarska, A. Filali-Mouhim, D. M. Provencher, A. M. Mes-Masson, and P. N. Tonin, Influence of monolayer, spheroid, and tumor growth conditions on chromosome 3 gene expression in tumorigenic epithelial ovarian cancer cell lines, *BMC Medical Genomics*, 1, 34, August 7, 2008.

12. M. Zietarska, C. M. Maugard, A. Filali-Mouhim, M. Alam-Fahmy, P. N. Tonin, D. M. Provencher, and A. M. Mes-Masson, Molecular description of a 3D *in vitro* model for the study of epithelial ovarian cancer (EOC), *Molecular Carcinogenesis*, 46, 872–885, October 2007.

13. T. T. Chang and M. Hughes-Fulford, Monolayer and spheroid culture of human liver hepatocellular carcinoma cell line cells demonstrate distinct global gene expression patterns and functional phenotypes, *Tissue Engineering Part A*, 15, 559–567, March 2009.

14. M. Shimada, Y. Yamashita, S. Tanaka, K. Shirabe, K. Nakazawa, H. Ijima, R. Sakiyama, J. Fukuda, K. Funatsu, and K. Sugimachi, Characteristic gene expression induced by polyurethane foam/spheroid culture of hepatoma cell line, Hep G2 as a promising cell source for bioartificial liver, *Hepato-Gastroenterology*, 54, 814–820, April–May 2007.

15. Y. Yamashita, M. Shimada, N. Harimoto, S. Tanaka, K. Shirabe, H. Ijima, K. Nakazawa, J. Fukuda, K. Funatsu, and Y. Maehara, cDNA microarray analysis in hepatocyte differentiation in Huh 7 cells, *Cell Transplantation*, 13, 793–799, 2004.

16. L. Gaedtke, L. Thoenes, C. Culmsee, B. Mayer, and E. Wagner, Proteomic analysis reveals differences in protein expression in spheroid versus monolayer cultures of low-passage colon carcinoma cells, *Journal of Proteome Research*, 6, 4111–4118, November 2007.

17. A. Frankel, R. Buckman, and R. S. Kerbel, Abrogation of taxol-induced G(2)-M arrest and apoptosis in human ovarian cancer cells grown as multicellular tumor spheroids, *Cancer Research*, 57, 2388–2393, June 15, 1997.

18. L. A. Hazlehurst, T. H. Landowski, and W. S. Dalton, Role of the tumor microenvironment in mediating de novo resistance to drugs and physiological mediators of cell death, *Oncogene*, 22, 7396–7402, October 20, 2003.

19. I. Serebriiskii, R. Castello-Cros, A. Lamb, E. A. Golemis, and E. Cukierman, Fibroblast-derived 3D matrix differentially regulates the growth and drug-responsiveness of human cancer cells, *Matrix Biology*, 27, 573–585, July 2008.

20. L. David, V. Dulong, D. Le Cerf, L. Cazin, M. Lamacz, and J. P. Vannier, Hyaluronan hydrogel: An appropriate three-dimensional model for evaluation of anticancer drug sensitivity, *Acta Biomaterialia*, 4, 256–263, March 2008.

21. A. Bullen, Microscopic imaging techniques for drug discovery, *Nature Reviews Drug Discovery*, 7, 54–67, January 2008.

22. K. Konig, Multiphoton microscopy in life sciences, *Journal of Microscopy-Oxford*, 200, 83–104, November 2000.

23. C. L. Evans and X. S. Xie, Coherent anti-stokes raman scattering microscopy: chemical imaging for biology and medicine, *Annual Review of Analytical Chemistry*, 1, 883–909, 2008.

24. R. K. K. Wang, Signal degradation by coherence tomography multiple scattering in optical of dense tissue: A Monte Carlo study towards optical clearing of biotissues, *Physics in Medicine and Biology*, 47, 2281–2299, July 7, 2002.

25. M. J. Cole, J. Siegel, S. E. D. Webb, R. Jones, K. Dowling, M. J. Dayel, D. Parsons-Karavassilis, P. M. W. French, M. J. Lever, L. O. D. Sucharov, M. A. A. Neil, R. Juskaitis, and T. Wilson, Time-domain whole-field fluorescence lifetime imaging with optical sectioning, *Journal of Microscopy-Oxford*, 203, 246–257, September 2001.

26. M. Salvador, J. Prauzner, S. Kober, K. Meerholz, J. J. Turek, K. Jeong, and D. D. Nolte, Three-dimensional holographic imaging of living tissue using a highly sensitive photorefractive polymer device, *Optics Express*, 17, 11834–11849, July 6, 2009.

27. B. J. Berne and R. Pecora, *Dynamic Light Scattering: With Applications to Chemistry, Biology, and Physics*. New York: Dover, 2000.

28. D. A. Weitz and D. J. Pine, Chapter. In *Dynamic Light Scattering: The Method and Some Applications*, Vol. 49, W. Brown, Ed. Oxford, U.K.: Oxford University Press, pp. 652–720, 1993.

29. V. A. Bloomfield, Quasi-elastic light-scattering applications in biochemistry and biology, *Annual Review of Biophysics and Bioengineering*, 10, 421–450, 1981.

30. T. G. Mason and D. A. Weitz, Optical measurements of frequency-dependent linear viscoelastic moduli of complex fluids, *Physical Review Letters*, 74, 1250–1253, February 13, 1995.

31. C. Joo, C. L. Evans, T. Stepinac, T. Hasan, and J. F. de Boer, Diffusive and directional intracellular dynamics measured by field-based dynamic light scattering, *Optics Express*, 18, 2858–2871, February 1, 2010.

32. M. Suissa, C. Place, E. Goillot, and E. Freyssingeas, Internal dynamics of a living cell nucleus investigated by dynamic light scattering, *European Physical Journal E*, 26, 435–448, August 2008.

33. L. Kramer, Theory of light scattering from fluctuations of membranes and monolayers, *Journal of Chemical Physics*, 55, 2097, 1971.

34. R. Hirn, T. M. Bayer, J. O. Radler, and E. Sackmann, Collective membrane motions of high and low amplitude, studied by dynamic light scattering and micro-interferometry, *Faraday Discussions*, 111, 17–30, 1998.

35. M. A. Haidekker, H. Y. Stevens, and J. A. Frangos, Cell membrane fluidity changes and membrane undulations observed using a laser scattering technique, *Annals of Biomedical Engineering*, 32, 531–536, April 2004.

36. M. S. Amin, Y. Park, N. Lue, R. R. Dasari, K. Badizadegan, M. S. Feld, and G. Popescu, Microrheology of red blood cell membranes using dynamic scattering microscopy, *Optics Express*, 15, 17001–17009, December 10, 2007.

37. R. B. Tishler and F. D. Carlson, A study of the dynamic properties of the human red-blood-cell membrane using quasi-elastic light-scattering spectroscopy, *Biophysical Journal*, 65, 2586–2600, December 1993.

38. A. Wax, C. H. Yang, R. R. Dasari, and M. S. Feld, Path-length-resolved dynamic light scattering: Modeling the transition from single to diffusive scattering, *Applied Optics*, 40, 4222–4227, August 20, 2001.

39. R. Carminati, R. Elaloufi, and J. J. Greffet, Beyond the diffusing-wave spectroscopy model for the temporal fluctuations of scattered light, *Physical Review Letters*, 92, 213903-1–213903-4, May 28, 2004.

40. P. A. Lemieux, M. U. Vera, and D. J. Durian, Diffusing-light spectroscopies beyond the diffusion limit: The role of ballistic transport and anisotropic scattering, *Physical Review E*, 57, 4498–4515, April 1998.

41. D. J. Pine, D. A. Weitz, J. X. Zhu, D. J. Durian, A. Yodh, and M. Kao, Diffusing-wave spectroscopy and interferometry, *Macromolecular Symposia*, 79, 31–44, March 1994.

42. D. A. Boas, L. E. Campbell, and A. G. Yodh, Scattering and imaging with diffusing temporal field correlations, *Physical Review Letters*, 75, 1855–1858, August 28, 1995.

43. D. A. Boas and A. K. Dunn, Laser speckle contrast imaging in biomedical optics, *Journal of Biomedical Optics*, 15, 011109-1–011109-12, January–February 2010.

44. T. Durduran, R. Choe, W. B. Baker, and A. G. Yodh, Diffuse optics for tissue monitoring and tomography, *Reports on Progress in Physics*, 73, 076701, July 2010.

45. X. L. Nan, P. A. Sims, and X. S. Xie, Organelle tracking in a living cell with microsecond time resolution and nanometer spatial precision, *Chemphyschem*, 9, 707–712, April 4, 2008.

46. K. J. Karnaky, L. T. Garretson, and R. G. Oneil, Video-enhanced microscopy of organelle movement in an intact epithelium, *Journal of Morphology*, 213, 21–31, July 1992.

47. N. A. Brazhe, A. R. Brazhe, A. N. Pavlov, L. A. Erokhova, A. I. Yusipovich, G. V. Maksimov, E. Mosekilde, and O. V. Sosnovtseva, Unraveling cell processes: Interference imaging interwoven with data analysis, *Journal of Biological Physics*, 32, 191–208, October 2006.

48. B. Trinczek, A. Ebneth, and E. Mandelkow, Tau regulates the attachment/detachment but not the speed of motors in microtubule-dependent transport of single vesicles and organelles, *Journal of Cell Science*, 112, 2355–2367, July 1999.

49. F. Brochard and J. F. Lennon, Frequency spectrum of flicker phenomenon in erythrocytes, *Journal De Physique*, 36, 1035–1047, 1975.

50. H. Strey and M. Peterson, Measurement of erythrocyte-membrane elasticity by flicker eigenmode decomposition, *Biophysical Journal*, 69, 478–488, August 1995.

51. A. Zilker, M. Ziegler, and E. Sackmann, Spectral-analysis of erythrocyte flickering in the 0.3-4-Mu-M-1 regime by microinterferometry combined with fast image-processing, *Physical Review A*, 46, 7998–8002, December 15, 1992.

52. M. A. Peterson, H. Strey, and E. Sackmann, Theoretical and phase-contrast microscopic eigenmode analysis of erythrocyte flicker—Amplitudes, *Journal De Physique Ii*, 2, 1273–1285, May 1992.

53. Y. Z. Yoon, H. Hong, A. Brown, D. C. Kim, D. J. Kang, V. L. Lew, and P. Cicuta, Flickering analysis of erythrocyte mechanical properties: Dependence on oxygenation level, cell shape, and hydration level, *Biophysical Journal*, 97, 1606–1615, September 16, 2009.

54. J. Evans, W. Gratzer, N. Mohandas, K. Parker, and J. Sleep, Fluctuations of the red blood cell membrane: Relation to mechanical properties and lack of ATP dependence, *Biophysical Journal*, 94, 4134–4144, May 15, 2008.

55. G. Maret and P. E. Wolf, Multiple light-scattering from disordered media—The effect of brownian-motion of scatterers, *Zeitschrift Fur Physik B-Condensed Matter*, 65, 409–413, 1987.

56. M. J. Stephen, Temporal fluctuations in wave-propagation in random-media, *Physical Review B*, 37, 1–5, January 1, 1988.

57. D. J. Pine, D. A. Weitz, P. M. Chaikin, and E. Herbolzheimer, Diffusing-wave spectroscopy, *Physical Review Letters*, 60, 1134–1137, March 21, 1988.

58. G. Maret, Diffusing-wave spectroscopy, *Current Opinion in Colloid and Interface Science*, 2, 251–257, June 1997.

59. B. J. Ackerson, R. L. Dougherty, N. M. Reguigui, and U. Nobbmann, Correlation transfer—Application of radiative-transfer solution methods to photon-correlation problems, *Journal of Thermophysics and Heat Transfer*, 6, 577–588, October–December 1992.

60. R. L. Dougherty, B. J. Ackerson, N. M. Reguigui, F. Dorrinowkoorani, and U. Nobbmann, Correlation transfer—Development and application, *Journal of Quantitative Spectroscopy and Radiative Transfer*, 52, 713–727, December 1994.

61. D. A. Boas and A. G. Yodh, Spatially varying dynamical properties of turbid media probed with diffusing temporal light correlation, *Journal of the Optical Society of America A-Optics Image Science and Vision*, 14, 192–215, January 1997.

62. J. Li, G. Dietsche, D. Iftime, S. E. Skipetrov, G. Maret, T. Elbert, B. Rockstroh, and T. Gisler, Noninvasive detection of functional brain activity with near-infrared diffusing-wave spectroscopy, *Journal of Biomedical Optics*, 10, 044002-1–044002-12, July–August 2005.

63. D. A. Zimnyakov and V. V. Tuchin, Optical tomography of tissues, *Quantum Electronics*, 32, 849–867, October 2002.

64. C. Menon, G. M. Polin, I. Prabakaran, A. Hsi, C. Cheung, J. P. Culver, J. F. Pingpank, C. S. Sehgal, A. G. Yodh, D. G. Buerk, and D. L. Fraker, An integrated approach to measuring tumor oxygen status using human melanoma xenografts as a model, *Cancer Research*, 63, 7232–7240, November 2003.

65. G. Q. Yu, T. Durduran, C. Zhou, H. W. Wang, M. E. Putt, H. M. Saunders, C. M. Sehgal, E. Glatstein, A. G. Yodh, and T. M. Busch, Noninvasive monitoring of murine tumor blood flow during and after photodynamic therapy provides early assessment of therapeutic efficacy, *Clinical Cancer Research*, 11, 3543–3552, May 2005.

66. E. M. Buckley, N. M. Cook, T. Durduran, M. N. Kim, C. Zhou, R. Choe, G. Q. Yu et al., Cerebral hemodynamics in preterm infants during positional intervention measured with diffuse correlation spectroscopy and transcranial Doppler ultrasound, *Optics Express*, 17, 12571–12581, July 2009.

67. A. K. Dunn, T. Bolay, M. A. Moskowitz, and D. A. Boas, Dynamic imaging of cerebral blood flow using laser speckle, *Journal of Cerebral Blood Flow and Metabolism*, 21, 195–201, March 2001.

68. J. M. Schmitt, S. H. Xiang, and K. M. Yung, Speckle in optical coherence tomography, *Journal of Biomedical Optics*, 4, 95–105, January 1999.

69. N. Iftimia, B. E. Bouma, and G. J. Tearney, Speckle reduction in optical coherence tomography by path length encoded angular compounding, *Journal of Biomedical Optics*, 8, 260–263, April 2003.

70. M. Bashkansky and J. Reintjes, Statistics and reduction of speckle in optical coherence tomography, *Optics Letters*, 25, 545–547, April 15, 2000.
71. G. Farhat , A. Mariampillai, V. X. D. Yang, G. J. Czarnota, and M. C. Kolios, Detecting apoptosis using dynamic light scattering with optical coherence tomography, *Journal of Biomedical Optics*, 16, 070505, July 2011.
72. R. A. Leitgeb, L. Schmetterer, W. Drexler, A. F. Fercher, R. J. Zawadzki, and T. Bajraszewski, Real-time assessment of retinal blood flow with ultrafast acquisition by color Doppler Fourier domain optical coherence tomography, *Optics Express*, 11, 3116–3121, November 17, 2003.
73. A. M. Rollins, S. Yazdanfar, J. K. Barton, and J. A. Izatt, Real-time *in vivo* colors Doppler optical coherence tomography, *Journal of Biomedical Optics*, 7, 123–129, January 2002.
74. P. Yu, L. Peng, M. Mustata, J. J. Turek, M. R. Melloch, and D. D. Nolte, Time-dependent speckle in holographic optical coherence imaging and the state of health of tumor tissue, *Optics Letters*, 29, 68–70, 2004.
75. K. Jeong, J. J. Turek, and D. D. Nolte, Imaging motility contrast in digital holography of tissue response to cytoskeletal anti-cancer drugs, *Optics Express*, 15, 14057–14064, 2007.
76. Y. Pan, E. Lankenau, J. Welzel, R. Birngruber, and R. Engelhardt, Optical coherence gated imaging of biological tissues, *IEEE Journal of Selected Topics in Quantum Electronics*, 2, 1029, 1996.
77. S. C. W. Hyde, R. Jones, N. P. Barry, J. C. Dainty, P. M. W. French, K. M. Kwolek, D. D. Nolte, and M. R. Melloch, Depth-resolved holography through turbid media using photorefraction, *IEEE Journal of Selected Topics in Quantum Electronics*, 2, 965–975, December 1996.
78. P. Yu, M. Mustata, J. J. Turek, P. M. W. French, M. R. Melloch, and D. D. Nolte, Holographic optical coherence imaging of tumor spheroids, *Applied Physics Letters*, 83, 575–577, July 21, 2003.
79. M. Laubscher, M. Ducros, B. Karamata, T. Lasser, and R. Salathe, Video-rate three-dimensional optical coherence tomography, *Optics Express*, 10, 429–435, May 6, 2002.
80. A. Dubois, K. Grieve, G. Moneron, R. Lecaque, L. Vabre, and C. Boccara, Ultrahigh-resolution full-field optical coherence tomography, *Applied Optics*, 43, 2874–2883, May 10, 2004.
81. B. Karamata, M. Leutenegger, M. Laubscher, S. Bourquin, T. Lasser, and P. Lambelet, Multiple scattering in optical coherence tomography. II. Experimental and theoretical investigation of cross talk in wide-field optical coherence tomography, *Journal of the Optical Society of America A-Optics Image Science and Vision*, 22, 1380–1388, July 2005.
82. S. C. W. Hyde, N. P. Barry, R. Jones, J. C. Dainty, and P. M. W. French, Sub-100 micron depth-resolved holographic imaging through scattering media in the near-infrared, *Optics Letters*, 20, 2330–2332, 1995.
83. D. D. Nolte, Semi-insulating semiconductor heterostructures: Optoelectronic properties and applications, *Journal of Applied Physics*, 85, 6259–6289, 1999.
84. I. Yamaguchi and T. Zhang, Phase-shifting digital holography, *Optics Letters,* 22, 1268–1270, August 15, 1997.
85. E. Cuche, F. Bevilacqua, and C. Depeursinge, Digital holography for quantitative phase-contrast imaging, *Optics Letters*, 24, 291–293, March 1, 1999.

86. F. Dubois, L. Joannes, and J. C. Legros, Improved three-dimensional imaging with a digital holography microscope with a source of partial spatial coherence, *Applied Optics*, 38, 7085–7094, December 1, 1999.

87. T. C. Poon, T. Yatagai, and W. Juptner, Digital holography—Coherent optics of the 21st century: Introduction, *Applied Optics*, 45, 821–821, February 10, 2006.

88. K. Jeong, J. J. Turek, and D. D. Nolte, Fourier-domain digital holographic optical coherence imaging of living tissue, *Applied Optics*, 46, 4999–5008, 2007.

89. G. Indebetouw and P. Klysubun, Imaging through scattering media with depth resolution by use of low-coherence gating in spatiotemporal digital holography, *Optics Letters*, 25, 212–214, February 15, 2000.

90. D. D. Nolte, *Optical Interferometry for Biology and Medicine*. New York: Springer, 2011.

91. S. H. Chen and F. R. Hallett, Determination of motile behavior of prokaryotic and eukaryotic cells by quasi-elastic light-scattering, *Quarterly Reviews of Biophysics*, 15, 131–222, 1982.

92. A. Palmer, T. G. Mason, J. Y. Xu, S. C. Kuo, and D. Wirtz, Diffusing wave spectroscopy microrheology of actin filament networks, *Biophysical Journal*, 76, 1063–1071, February 1999.

93. M. Heckmeier, S. E. Skipetrov, G. Maret, and R. Maynard, Imaging of dynamic heterogeneities in multiple-scattering media, *Journal of the Optical Society of America a-Optics Image Science and Vision*, 14, 185–191, January 1997.

94. K. K. Bizheva, A. M. Siegel, and D. A. Boas, Path-length-resolved dynamic light scattering in highly scattering random media: The transition to diffusing wave spectroscopy, *Physical Review E*, 58, 7664–7667, December 1998.

95. K. Jeong, J. J. Turek, and D. D. Nolte, Speckle fluctuation spectroscopy of intracellular motion in living tissue using coherence-domain digital holography, *Journal of Biomedical Optics*, 15, 030514, May–June 2010.

96. D. D. Nolte, R. An, J. J. Turek, and K. Jeong, Holographic tissue dynamics spectroscopy, *Journal of Biomedical Optics*, 16, 087004, 2011.

97. D. D. Nolte , R. An, J. J. Turek, and K. Jeong, Tissue dynamics spectroscopy for three-dimensional tissue-based drug screening, *Journal of Laboratory Automation*, 16, 431–442, 2011.

98. Y. Feng, T. J. Mitchison, A. Bender, D. W. Young, and J. A. Tallarico, Multiparameter phenotypic profiling: Using cellular effects to characterize small-molecule compounds, *Nature Reviews Drug Discovery*, 8, 567–578, July 2009.

99. D. Thomas, J. Cortes, F. Giles, S. Faderl, S. O'Brien, M. B. Rios, J.-P. Issa et al., Combination of hyper-CVAD with imatinib mesylate (STI571) for Philadelphia (PH)-positive adult acute lymphocytic leukemia or chronic myelogenous leukemia in lymphoid blast phase, *Blood*, 98, 803a, 2001.

100. H. Kantarjian, C. Sawyers, A. Hochhaus, F. Guilhot, C. Schiffer, C. Gambacorti-Passerini, D. Niederwieser et al., Hematologic and cytogenetic responses to imatinib mesylate in chronic myelogenous leukemia, *New England Journal of Medicine*, 346, 645–652, 2002.

101. P. J. Keller, F. Pampaloni, and E. H. K. Stelzer, Life sciences require the third dimension, *Current Opinion in Cell Biology*, 18, 117–124, February 2006.

102. E. Cukierman, R. Pankov, D. R. Stevens, and K. M. Yamada, Taking cell-matrix adhesions to the third dimension, *Science*, 294, 1708–1712, November 23, 2001.

# 6

## Lasers in Medicine: Innovations and Applications

Thomas D. Weldon

**CONTENT**

The ability to use technology and tools for the betterment of humankind is the hallmark differentiating our species from all others on the planet. In the development of tools for use in the treatment of disease, very little is truly unique. Medical products, or devices, typically are the combination of existing technologies, brought together to solve specific clinical unmet needs. Lasers are an excellent example of a technology that has been adapted for medical use in order to solve a variety of medical problems (Figure 6.1). Laser energy is efficient and controllable and can be transmitted through tiny optical fibers, making it an ideal energy source for use with catheter-based devices (Figure 6.2).

I have been directly involved with starting dozens of medical device companies and have contributed to the development of nearly as many medical products. Two of the more interesting applications of laser technology in clinical practice I have been involved with include the treatment of atrial fibrillation (AF) and the treatment of severe acne.

The human heart pumps blood by generating an electrical signal from its natural pacemaker, which travels through fibers in the cardiac tissue, finally reaching the apex of the heart, causing contractions of the cardiac muscle. In patients who suffer from AF, the atrial chambers of the heart rapidly quiver (*fibrillate*), instead of contracting in a controlled, healthy manner. The electrical signal travels through the proper pathways to the apex of the heart, but an extra signal is generated, predominantly in the pulmonary veins, directly reaching the tissue of the atrium, causing it to beat at much higher rates. This defeats its purpose as an antechamber and eventually reduces cardiac output. This condition was first treated with drugs and/or surgery, but drugs were ineffective and surgery involved serious safety issues.

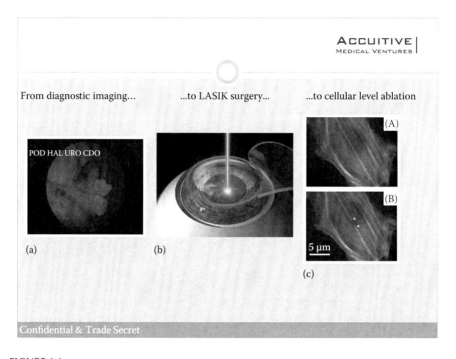

**FIGURE 6.1**
**(See color insert.)** Medical application examples. (a) Compounds find the tumor and a laser is used to activate fluorescence. (From Peng, Q. et al., *Rep. Prog. Phys.*, 71, 056701, 2008.) (b) Correct vision by vaporizing tissue to permanently change shape of the cornea. (From www.advanced-visionnetwork.com.) (c) Ablation of actin filaments inside a cell with femtosecond pulsed laser, (A) before laser and (B) after laser. (From Peng, Q. et al., *Rep. Prog. Phys.*, 71, 056701, 2008.)

The development of devices to treat AF rapidly evolved, and catheter-based systems utilizing radio-frequency (RF) energy became the standard of care. These devices are inserted through a vein in the leg and moved up into the heart. The RF energy is deployed to create a burn, or lesion, on the inside of the heart tissue, thus making it impossible for an electrical signal to cross the lesion. The lesions created by the catheter are, however, small round dots. Literally connecting the dots to wall off the errant electrical signal is very time-consuming and often unsuccessful if all the dots do not result in a completely continuous line.

CardioFocus, a company in which our venture capital (VC) firm has invested, was created both to address the limitations of RF catheters by employing a laser to reduce the total treatment time and to increase the efficacy of the procedure (Figure 6.3). The CardioFocus catheter is serially positioned into each of the pulmonary veins, and a compliant balloon is inflated so that it is pressing circumferentially against the pulmonary vein

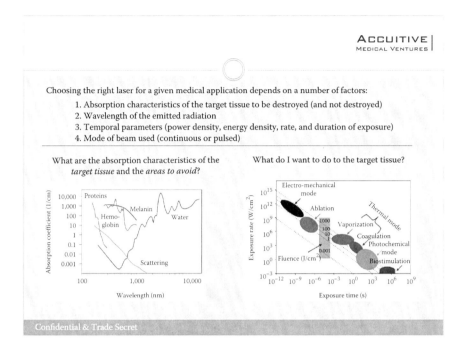

**FIGURE 6.2**
**(See color insert.)** Selecting a medical laser. (From Peng, Q. et al., *Rep. Prog. Phys.*, 71, 056701, 2008.)

wall. A channel inside the catheter contains fiber-optic strands, and the distal tip of this cable is perpendicular to the center line of the catheter inside the balloon at the end of the catheter system. When the laser is fired through the fiber-optic cable, the tip is rotated inside the balloon in separate arcs, eventually creating a lesion that is a circle around the diameter of the pulmonary vein. This lesion then blocks the errant electrical signal from reaching the tissue of the atrium, thus curing AF (Figure 6.4).

A second laser-based company, Sebacia, was started at The Innovation Factory to address the needs of patients with severe acne. The condition is caused by the sebaceous glands producing an excess of sebum, clogging up the pores and causing a blemish on the surface of the skin. The current treatment gold standard is to take the drug Accutane. It is effective, but has several serious side effects. Sebacia set out to develop a nonsystemic approach to the treatment of acne. The first approach proposed by Sebacia was to work a photosensitive drug into the skin and then activate the drug using laser light. To avoid the need for lengthy, expensive, and complex Food and Drug Administration (FDA) approval of a new drug, an innovative device treatment was preferred. It was suggested that the development focus shift to a

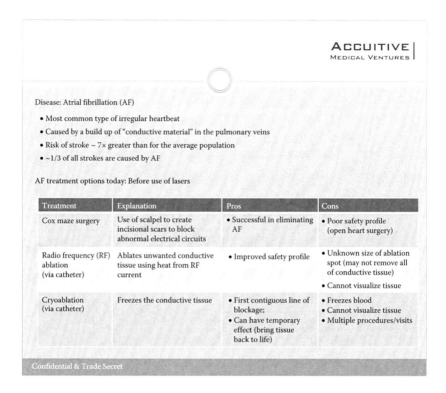

**FIGURE 6.3**
Cardiology example—atrial fibrillation.

device-based approach, with no biologic activity from a drug or compound contributing to the clinical outcome (Figure 6.5).

A solution infused with gold microshells is worked into the skin area to be treated. The shells are worked down into the pores and near the sebaceous glands. A laser, tuned to the gold microshells, is fired onto the surface of the skin. The laser energy is not absorbed by the skin and passes through the upper layers to reach the microshells. The shells become excited by the laser energy and radiate heat, which in turn kills the sebaceous glands. This potentially represents a cure for acne (Figure 6.6).

Both of the VC-funded companies described earlier have employed lasers in creative ways to address poorly met clinical needs. The use of lasers in medical device applications will continue to evolve, because of the unique advantages of the energy source itself and the precision and flexibility of its delivery. Over time, hopefully, many other poorly met clinical needs can be better addressed using creative approaches, such as the examples described earlier.

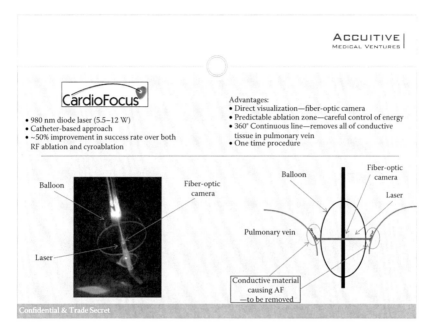

**FIGURE 6.4**
Laser to remove conductive tissue.

**FIGURE 6.5**
Dermatology example—acne.

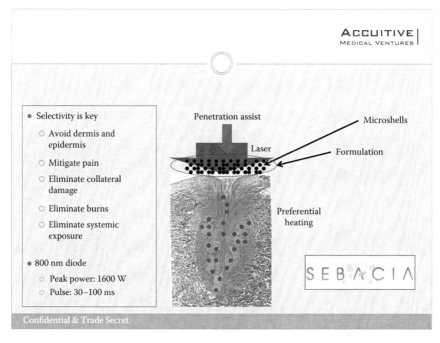

**FIGURE 6.6**
Laser to deliver "Accutane-like" results.

# Reference

Peng, Q. et al. (2008) Lasers in medicine, *Rep. Prog. Phys.*, 71, 056701.

# 7

## Shaping Ultrafast Laser Fields for Photonic Signal Processing

Andrew M. Weiner

### CONTENTS

### Introduction

Laser and photonics technologies are capable of very high speeds and band-widths, far beyond those available with state-of-the-art electronics such as those used in computers and wireless communications equipment. This capability is developed and exploited in two main technology areas: light-wave communications and ultrafast optics.

Lightwave refers to the use of optical signals—that is, light—to carry information, principally over glass fiber-optic cables. The highly purified glass used for fiber optics exhibits extremely low optical loss, transmitting light for up to 50 km with only a factor of 10 attenuation. Furthermore, the loss can be compensated using in-line, optical fiber amplifiers, enabling signal transmission over transoceanic distances. In contrast, attenuation over electrical cables is many orders of magnitude stronger for high-frequency electrical signals; wired transmission of high-speed electrical data in the many-gigahertz (GHz) range is practically ruled out for any substantial distance due to excessive electrical loss of all known materials. Furthermore, due to the ability to modulate light at very high speeds and then propagate with low distortion over broad bandwidth, transmission of data at terabit/s (Tb/s or $10^{12}$ bit/s) rates can be achieved. These unique capabilities make lightwave the primary information carrier enabling the high-speed Internet.

Optical technologies such as mode-locked lasers provide signals that are even faster, such as visible light pulses with durations as short as a couple of femtoseconds (1 femtosecond, or 1 fs, is equal to $10^{-15}$ s). The shortest of such pulses have reached the ultimate limit to the pulse duration of a single optical cycle; in this limit, the light field contains frequency components spanning the visible spectral range and appears white in color. Texts providing comprehensive treatments of ultrafast optics technology may be found, for example, in Diels and Rudolph (2006) and Weiner (2009). Even shorter pulses in the attosecond regime are possible when such fields interact with matter in a strongly nonlinear fashion to produce light in the deep ultraviolet or x-ray spectral region (1 attosecond, or 1 as, is equal to $10^{-18}$ s). Pulsed optical fields in the femtosecond and attosecond regimes represent time scales much faster than available from any other known technology and hence provide many new opportunities. The application space for ultrashort pulses is now extremely broad, including but not limited to high-field laser–matter interactions, ultrafast time-resolved spectroscopy, high-precision frequency metrology and development of optical clocks, machining and materials processing, nonlinear microscopy, optical communications, and radio-frequency (RF) signal processing. Within the world of science, applications abound. In time-resolved spectroscopy, for example, such fields act as the world's shortest strobe pulses, enabling stop-action measurements of ultrafast motions within semiconductors and molecules, including the initial steps in photochemical reactions. In another aspect, the peak power of a signal equals its energy divided by its duration. Because the durations are so short, femtosecond pulses can reach incredibly high peak powers, enabling investigation of matter excited to extreme conditions otherwise inaccessible under controlled laboratory conditions.

Within an engineering context, one prominent area of application is communications. In addition to the ability to generate ultrashort, broad bandwidth signals, techniques for precision manipulation of such signals are also very important. In this chapter, I provide an overview of optical pulse-shaping methods that enable programmable reshaping of ultrafast laser pulses, or generation of arbitrary optical waveforms, according to user specifications. These techniques enable precision control of ultrafast optical fields at time scales far faster than available with high-speed electronics, thus circumventing so-called electronic bottlenecks. In particular, in the "Optical Pulse Shaping" section, I will first introduce the field of femtosecond pulse shaping, including some examples of applications in ultrafast optical science. In the "Applications in Fiber Optics" section, I will describe some experiments relevant to optical fiber communications and highlight the use of pulse shaping to overcome distortions typically encountered by short-pulse signals. In the "Applications in the Ultrabroadband Radio-Frequency Photonics" section, I will discuss the use of pulse shaping in hybrid optical–electrical systems to generate arbitrary short-pulse RF electrical signals with bandwidths difficult to achieve using electronic technologies alone and illustrate the application to compensation of signal distortions encountered in wireless systems. In the "Conclusion" section, I will conclude.

## Optical Pulse Shaping

Ultrafast laser systems now routinely generate pulses of light with durations deep into the femtosecond range, much shorter than available in any other engineered system. Although this opens up many new opportunities, it also raises new technical challenges. As one example, reshaping ultrashort pulses produced by femtosecond lasers into user-specified waveforms for specific applications becomes difficult, because the relevant time scale is too fast for direct modulation under electronic control. Femtosecond pulse shaping refers to a process, implemented using special optical hardware, of taking an ultrafast input optical signal and modifying it to produce a desired output signal with features controllable on an ultrafast time scale (Weiner 2000, 2009, 2011). The process can have applications in a wide range of areas—ranging from high-speed optical telecommunications to ultrafast optical science, nonlinear microscopy, life sciences, and engineering applications. Future potential uses of interest include ultrabroadband RF photonics, single-shot biochemical detection, optical particle acceleration, and even quantum computing.

Pulse shaping requires the convolution of an input signal with a filter function, which is most commonly implemented in the optical frequency domain using a so-called Fourier transform pulse-shaping approach (Weiner 2000, 2009, 2011). As illustrated in Figure 7.1, the different optical frequencies

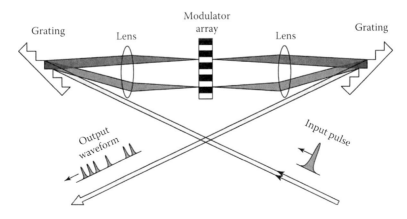

**FIGURE 7.1**
**(See color insert.)** Basic setup for Fourier transform optical pulse shaping.

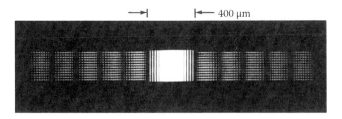

**FIGURE 7.2**
Intensity mask used for generation of ultrafast square pulse. (From Weiner, A.M. et al., *J. Opt. Soc. Am. B*, 5, 1563, 1988.)

(or colors) that make up the input field are first separated using well-known optical components such as a diffraction grating and lens. At the back focal plane of the first lens, the amplitude and phase of each frequency component may be independently adjusted using a complex spatial mask. After passing through another lens and grating, the different optical frequencies are reassembled into a single spatial beam. The resulting waveform in the time domain is given by the inverse Fourier transform of the complex pattern transferred from the spatial mask onto the optical spectrum. In the early days of pulse shaping, masks were fabricated in advance using microlithographic patterning techniques to place a fixed amplitude and phase modulation onto the expected range of optical input frequencies (Weiner et al. 1988). An image of one such fixed pulse-shaping mask is presented in Figure 7.2. However, for the greatest versatility, it is quite useful to have the ability to program the Fourier pulse-shaping waveform dynamically. The technology capable of achieving this functionality is known today as the spatial light modulator (SLM). Commonly used types of SLMs include liquid-crystal SLMs, acousto-optic SLMs, and deformable mirrors and micromirror arrays based on microelectromechanical systems (MEMS) technology (Weiner 2000, 2011). SLMs used in pulse shaping are related to computer and other display technologies; a difference is that computer displays usually use incoherent light sources, whereas SLMs are usually optimized for use with coherent (i.e., laser) light. The Fourier transform pulse shaper concept exploits parallel manipulation of optical frequencies to overcome severe speed limitations of (opto)electronic modulator technologies, resulting in the generation of complex photonic signals modulated on a femtosecond time scale according to user specification. Thus, waveforms with effective serial modulation bandwidths as high as hundreds of terahertz can be generated and manipulated, without requiring ultrafast modulators.

A detailed theoretical analysis of pulse shaping is given in Weiner (2009). Here, we will confine our discussion to a simple level, for which we may envision pulse-shaping operation as illustrated in Figure 7.3. An input pulse, with complex amplitude functions in the time and frequency domains denoted $e_{in}(t)$ and $E_{in}(\omega)$, respectively, passes through a pulse shaper with

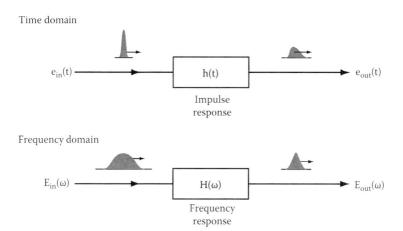

**FIGURE 7.3**
Pulse shaping viewed as a linear filter. In the time domain, the shaped output pulse is the convolution of the input pulse with the impulse response function of the filter. In the frequency domain, the complex spectrum at the output is equal to the product of the complex input spectrum and the frequency response function of the filter. These pictures are connected by a Fourier transform relationship.

spatial mask M(x). M(x) is taken to be complex to represent both intensity and phase filtering. The output field in the spectral domain is written

$$E_{out}(\omega) = M(\alpha\omega)E_{in}(\omega) \tag{7.1}$$

where $\alpha = \partial x / \partial \omega$ is the spatial dispersion at the masking plane. In this description of the pulse shaper, the output spectrum is given by the input spectrum multiplied by a complex frequency response function equal to a directly scaled version of the mask. The output field in time is obtained from the inverse Fourier transform of Equation 7.1:

$$e_{out}(t) = e_{in}(t) * m\left(\frac{t}{\alpha}\right) \tag{7.2}$$

where

$$m\left(\frac{t}{\alpha}\right) = \frac{1}{2\pi} \int M(\alpha\omega) e^{j\omega t} \, d\omega \tag{7.3}$$

and the "*" symbol represents convolution. The output field is given by the input field convolved with the impulse response function of the pulse shaper, which is equal to the inverse Fourier transform of the scaled masking function.

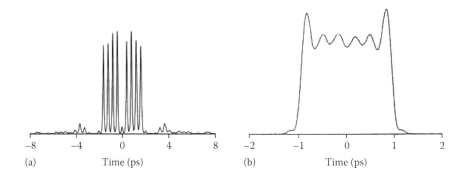

**FIGURE 7.4**
Pulse-shaping examples. (a) Femtosecond data packet. (b) Ultrafast square pulse.

Figure 7.4 shows two examples of shaped pulses generated early on in the author's laboratory. In each of these examples, experiments were performed using ~100 fs duration input pulses. The time-dependent intensity profiles of shaped pulses were measured by nonlinear mixing with unshaped reference pulses (directly from the laser) as a function of relative delay, a standard technique in ultrafast optics, which may be considered the optical equivalent of sampling oscilloscopes widely used for measurement of fast electronic signals. Figure 7.4b shows an ultrafast optical square pulse, in which the spectrum is patterned according to a sinc function (Weiner et al. 1988). The intensity profile exhibits an ~2-ps, approximately flat-topped region, with fast rise and fall times consistent with the original 100 fs pulse duration. The overshoot and ringing arise from truncation of the sinc-function spectrum at finite bandwidth and are expected. Figure 7.4a shows a femtosecond *data packet*, in which the single input pulse is reshaped into a sequence of evenly spaced pulses, with eight pulses on and one pulse missing (Weiner and Leaird 1990). The pulse spacing is approximately 400 fs, corresponding to peak data rate of 2.5 Tb/s, far beyond the modulation capabilities of electronic approaches. An interesting point in this example is that because the phase of individual pulses in the output waveform was not specified, it was possible to accomplish the pulse shaping using manipulation of spectral phase alone.

It is worth commenting on two contrasting control strategies for programmable pulse shaping, namely, open-loop control and feedback (adaptive) control, both of which are depicted in Figure 7.5. In the open-loop configuration, the desired output waveform is specified by the user, and reasonable knowledge of the input pulse is also usually available. Therefore, the desired transfer function is known, and one simply programs the pulse-shaping SLM to provide this transfer function. The examples shown earlier fall into this class.

The ability to program a pulse shaper under computer control leads also to an alternative adaptive control strategy (Baumert et al. 1997, Yelin et al. 1997,

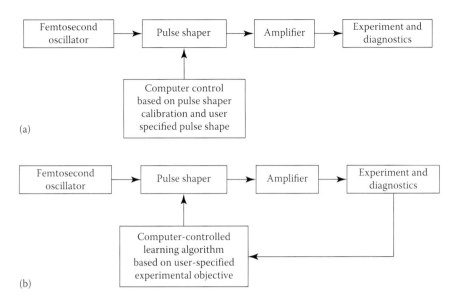

**FIGURE 7.5**
Control strategies for programmable pulse shaping. (a) Open-loop control. (b) Feedback or adaptive control.

Meshulach et al. 1998). In these experiments, one usually starts with a random spectral pattern programmed into the pulse shaper, which is updated iteratively according to a stochastic optimization algorithm based on the difference between a desired and measured experimental output. In this scheme, there is no need to explicitly program the pulse shaper. This adaptive control scheme is less intuitive but is often viewed as especially suitable for optimization of strongly nonlinear processes or for manipulation of quantum mechanical motions in systems, such as molecules, where knowledge of the dynamics is not sufficiently accurate (Judson and Rabitz 1992). In such cases, the adaptive control approach can be used to search for the laser waveform, which gives the best experimental result, as judged according to a user-defined metric. Such adaptive optimization schemes are well known to the industrial engineering community. Experimental examples are reviewed briefly in Weiner (2011) and include laser-controlled chemistry, (Bardeen et al. 1997, Assion et al. 1998, Brixner et al. 2001, Levis et al. 2001), selective enhancement of short-wavelength radiation from atoms driven by strong laser fields (Bartels et al. 2000), manipulation of energy flow in light-harvesting molecules (Herek et al. 2002), and spatially selective excitation of subwavelength metallic nanostructures (Aeschlimann et al. 2007).

As one relatively recent example, the last experiment named earlier, involving pulse-shaping control of nanooptical fields, is sketched in Figure 7.6 (Aeschlimann et al. 2007). Here, a submicron scale silver pattern fabricated onto a substrate serves as a nanoantenna that enhances the

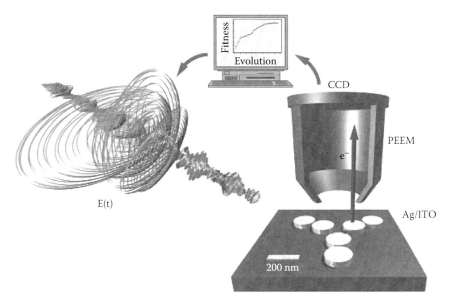

**FIGURE 7.6**
Conceptual view of experiments demonstrating pulse-shaping control of nanooptical fields on metallic nanostructures. (From Aeschlimann, M. et al., *Nature*, 446(7133), 301, 2007.)

collection of incident light. The nanoantenna has spatial features smaller than the optical wavelength, which represents the usual limit to which light can be focused. A generalized pulse shaper provides light pulses that are temporally shaped not only in the usual phase and amplitude, but also in polarization (orientation of the laser beam's electric field). The resulting excitation of the nanoantenna is observed through photoemission electron microscopy, that is, by imaging the electrons photoemitted in response to the illuminating light in an electron microscope. Because the spatial resolution of the electron microscope is not limited by the optical wavelength, it can therefore reveal the region of the nanoantenna from which photoemitted electrons originate. Furthermore, because the experiment relies on a nonlinear photoemission process, imaging the photoemitted electrons provides a kind of subwavelength image of the peak optical intensity captured at different locations on the nanoantenna. The electron microscope images are fed into a computer running an adaptation algorithm that seeks to vary the shape of the incident laser field in order to optimize the yield of electrons emitted from specific subregions of the nanoantenna. Remarkably, this procedure is successful, arriving at different laser waveforms that respectively peak up the electron yield (hence, optical intensity) either from the base or from the upper branches of the Y-shaped nanostructure. An important conclusion emerging from these experiments is that spectral–temporal shaping of a far-field waveform can affect subwavelength spatial degrees of freedom in the optical near-field. In the long term, such studies may enhance the ability to address electronic devices with deep-subwavelength size scales using optical means. This in turn would have implications for the use of light to circumvent serious communications bottlenecks affecting today's high-speed electronic integrated circuit technologies.

## Applications in Fiber Optics

Figure 7.7 shows an end-on view of a strand of optical fiber. The fiber is made of silica glass that has been purified to minimize the absorption of light. The central region, less than 10 µm in diameter, is doped in order to increase its refractive index, resulting in a waveguide structure that very effectively confines light within this central core. Figure 7.8 shows the transmission loss of modern optical fibers as a function of laser wavelength (Essiambre et al. 2010). The minimum transmission loss is below 0.2 dB attenuation per kilometer, meaning that more than 10% of the light remains even after 50 km. This is in sharp contrast to electrical cables, which suffer transmission losses in the hundreds of dB per kilometer at GHz frequencies. Furthermore, the loss remains low for bandwidths of tens of THz (at wavelengths around the ~1550 nm loss minimum, the conversion from bandwidth in wavelength units to bandwidth in frequency units is 8 nm/THz). In most systems, the bandwidth is limited by that of in-line erbium-doped fiber amplifiers (EDFAs) to about 5–10 THz, depending on the amplifier details. Nevertheless, the bandwidth

**FIGURE 7.7**
End-on view of a single-mode optical fiber. The bright area in the center is the waveguide core, less than 10 µm in diameter, in which light propagates with low loss.

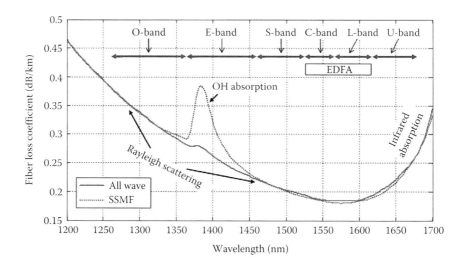

**FIGURE 7.8**
**(See color insert.)** Transmission loss of modern glass fibers (dB/km vs. wavelength). Different frequency bands considered for fiber-optic transmission are noted, as is the operating band for the dominant optical amplifier technology (EDFAs). The principal physical mechanisms defining the fiber loss are also indicated. (From Essiambre, R.-J. et al., *J. Lightwave Technol.*, **28**(4), 662, 2010.)

remaining still far exceeds that of other approaches. The combination of very low propagation loss and very high bandwidth has made lightwave transmission using fiber-optic cables a key enabler for today's high-speed communications systems operating over long distances at low cost. For a detailed discussion, refer, for example, to Agrawal (2010).

Because the bandwidth capacity of fiber optics is very large and beyond that accessible with electronic processors, strategies are needed to access the bandwidth. In the most common strategy, known as wavelength-division multiplexing (WDM, Figure 7.9a), the fiber bandwidth is divided into different frequency channels. Different, independently modulated laser sources producing different frequencies (or wavelengths or colors) of light carry the data in the different channels. Individual lasers operate at tens of giga-symbols/s, compatible with electronic processing. However, by having up to hundreds of lasers operating in different wavelength channels, overall capacities of Tb/s are achieved. Approaches inspired by pulse shaping allow manipulation of different wavelengths on a channel by channel basis. In a second strategy, known as time-division multiplexing (TDM, Figure 7.9b), the modulation rate of individual channels becomes much faster, and accordingly, the bandwidth of the individual channels becomes much larger. In the limit of very short pulses and very fast modulation, only a single channel remains. Although this TDM approach reduces the need for large inventories of lasers controlled to operate at many different wavelengths, it creates

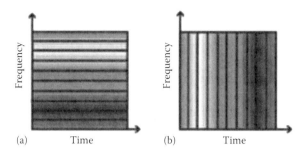

**FIGURE 7.9**
Common schemes for using fiber bandwidth. (a) WDM. (b) TDM.

new challenges, especially for handling severe distortions accompanying ultrashort pulse propagation in long lengths of fiber. Perhaps for this reason, WDM remains the dominant approach in today's commercial lightwave systems. TDM has been the subject of substantial research, and impressive transmission experiments demonstrating bit rates exceeding 1 Tb/s have been reported (e.g., Yamamoto et al. 2003). Pulse shaping has been applied to compensate waveform distortions in such contexts, described later in this section.

As sketched in Figure 7.10, we may generalize our picture of pulse shaping to consider various forms of broadband inputs, including not only ultrashort pulses but also continuous-wave lasers that are modulated with data and multiple wavelength sources. For applications in WDM, manipulation is performed on a wavelength-by-wavelength basis, with no concern for phase between channels. This is in sharp contrast to shaping of ultrashort pulses, where phase coherence across the entire spectrum is an essential ingredient. Pulse-shaping arrangements have been mostly applied for manipulating

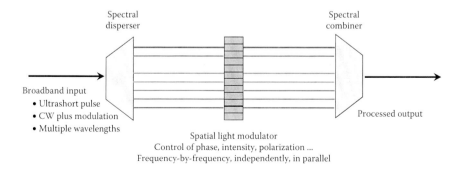

**FIGURE 7.10**
**(See color insert.)** Generalized view of pulse shaping, including not only ultrashort pulses but also continuous-wave lasers that are modulated with data and multiple wavelength sources. In the optical communications community, signal manipulation based on generalized pulse-shaping geometries is often termed dynamic wavelength processing.

the power spectrum of WDM signals, for example, to correct for the effect of wavelength-dependent optical amplification or to achieve wavelength-selective optical switching (where a switch handling multiple wavelengths may be programmed to impose different, independent switching operations on different wavelengths). Recently, increasing attention is being given to manipulation of spectral phase within individual wavelength channels as well. Within the lightwave communications area, pulse-shaping arrangements and their adaptations are often now referred to by terms such as *dynamic wavelength processors* and *dynamic spectral equalizers*. Although not shown in the figure, for applications such as wavelength-selective switching, the apparatus may be modified to encompass multiple input and output fibers. A common theme is that in all cases, one manipulates light frequency by frequency, independently and in parallel.

An application known as dynamic gain equalization is important because optical fiber amplifiers have a gain factor that varies with optical frequency. Repeated passage through such fiber amplifiers in an optical network results in a strong intensity difference between different WDM frequencies, which may seriously degrade data transmission. As an example, in one early pre-commercial experiment (Figure 7.11), 36 WDM channels spread over a frequency range of nearly 4 THz centered around 1.55 μm wavelength exhibited a spread of 7.1 dB between the intensity of each channel; the intensity spread was reduced by a factor of 10 under pulse-shaping action (Ford et al. 2004). An important point is that because this approach is programmable, such equalizers can track slow changes in the channel intensities. Pulse shapers have also been generalized to realize wavelength-selective switches, in which different WDM channels can be routed, programmably and independently, to different output fibers (Patel and Silberberg 1995, Ford et al. 1999,

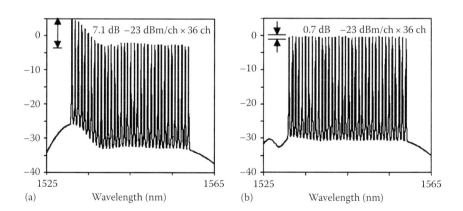

**FIGURE 7.11**

Spectral gain equalization using a pulse shaper actuated by an MEMS structure that acts as an intensity SLM. (a) Before equalization. (b) After equalization. (From Ford, J.E. et al., *IEEE J. Sel. Top. Quant. Electron.*, 10(3), 579, 2004.)

Roelens et al. 2008). Such capability allows network operators to implement different network connections simultaneously on a single fiber infrastructure by using wavelength as an independent variable.

We now illustrate application of spectral phase control. This can be used for compensation of chromatic dispersion, which broadens the durations of signals sent through fiber-optic links. Unless compensated, such broadening leads to intersymbol interference, which limits the bit rates of high-speed optical fiber communication links. Physically, chromatic dispersion represents frequency-dependent group velocity and frequency-dependent delay. A key relation, based on Fourier transform theory, is that the frequency-dependent delay $\tau(\omega)$ is fundamentally related to the derivative of the frequency-dependent phase $\psi(\omega)$, according to

$$\tau(\omega) = \frac{-\partial\psi(\omega)}{\partial\omega} \tag{7.4}$$

Hence, one can compensate for the effect of dispersion by programming a pulse shaper to impose a spectral phase function equal and opposite to that specified earlier.

A block diagram of experiments demonstrating such dispersion compensation is shown in Figure 7.12. Here, a programmable pulse shaper was used to complement fiber dispersion-compensation techniques in propagating sub-500 fs pulses over optical fiber links ranging from 3 km in early experiments to 50 km most recently (Chang et al. 1998, Shen and Weiner 1999, Jiang et al. 2005). In all cases, the link consisted of a length of standard single-mode fiber (SMF) concatenated to an approximately matching length of dispersion-compensating fiber (DCF). Since SMF and DCF have dispersion with opposite signs at the operating wavelength, the fiber lengths can be adjusted to cancel all of the lowest-order dispersion (i.e., phase varying quadratically with frequency). Remaining pulse distortions, arising, for example, due to uncompensated cubic spectral phase (dispersion slope in the terminology of fiber optics), are corrected by using the pulse shaper as

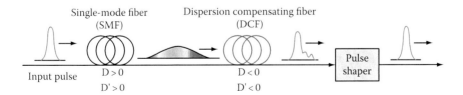

**FIGURE 7.12**
Schematic view of fiber dispersion compensation using a pulse shaper as a programmable spectral phase equalizer. Similar schemes involving pulse shapers are used to compensate residual spectral phase in chirped pulse amplifier systems and few cycle pulse compression experiments.

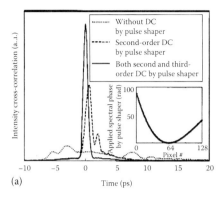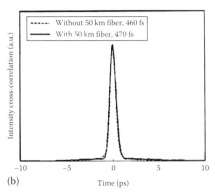

**FIGURE 7.13**

Data from experiments in which <500 fs pulses are transmitted through 50 km of standard SMF, with subsequent compression using DCF. (a) Intensity profiles of pulses after various degrees of further compression using the pulse shaper for spectral phase equalization. Inset: applied spectral phase profile for full compression. (b) Comparison of pulses before and after transmission. With spectral phase equalization, the pulse duration is preserved.

a spectral phase equalizer. In the experiments in Jiang et al. (2005), ~460 fs input pulses at 1542 nm center wavelength are first broadened 10,000 times to ~5 ns in propagating through 50 km of SMF, then recompressed by the DCF to ~14 ps, as shown in Figure 7.13. Although compression by the DCF is roughly 99.7% effective, the residual pulse broadening is still a factor of 30 relative to the bandwidth limit. Most of this broadening is due to a mismatch in fiber lengths, amounting to 120 m of SMF. This can be compensated by programming the pulse shaper for an appropriate quadratic spectral phase, which further compresses the pulse to within a factor of 2 of the original duration. Compensation of the residual dispersion is achieved by programming the shaper to superimpose an appropriate additional cubic phase variation. An unwrapped view of the total applied phase, comprising almost 100 rad of phase variation, is shown in the inset to Figure 7.13a. This leads to a completely recompressed pulse with essentially the original pulse duration (~470 fs) and no observable distortion (Figure 7.13b). Thus, in these experiments, all-fiber techniques are used for coarse dispersion compensation, while a programmable pulse shaper is used as a spectral phase equalizer to fine-tune away any remaining dispersion.

In a completely analogous fashion, programmable pulse shapers are now used extensively in compensation of residual dispersion in femtosecond amplifier systems and in generation or compression of pulses at durations comprising only a few optical cycles, approaching fundamental limits.

It is worth noting that although the unwrapped spectral phase pictured in Figure 7.13 is quite large, liquid-crystal SLMs are in fact limited to relatively small phase shifts, typically a few π. The actual phase function is applied with phase wraps that occur modulo 2π, which fundamentally does not change the pulse-shaping operation. For discussion of practical limitations, see, for example, Weiner et al. (1992) and Weiner (2009).

In addition to experiments with ultrashort pulses, tunable dispersion compensation of conventional laser sources modulated with data in the range 10–40 Gb/s has also been demonstrated via pulse shaping (Ooi et al. 2002, Sano et al. 2003, Lee et al. 2006, Marom et al. 2006). Applications of such tunable dispersion compensation include replacing various lengths of conventional DCFs in lower data rate systems and compensating residual dispersion after the DCF to meet the tight dispersion tolerances required in some higher data rate systems. It is worth noting that substantial success has also been achieved recently in compensating dispersion electronically (via digital signal processing of the received signal) as an alternate to hardware optical compensators such as described earlier.

## Applications in the Ultrabroadband Radio-Frequency Photonics

In the area of RF photonics, one seeks to use photonics technologies to augment the capabilities of RF electrical or wireless systems (Minasian 2006, Capmany and Novak 2007, Yao 2009). Popular applications that have been explored include fiber transmission of high-frequency RF signals over distances that would be impossible over electrical cables, optical control of phased-array radars, and photonic implementation of RF filters. In a relatively new subfield of RF photonics, which I have termed ultrabroadband RF photonics, one seeks to exploit the high-speed capabilities of photonic approaches to achieve new functionality for generation and manipulation of ultrawideband (UWB) (often pulsed) electrical signals, with instantaneous bandwidths beyond those which can be handled by conventional electronic solutions. In the following, I describe examples of ultrabroadband RF arbitrary waveform generation as well as ultrabroadband RF pulse compression through phase compensation. Manipulation of broadband RF phase was virtually unexplored in RF photonics until relatively recently.

UWB wireless, a short-distance communications modality introduced into practice about a decade ago, furnishes a nice application example (Reed 2005). As defined by the Federal Communications Commission (FCC), UWB covers frequencies from 3.1 to 10.6 GHz (see Figure 7.14a). What is novel about UWB is the wide frequency band. Conventional RF applications involve much narrower bands. Furthermore, frequency bands are subject to license, and literally dozens of licensed narrow frequency bands already exist in the 3.1–10.6 GHz UWB band. Also, in UWB, the wide frequency band permits the use of very short RF pulses, termed impulse radio (Figure 7.14b). The use of short pulses and high bandwidth provides a number of potentially attractive features—both for commercial systems and for defense applications: high

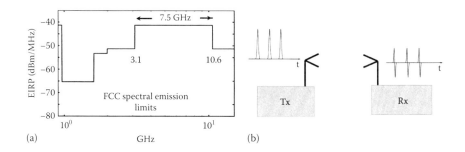

**FIGURE 7.14**
(a) Spectral emission limits, as defined by the FCC, for UWB wireless. EIRP: equivalent isotropic radiated power. (b) Schematic view of impulse radio, in which the large bandwidth available through UWB supports the transmission of pulses.

time resolution (for RF localization applications), high data rate, resistance to multipath interference, and license-free overlay with conventional narrowband services. A constraint is that the power spectral density of UWB transmitters is regulated to low levels to avoid interference with preexisting narrowband services occupying the same band. This limits UWB systems to short distances. Another potential advantage that is less commonly cited is low probability of intercept, which may be of interest for secure communication applications. Practically, electronics operating speeds originally limited pulsed UWB systems to only about 1.5 GHz of instantaneous bandwidth, that is, only 20% of the full UWB band (Reed 2005). Photonics-enabled approaches allow generation of ultrabroadband RF signals with instantaneous bandwidth spanning the UWB band (and well beyond). This enabled experiments that were not possible at the time with available electronic technologies; examples are discussed later in this section. Although electronic waveform generators have recently improved to cover the UWB band, photonics approaches scale at least an order of magnitude beyond.

Photonics also bring new ways of thinking into RF practice. Conventional RF systems are usually designed to operate at low instantaneous bandwidth with signals possessing a well-defined center frequency (that may however be tuned slowly over a large range). Furthermore, although such systems generally employ coherent, phase-based signaling schemes within their narrow instantaneous frequency band, control of phase across large-frequency bands is usually not considered. On the other hand, optical systems can be extremely broadband, generating pulses with durations down to femtoseconds and with terahertz instantaneous bandwidths. Such ultrashort pulse optical systems are fundamentally time domain in nature, and control of phase over the full bandwidth is crucial and commonly practiced—as in the fiber dispersion-compensation experiments described earlier. A little later in this section, we will discuss the extension of such ideas to the RF domain for compensation of dispersion in commercial antennas. Another point

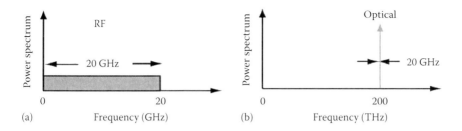

**FIGURE 7.15**

(a) Representation of electrical signal with large fractional bandwidth. (b) Representation of equivalent bandwidth placed on an optical carrier, now comprising very small fractional bandwidth.

empowering optics is its high center frequencies. At wavelengths around 1.5 μm employed for telecommunications, the optical frequency is $2 \times 10^{14}$ Hz. As per the illustration of Figure 7.15, consider a signal that would normally be considered to have a very large instantaneous bandwidth (say 20 GHz). In the electrical domain, this signal might span frequencies from 0 to 20 GHz; such a signal would have multiple octave fractional bandwidth and would be very hard to deal with. However, converted to the optical domain, such a signal would constitute only a very small 0.01% fractional bandwidth, allowing propagation or processing with minimal difficulty. Bandwidths that are large and challenging to work with in the RF domain are relatively small and much easier to handle when translated into the optical regime.

Optical pulse-shaping technology can be directly exploited to realize arbitrary waveform generation capability for UWB RF electrical signals. As depicted in Figure 7.16a, an ultrashort optical input pulse is first shaped as desired, then directed to a fast optical-to-electrical converter (O/E). By controlling the optical excitation waveform onto the O/E, programmable

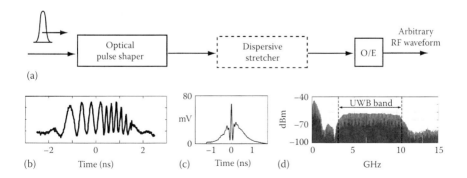

**FIGURE 7.16**

(a) Photonics-enabled RF arbitrary waveform generator, including optional dispersive fiber stretcher. (b) 1.2-2.4-4.9 GHz chirp waveform. (c) Impulsive waveform with fine structure designed to achieve (d) flat RF power spectrum.

cycle-by-cycle synthesis of burst RF waveforms can be achieved. In some cases, an optical fiber acts as a dispersive stretcher, or a frequency-to-time converter, in order to match the optical waveform duration to the desired electrical waveform duration (Chou et al. 2003, Lin et al. 2005). Different choices of pulse shaper configurations, coupled with different choices of O/E converter technologies, have allowed demonstrations of waveform generation from the GHz to the THz. Hence, this approach is scalable over several orders of magnitude in RF frequency. An in-depth discussion of ultrabroadband RF waveform generation is presented in McKinney and Weiner (2013).

Figure 7.16b through d shows examples of subnanosecond RF waveform burst waveforms approximately within the 3.1–10.6 GHz band allocated by the FCC for UWB wireless communications. These waveforms were generated using the modified generator approach, in which the optical power spectrum of a short optical pulse is tailored using a pulse shaper, followed by optical frequency-to-time conversion in a dispersive medium (Chou et al. 2003, Lin et al. 2005). Figure 7.16b shows a signal with abrupt frequency hops inserted on a cycle-by-cycle basis (Lin et al. 2005), while Figure 7.16c and d shows the temporal and spectral profiles of a modulated impulsive signal designed to yield nearly flat RF power spectrum over the full 3.1–10.6 GHz band (McKinney et al. 2006). The flat power spectrum allows maximized transmitted pulse energy, hence increasing range and signal-to-noise, in a radar or wireless communication system with constrained peak power spectral density. Such waveforms, generated using photonics means more than 7 years ago, substantially exceeded the bandwidth of the electronic waveform generators then available and could only be generated via photonics. Although electronic solutions capable of generating such signals have now become available, photonics scales to much higher frequencies. Examples of waveforms generated using a pulse shaper followed directly by an O/E (no frequency-to-time mapping) are shown in Figure 7.17. Figure 7.17a shows a burst millimeter-wave signal consisting of 3 cycles at 48 GHz changing abruptly, on a cycle-by-cycle basis, to 2 cycles at 24 GHz (McKinney et al. 2002). Also shown is the driving optical waveform; here, the 60 GHz bandwidth photodiode acts as a low-pass filter, smoothing the optical pulse train into the desired smooth RF electrical waveform. Figure 7.17b shows a periodic millimeter-wave signal with abrupt phase modulations at ~50 GHz (McKinney et al. 2003). Waveforms such as these are available only via photonic solutions and open up new possibilities for impulsive radar, electronic warfare, RF sensing, and secure wireless communications. Figure 7.17c shows a final example in a still higher frequency (THz) regime (Liu et al. 1996). In this case, the O/E is a special photoconductive antenna device. The waveform pair demonstrates the ability to intentionally inject an abrupt phase shift into a THz signal, again something not possible by any other current means.

As one example of the new opportunities enabled, here we discuss the use of such electrical waveform shaping to allow compensation of dispersive effects arising from the antennas used in UWB wireless links. This

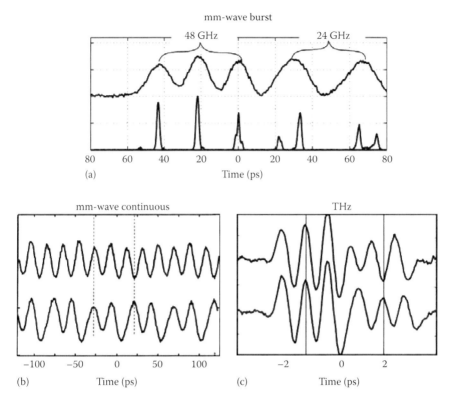

**FIGURE 7.17**
Photonically generated RF waveforms at higher frequencies and bandwidths. (a) 48/24 GHz burst waveform. (b) Periodic and shaped waveforms at 50 GHz: (top) unmodulated sinusoid; (bottom) with abrupt phase modulation. (c) THz waveforms with abrupt phase modulation. In (b) and (c), vertical lines are to visualize the phase modulation.

work is closely analogous to the optical dispersion-compensation experiments described earlier and constitutes the first hardware implementation of dispersion compensation in the UWB electrical domain. Antenna dispersion, associated with a frequency-dependent phase response and hence frequency-dependent delay, is an important issue in UWB, as many common antennas that have been optimized for broadband amplitude response do not exhibit linear spectral phase. A related point is that although RF components, including antennas, may be designed for systems that are tunable over a broad frequency range (e.g., 2–18 GHz), they are often not designed for impulsive signals with broad instantaneous bandwidth. The concept of optimizing the antenna feed voltage to obtain desirable temporal properties in the received waveforms, such as peak amplitude or minimal duration, had been explored theoretically (Pozar 2003), but had not been tested experimentally due to lack of waveform generation capability. Photonic waveform

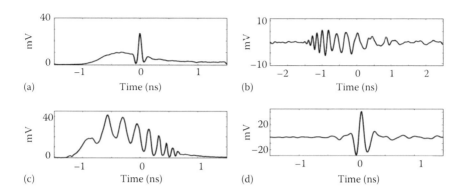

**FIGURE 7.18**

Precompensation of broadband dispersive antenna link. (a) Impulse drive waveform and (b) resulting received signal showing strong pulse distortion. (c) Predistorted drive waveform, resulting in (d) compressed output.

generator technology enabled our group to publish the first such experiments (McKinney and Weiner 2006, McKinney et al. 2008).

An example is shown in Figure 7.18. First, an ~200 ps RF impulse excites a commercial log-periodic antenna for transmission over a short line-of-sight wireless link. The log-periodic antenna, an example of a so-called frequency-independent antenna, exhibits strong dispersion, leading to a 10-fold broadening in the received signal and an obvious frequency sweep. We then generate an intentionally predistorted electrical drive waveform, with frequency-dependent delay opposite to that of the dispersive antenna link. The output signal is now recompressed to nearly the original 200 ps duration. Furthermore, the received peak power, normalized to the peak drive power, is increased nearly 20-fold (13 dB). Such pulse compression experiments, though commonplace in optical systems, are unprecedented in RF systems with the ~10 GHz instantaneous bandwidths considered here. The ability to precompensate antenna distortions could significantly extend the choices of antenna structures that can be applied to UWB wireless.

A final point is that in the experiments earlier, the antennas are located only 1–2 m apart, and the signal propagates over a direct wireless path from the transmitter to the receiver. For real applications in indoor wireless, however, the channel is not so simple. In particular, transmit and receive antennas may be located in different rooms with no direct path between them. In such non-line-of-sight channels, the input signal experiences strong multiple scattering before reaching the receiver. As a result, a short-pulse transmit signal suffers a complex signal distortion, reaching the receiver as a series of echoes arriving randomly within a certain time spread. This is known as multipath distortion. In the last couple of years, we have been studying measurement and compensation of multipath distortion in ultra-broadband RF propagation using both electronic (Dezfooliyan and Weiner

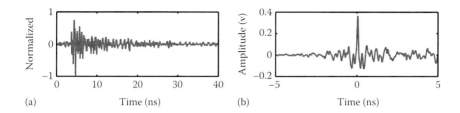

**FIGURE 7.19**
Recent experimental results on ultrabroadband wireless propagation in indoor environment showing strong multipath distortion. (a) Channel impulse response. (b) Received waveform after precompensation using a photonically shaped RF drive signal.

2012) and photonic waveform generator approaches. An example using an advanced photonic waveform generator, which provides greater bandwidth, is shown in Figure 7.19 (Dezfooliyan and Weiner 2013). Figure 7.19a shows the measured impulse response between 2 antennas located about 10 m apart and separated by a cement wall. The response consists of a series of echoes spread over ~ 30 ns, a delay spread that is more than 1000 times larger than the ~20 ps fundamental time resolution corresponding to the 2–18 GHz transmit bandwidth. The impulse response is much less regular than that from the line-of-sight antenna link of Figure 7.18b and is spread over substantially longer time. Generating waveforms suitable for compensation of such a complicated distortion required advances in our photonic waveform generators, which we have recently achieved. As a result, we have been able to demonstrate compensation of the multipath distortion as in Figure 7.19b. Here, a strong central peak is obtained with duration below 80 ps, substantially below the ~30 ns delay spread of the uncompensated impulse response. By mitigating the severe multipath distortion, it may become possible to achieve higher communication rates for data-hungry applications such as multimedia wireless interconnects. Furthermore, since multipath distortion signatures such as that in Figure 7.19a are specific to the precise antenna locations, one may selectively compensate the response seen by a desired receive antenna, while leaving the response at a nearby receive antenna nearly uncompensated. This behavior, under current investigation in our laboratory, leads to new opportunities for covert or private wireless communication.

# Conclusion

In summary, I have first introduced an approach for shaping of ultrafast laser fields that enables the generation of nearly arbitrary optical waveforms on time scales too fast for direct modulation under electronic control.

This technology is now widely employed in laboratories around the world, with applications in diverse areas of ultrafast science and technology. Generalizations of pulse-shaping technology have led to modules deployed in lightwave communications networks. In this article, I have focused on research in photonic signal processing directed at transmission or manipulation of information. These activities are relevant to either fiber-optic communications or RF electrical signal generation and wireless systems. We may anticipate that the ability to programmably generate waveforms on an ultrafast time scale and according to user specification will continue to lead to fascinating new opportunities, both in optical science and in engineering applications.

## Acknowledgments

I gratefully acknowledge Dr. Dan Leaird for providing the fiber photo in Figure 7.7. This work was supported in part by the National Science Foundation (grant number ECCS-1102110) and the Naval Postgraduate School (grant number N00244-09-1-0068) under the National Security Science and Engineering Faculty Fellowship program. (Any opinions, findings, and conclusions or recommendations expressed in this publication are those of the authors and do not necessarily reflect the views of the sponsors.)

## References

Aeschlimann, M. et al. (2007). Adaptive subwavelength control of nano-optical fields. *Nature* **446**(7133): 301–304.

Agrawal, G. P. (2010). *Fiber-Optic Communication Systems*, Wiley, Hoboken, NJ.

Assion, A. et al. (1998). Control of chemical reactions by feedback-optimized phase-shaped femtosecond laser pulses. *Science* **282**: 919–922.

Bardeen, C. J. et al. (1997). Feedback quantum control of molecular electronic population transfer. *Chemical Physics Letters* **280**: 151–158.

Bartels, R. et al. (2000). Shaped-pulse optimization of coherent emission of high-harmonic soft x-rays. *Nature* **406**: 164–166.

Baumert, T. et al. (1997). Femtosecond pulse shaping by an evolutionary algorithm with feedback. *Applied Physics B* **65**: 779–782.

Brixner, T. et al. (2001). Photoselective adaptive femtosecond quantum control in the liquid phase. *Nature* **414**(6859): 57–60.

Capmany, J. and D. Novak (2007). Microwave photonics combines two worlds. *Nature Photonics* **1**(6): 319–330.

Chang, C.-C. et al. (1998). Dispersion-free fiber transmission for femtosecond pulses using a dispersion-compensating fiber and a programmable pulse shaper. *Optics Letters* **23**: 283–285.

Chou, J. et al. (2003). Adaptive RF-photonic arbitrary waveform generator. *IEEE Photonics Technology Letters* **15**(4): 581–583.

Dezfooliyan, A. and A. M. Weiner (2012). Experimental investigation of UWB impulse response and time reversal technique up to 12 GHz: Omnidirectional and directional antennas. *IEEE Transactions on Antennas and Propagation* **60**(7): 3407–3415.

Dezfooliyan , A. and A. M. Weiner (2013). Temporal focusing of ultrabroadband wireless signals using photonic radio frequency arbitrary waveform generation. *Optical Fiber Communication Conference*, Optical Society of America, Anaheim, CA.

Diels, J. C. and W. Rudolph (2006). *Ultrashort Laser Pulse Phenomena*, Academic Press, San Diego, CA.

Essiambre, R.-J. et al. (2010). Capacity limits of optical fiber networks. *Journal of Lightwave Technology* **28**(4): 662–701.

Ford, J. E. et al. (1999). Wavelength add-drop switching using tilting micromirrors. *Journal of Lightwave Technology* **17**(5): 904–911.

Ford, J. E. et al. (2004). Interference-based micromechanical spectral equalizers. *IEEE Journal of Selected Topics in Quantum Electronics* **10**(3): 579–587.

Herek, J. L. et al. (2002). Quantum control of energy flow in light harvesting. *Nature* **417**(6888): 533–535.

Jiang, Z. et al. (2005). Fully dispersion-compensated ~500 fs pulse transmission over 50 km single-mode fiber. *Optics Letters* **30**(12): 1449–1451.

Judson, R. S. and H. Rabitz (1992). Teaching lasers to control molecules. *Physical Review Letters* **68**: 1500.

Lee, G. H. et al. (2006). Optical dispersion compensator with > 4000-ps/nm tuning range using a virtually imaged phased array (VIPA) and spatial light modulator (SLM). *IEEE Photonics Technology Letters* **18**(17–20): 1819–1821.

Levis, R. J. et al. (2001). Selective bond dissociation and rearrangement with optimally tailored, strong-field laser pulses. *Science* **292**: 709–713.

Lin, I. S. et al. (2005). Photonic synthesis of broadband microwave arbitrary waveforms applicable to ultra-wideband communication. *IEEE Microwave and Wireless Components Letters* **15**: 226–228.

Liu, Y. et al. (1996). Terahertz waveform synthesis via optical pulse shaping. *IEEE Journal of Selected Topics in Quantum Electronics* **2**: 709–719.

Marom, D. M. et al. (2006). Compact colorless tunable dispersion compensator with 1000-ps/nm tuning range for 40-Gb/s data rates. *Journal of Lightwave Technology* **24**(1): 237–241.

McKinney, J. D. et al. (2002). Millimeter-wave arbitrary waveform generation with a direct space-to-time pulse shaper. *Optics Letters* **27**: 1345–1347.

McKinney, J. D. et al. (2003). Photonically assisted generation of arbitrary millimeter-wave and microwave electromagnetic waveforms via direct space-to-time optical pulse shaping. *Journal of Lightwave Technology* **21**(12): 3020–3028.

McKinney, J. D. et al. (2006). Shaping the power spectrum of ultra-wideband radio-frequency signals. *IEEE Transactions on Microwave Theory and Techniques* **54**(12): 4247–4255.

McKinney, J. D. et al. (2008). Dispersion limitations of ultra-wideband wireless links and their compensation via photonically enabled arbitrary waveform generation. *IEEE Transactions on Microwave Theory and Techniques* **56**(3): 710–719.

McKinney, J. D. and A. M. Weiner (2006). Compensation of the effects of antenna dispersion on UWB waveforms via optical pulse-shaping techniques. *IEEE Transactions on Microwave Theory and Techniques* **54**(4): 1681–1686.

McKinney, J. D. and A. M. Weiner (2013). Photonic synthesis of ultrabroadband arbitrary electromagnetic waveforms. *Microwave Photonics*. C. H. Lee. (ed.), CRC Press, Boca Raton, FL.

Meshulach, D. et al. (1998). Adaptive real-time femtosecond pulse shaping. *Journal of the Optical Society of America B* **15**: 1615–1619.

Minasian, R. A. (2006). Photonic signal processing of microwave signals. *IEEE Transactions on Microwave Theory and Techniques* **54**(2): 832–846.

Ooi, H. et al. (2002). 40-Gb/s WDM transmission with virtually imaged phased array (VIPA) variable dispersion compensators. *Journal of Lightwave Technology* **20**(12): 2196–2203.

Patel, J. S. and Y. Silberberg (1995). Liquid crystal and grating based multiple-wavelength cross-connect switch. *IEEE Photonics Technology Letters* **7**: 514–516.

Pozar, D. M. (2003). Waveform optimizations for ultrawideband radio systems. *IEEE Transactions on Antennas and Propagation* **51**(9): 2335–2345.

Reed, J. H. (2005). *An Introduction to Ultra Wideband Communication Systems*, Prentice Hall, Upper Saddle River, NJ.

Roelens, N. A. F. et al. (2008). Dispersion trimming in a reconfigurable wavelength selective switch. *Journal of Lightwave Technology* **26**(1–4): 73–78.

Sano, T. et al. (2003). Novel multichannel tunable chromatic dispersion compensator based on MEMS and diffraction grating. *IEEE Photonics Technology Letters* **15**(8): 1109–1110.

Shen, S. and A. M. Weiner (1999). Complete dispersion compensation for 400-fs pulse transmission over 10-km fiber link using dispersion compensating fiber and spectral phase equalizer. *IEEE Photonics Technology Letters* **11**: 827–829.

Weiner, A. M. (2000). Femtosecond pulse shaping using spatial light modulators. *Review of Scientific Instruments* **71**: 1929–1960.

Weiner, A. M. (2009). *Ultrafast Optics*, Wiley, Hoboken, NJ.

Weiner, A. M. (2011). Ultrafast optical pulse shaping: A tutorial review. *Optics Communications* **284**(15): 3669–3692.

Weiner, A. M. and D. E. Leaird (1990). Generation of terahertz-rate trains of femtosecond pulses by phase-only filtering. *Optics Letters* **15**: 51–53.

Weiner, A. M. et al. (1988). High-resolution femtosecond pulse shaping. *Journal of the Optical Society of America B* **5**: 1563–1572.

Weiner, A. M. et al. (1992). Programmable shaping of femtosecond pulses by use of a 128-element liquid-crystal phase modulator. *IEEE Journal of Quantum Electronics* **28**: 908–920.

Yamamoto, T. et al. (2003). 1.28 Tbit/s-70-km OTDM femtosecond-pulse transmission using third- and fourth-order simultaneous dispersion compensation with a phase modulator. *Electronics and Communications in Japan Part I-Communications* **86**(3): 68–79.

Yao, J. P. (2009). Microwave photonics. *Journal of Lightwave Technology* **27**(1–4): 314–335.

Yelin, D. et al. (1997). Adaptive femtosecond pulse compression. *Optics Letters* **22**(23): 1793–1795.

# 8

## Terahertz Wave Air Photonics: Bridging the Gap and Beyond

Xi-Cheng Zhang

### CONTENTS

Terahertz (THz) wave sensing and imaging have received a great deal of attention in the past decade owing to its significant scientific and technological potentials in various multidisciplinary fields. However, two major challenges were present in the development and application of remote open-air broadband THz spectroscopy in addressing the urgent needs of homeland security, the fields of astronomy, and environmental monitoring: (1) Due to the limited performance of existing emitters and sensors in the far-infrared

(IR) range, there was a THz gap, and (2) due to strong water vapor absorption, remote sensing with broadband THz waves was considered impossible. The requirement for on-site bias or forward collection of the optical signal in conventional THz detection techniques has inevitably prohibited their use in remote sensing. An "all-optical" technique of broadband THz wave detection has been recently developed to tighten the mentioned gap. The technique is based on coherent manipulation of the fluorescence emission from asymmetrically ionized gas plasma interacting with THz waves. Unique features of measured THz pulses at standoff distances (30 m) are minimal water vapor absorption and unlimited directionality for optical signal collection. This chapter presents an introduction to the THz wave air photonics technique, along with its applications in various fields such as biomedical imaging, explosive identification, pharmaceuticals, food safety, defense, and security.

## Overview

THz waves, also known as submillimeter radiation or T-rays, refer to the electromagnetic (EM) waves of frequencies 0.1–10 THz. These waves lie between the IR and microwave ranges of EM radiation. THz waves also exhibit some similar properties of the two, among which are penetrability into cloth, paper, cardboard, wood, plastic, and ceramic. THz waves can be radiated from both natural and artificial sources. Objects at room temperature (300 K) emit thermal energy in this range (6 THz). Half the luminosity and 98% of the photons released since the Big Bang fall into the submillimeter and far-IR regions. From gigahertz to THz frequencies, numerous organic molecules exhibit strong absorption and dispersion due to rotational and vibrational transitions. These transitions are specific to the targets and enable THz fingerprinting (Hauser and Dwek, 2001). THz waves have low photon energies (4 meV at 1 THz, one million times weaker than an x-ray photon at keV) and will not cause harmful photoionization in biological tissues. Thus, THz wave could be a superior means for sensing and imaging purposes.

Photoconductive antennas (PCAs) and electro-optic (EO) sampling have been widely used in recent years for detection of broadband THz radiation in various applications, for example, biomedical imaging, nondestructive inspection, and material characterization (Ferguson and Zhang, 2002; Mittleman, 2003; Tonouchi, 2007). THz wave generation and detection method has been developed over the recent decades, offering a powerful yet safe approach toward a large range of phenomena. Figure 8.1 shows examples of the THz sensing and imaging systems (courtesy of Zomega). THz radiation

 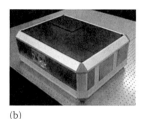 

(a)  (b)  (c)

**FIGURE 8.1**

(a) Mini-Z THz TDS: The most compact, fully integrated THz TDS, weighing in at less than 5 lb with true turnkey operation. Produces and measures pulsed THz waves from 0.1 to 4.0 THz using TDS techniques in both transmission and reflection geometries with a waveform measurement of up to 500 Hz. (b) Z-series THz TDS: Produces and measures pulsed THz waves from 0.1 to 3.5 THz using TDS techniques in both transmission and reflection geometries with a spectral resolution of 5 GHz. (c) ZAP THz TDS: ZAP is a THz TDS that uses air plasma to generate and detect ultrabroadband THz from 0.3 to 20 THz. Coherent detection over the entire THz spectrum and into the far-IR, with electric fields in excess of 100 kV cm$^{-1}$ unlocks a new field of nonlinear THz spectroscopy, for the first time available in a commercial package.

generated from air plasma offers much broader bandwidth than PCA or EO rectification, since plasmas are phonon-free and their dispersion is negligible. This is a significant property in meeting the merging requirements of homeland security and environmental science for broadband remote THz spectroscopy. Focusing two-color optical beams is a potential solution to the remote THz wave generation. However, due to the strong absorption of ambient water vapor in the THz band and the difficulties inherent to remote optical signal collection, realization of broadband THz remote sensing is even more challenging. In THz wave detection, collection of the second-harmonic beam is difficult at standoff distances due to weak scattering (Liu et al., 2010). Furthermore, application of biased PCA or EO crystal for THz wave remote sensing is impractical.

The previously developed technique of THz air-biased coherent detection (ABCD) provides ultrabroadband sensing capability. Figure 8.2 shows the THz ABCD detector module, the Zomega Air Photonics (ZAP)-BDP, commercialized by Zomega Terahertz Corporation. The THz ABCD technique, however, requires electrodes or wires near the target to provide local oscillator for heterodyne detection. Hence, it cannot be used in remote standoff measurement applications (Zhang and Xu, 2009). Recent studies have involved innovative mixes of ultrafast laser and THz technologies, namely, "THz wave air photonics," to enhance broadband THz remote sensing. Examples include radiation-enhanced emission of fluorescence (REEF) and THz-enhanced acoustics (TEA) (Zhang and Xu, 2009). The ability to extend THz to remote sensing scenarios is especially relevant to standoff detection of explosives and hazardous agents and environmental monitoring applications.

**FIGURE 8.2**
The THz ABCD detector module, commercialized by Zomega, the ZAP-BDP. The detection module implements the detection part of the system in a convenient and relatively easy to setup package.

## THz Wave Air Photonics

### Generation in Plasma

THz gas photonics research was initiated by focusing on using gas plasmas as THz emitters. The idea is to frequently double an intense ultrafast laser pulse, with a center wavelength of 800 nm, by a nonlinear crystal, for example, barium borate (BBO). The gas is then ionized by mixing 800 and 400 nm pulses, creating a filament of plasma to emit intense and broadband THz radiation. Perturbative four-wave mixing is a simplified version of the generation mechanism, that is,

$$E_{THz} = \chi^{(3)} E_\omega E_\omega E_{2\omega}^* + c.c. \tag{8.1}$$

where $\chi^{(3)}$ denotes the effective third-order nonlinear susceptibility of the gas. The two fundamental photons with frequency of $\omega$ are mixed with a second-harmonic photon of frequency $2\omega$, through the effective third-order susceptibility of plasma. Thus, the result of the process will be a photon of frequency $\Omega = 2\omega - \omega - \omega$. The physical mechanism of the THz radiation is a two-step process of asymmetric ionization followed by collisions with

**FIGURE 8.3**
**(See color insert.)** Standoff THz generation with ~30 fs laser pulses.

neighboring atoms. In this mechanism, the plasma has to be generated for THz emission to occur. Hence, this physical requirement means that a powerful laser is necessary for this technique (see Figure 8.3).

### Detection in Plasma

The broadband generation mechanism can be inverted to realize broadband detection. In the four-wave mixing approach, two fundamental photons of frequency ω are coupled with a THz photon with frequency of Ω, generating a second-harmonic photon of frequency 2ω, that is,

$$E_{2\omega}^{\text{THz}} \propto \chi^{(3)} E_\omega E_\omega E_{\text{THz}}^{(*)} + c.c. \tag{8.2}$$

Detectors for the second harmonic, however, are only sensitive to the "intensity" and not the "electric field." Hence,

$$I_{2\omega} \propto \left(\chi^{(3)}\right)^2 I_\omega^2 E_{\text{THz}}^2. \tag{8.3}$$

With sufficiently high pulse energies at 800 nm, a 400 nm local oscillator could be generated within the plasma, allowing the THz to be detected quasi-coherently (ABCD was first called air-breakdown coherent detection).

However, this technique places strong limits on both the probe and THz power. As a modified alternative technique, the electrodes can be used to directly provide a local oscillator (ABCD now is called air-biased coherent detection). This heterodyne technique is known as ABCD and is the main principle corresponding to the ZAP coherent detector (Figure 8.2b). In the following are the general forms of the second-harmonic electric field (8.4) and the second-harmonic intensity (8.5):

$$E_{2\omega} \propto \chi^{(3)} E_{\omega} E_{\omega} \left( E_{\text{THz}} + E_{\text{LO}} \right) + c.c. \tag{8.4}$$

$$I_{2\omega} \propto \left( \chi^{(3)} \right)^2 I_{\omega}^2 \left( E_{\text{THz}} + E_{\text{LO}} \right)^2 = \left( \chi^{(3)} \right)^2 I_{\omega}^2 \left( E_{\text{THz}}^2 + 2E_{\text{THz}}E_{\text{LO}} + E_{\text{LO}}^2 \right), \tag{8.5}$$

where $F_{\text{LO}}$ is the electric field generated across the electrodes. The benefits inherent to the ABCD mechanism include

1. Coherency and ability to distinguish between THz electric fields of opposite sign
2. Consuming lower probe pulse energies due to the improved sensitivity by the oscillator
3. Ultrabroadband

The limiting factors for the bandwidth are the laser pulse width and the phase mismatch between the fundamental and the second-harmonic beams.

### Experimental Setup

The general features and characteristics of the THz ABCD setup (Figure 8.3) are as follows:

- The harmonic beam is produced from the fundamental beam using a BBO crystal.
- The fundamental beam wavelength is 800 nm, and the harmonic is 400 nm.
- Both beams create plasma, thus generating the THz pulse.
- The optical pump beams are blocked by a Si filter.
- The THz and probe pulses are focused together in a gas biased by a high-voltage modulator.
- The resulting second-harmonic (400 nm) light is detected by a photomultiplier tube (PMT).
- The high voltage is modulated at a frequency of one-half the laser repetition rate.
- The signal from the PMT is measured by a lock-in amplifier.

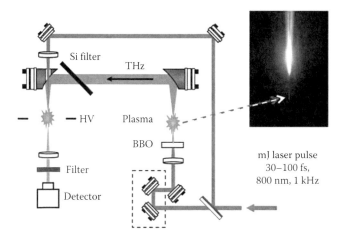

**FIGURE 8.4**
**(See color insert.)** The THz wave ABCD setup schema.

The minimum requirements for a laser system to make a THz ABCD system work are an amplified laser at 800 nm with a pulse energy of 1 mJ, a pulse duration of 40 fs, and a repetition rate between 1 and 10 kHz. The diagram of the THz ABCD setup is shown in Figure 8.4.

## All-Optical Standoff THz Wave Detection

Standoff broadband coherent THz wave detection has been recently realized by Liu et al. (2010). The idea is to probe the THz pulse with fully controllable two-color laser-induced gas plasmas in order to analyze the interaction by detecting the omnidirectional fluorescence emission. The high transparency of UV fluorescence in the atmosphere enhances circumvention of the sensing distance limitation that arises due to strong water vapor absorption in the THz region. The two-color laser field functions as a "remote optical modulator" for the THz REEF signal through coherent manipulation of the ionized electron drift velocity and subsequent collision-induced fluorescence emission. This technique circumvents the limitations of the on-site bias requirement, water vapor attenuation, and signal-collection direction at standoff distances and realizes the detection of broadband THz radiation from a distance of 10 m.

### Experimental Details

The THz REEF experiment with gas plasma excited by two-color laser fields, superposition of a linearly polarized fundamental pulse $E_\omega$, and its second-harmonic pulse $E_{2\omega}$ is conducted by propagating a 80 fs, 100 mJ, 800 nm laser pulse through a 250 mm type-I β-BBO. The relative phase, that is, $\phi_{\omega,2\omega}$, can be controlled by an in-line phase compensator consisting of an α-BBO time plate, a pair of fused silica wedges, and a dual-wavelength plate (DWP, ALPHALAS GmbH), with attosecond phase-control accuracy (Dai et al., 2009). The two-color laser beam is focused into air to generate plasma. The two optical intensities at the focus are $I_{\omega}$ (~$10^{13}-10^{14}$ W cm$^{-2}$) for the fundamental beam and $I_{2\omega}(= I_{\omega}/10)$ for the second-harmonic beam. The $E_{\omega}$ direction can be changed by rotating the DWP while keeping the $E_{2\omega}$ direction unchanged. To create an asymmetric electron drift velocity, $E_{\omega}$ and $E_{2\omega}$ are aligned parallel. The synthesized optical field $E_{Opt}$ can be expressed as follows:

$$E_{Opt} = E_{\omega}(t) + E_{2\omega}(t) = A_{\omega 0}(t)\cos(\omega t) + A_{2\omega 0}(t)\cos(2\omega t + \phi_{\omega,2\omega}), \quad (8.6)$$

where $A_{\omega 0}(t)$ and $A_{2\omega 0}(t)$, respectively, represent the envelopes of the fundamental (frequency ω) and the second-harmonic (frequency 2ω) pulses. For the conceptual demonstration of THz wave remote sensing, a single-cycle THz pulse $E_{THz}(t)$ with a peak field of 100 kV cm$^{-1}$ can be generated locally from a LiNbO$_3$ prism using an optical pulse with a tilted pulse front as the excitation (Yeh et al., 2007) and focused collinearly with the optical beam onto the plasma.

## Remote Sensing

Figure 8.5 shows the remote sensing mechanism. An experiment was conducted on THz wave remote sensing, using coherent manipulation of THz-wave-enhanced fluorescence from asymmetrically ionized gas. The two-color laser beam with parallel polarization was focused into air to generate plasma, with the relative phase being controlled by an in-line phase compensator (Dai et al., 2009). A single-cycle THz pulse with a peak field of 100 kV cm$^{-1}$ was focused collinearly with the optical beam onto the plasma. Figure 8.4 depicts a schematic of the experiment. A rotatable UV-enhanced concave mirror (M1) with a diameter of 200 mm and focal length of 500 mm was deployed to control the fluorescence emitted from the two-color laser-induced plasma. Then another UV plane mirror (M2) with a diameter of 75 mm was used to guide the emitted fluorescence into a PMT through a monochromator.

## Ionization Process

After passing the laser pulse in the laser-induced ionization processes, the electrons released from molecules or atoms get a constant drift velocity (Corkum et al., 1989), determined by the phase of the laser field at the birth of the free electron. Residual current density or asymmetric electron velocity distribution could remain in the plasma, ionized by single-color few-cycle pulses (Kresβ et al., 2006) or by two-color fields with optimized relative phase

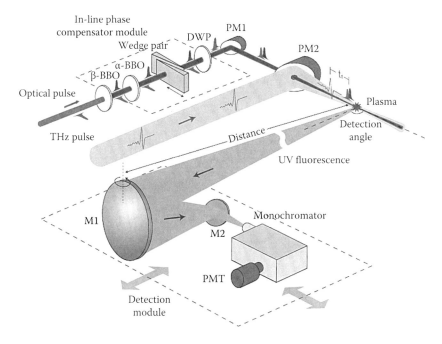

**FIGURE 8.5**
**(See color insert.)** Schematic of the THz wave remote sensing technique. (From Liu, J. et al., *Nat. Photonic.*, 4, 627, 2010.) The 2ω pulse is generated by passing the fundamental beam through a type-I β-BBO crystal. Both the fundamental and second-harmonic optical pulses are linearly polarized along a vertical direction. The relative phase change between the ω and 2ω pulses is tuned by the lateral translation of fused silica wedges in the optical beam path after the α-BBO. The two optical pulses are focused by a parabolic mirror (effective focal length, 150 mm) into air to generate plasma. The time delay $t_d$ is defined as the delay between the optical pulse peak and THz pulse peak. The fluorescence detection system consists of a UV concave mirror (M1; diameter of 200 mm and focal length of 500 mm), a UV plane mirror (M2), a monochromator, and a PMT. The distance of remote sensing is varied by moving the fluorescence detection system with respect to the plasma (DWP).

(Schumacher et al., 1994; Kim et al., 2008). Under irradiation from intense laser pulses, some of the excited electrons are trapped in high-lying states of atoms and molecules (Kulander et al., 1991; Talebpour et al., 1996a,b; Fedorov, 2006). Those trapped states have a large principal quantum number ($n \gg 1$) and are more easily ionized by collision with energetic electrons (Phelps, 1968; Azarm et al., 2008; Filin et al., 2009), as illustrated in Figure 8.6a. The interaction of laser-induced plasma with a THz wave increases the plasma temperature through electron acceleration. Electron impact produces more ionized gas species and subsequently generates more $N_2(C^3\Pi_u)$ through dissociative recombination (Xu et al., 2009). In the single-color, multi-cycle laser pulse excitation, THz REEF from nitrogen plasma quadratically depends on the THz field (Liu and Zhang, 2009). Similar phenomena were also observed in argon, krypton, and xenon gas plasmas.

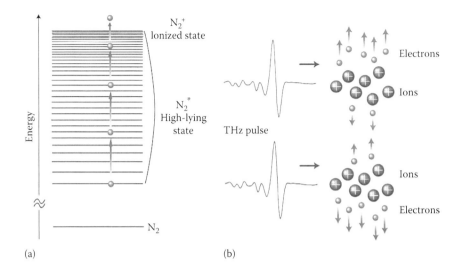

**FIGURE 8.6**
The THz-wave-assisted electron impact ionization of high-lying states in plasma. (From Liu, J. et al., *Nat. Photon.*, 4, 627, 2010.) (a) High-lying states can be ionized by a series of collisions with energetic electrons, (b) interaction between the THz pulse and the asymmetric photoelectron velocity distributions generated by two-color field ionization.

In contrast, the synthesized optical field of two-color pulses generates ionized electrons with an asymmetric drift velocity. The drift velocity distribution and electron trajectories can be controlled by the polarizations and relative phase of two optical fields $\phi_{\omega,2\omega}$ (Dai et al., 2009; Xu et al., 2009). After the passage of two-color pulses, the electric field of a single-cycle THz pulse applied to the laser-induced plasma alters the ionized electron momentum by acceleration or deceleration, depending on the initial velocity $v(0)$ of the electron (Figure 8.6b). Since both the amplitude and direction of the THz field affect plasma fluorescence, the THz waveform information is encoded into a change of fluorescence at a different time delay $t_d$ between the THz pulse and the optical pulses. We demonstrate that a THz waveform can be retrieved by measuring time-dependent fluorescence emission when $v(0)$ is aligned both parallel and antiparallel to $E_{THz}(t)$.

## THz Wave Time-Domain Spectroscopy

THz time-domain spectroscopy (TDS) is the most widely used technique to characterize the spectral properties of materials in the THz band. In particular, most materials, especially organic ones, have their vibration and rotational modes in the THz band. Therefore, THz TDS is poised as a suitable technique to study such modes and get an understanding of molecular structure and dynamics. THz TDS, in contrast with other alternative techniques such as Raman and Fourier transform infrared (FTIR) spectroscopy that measure "energy," measures "electric field." This feature makes THz TDS a coherent detection technique that allows recovering both phase and amplitude data. This coherence allows measuring the real and imaginary part of the dielectric constant of a material in the THz band directly. On the other hand, FTIR and Raman require the use of models and indirect computation (e.g., the Kramers–Kronig) for retrieving the complex dielectric constant. Accordingly, the unique features of THz TDS are as follows:

- *Fingerprint*: The rotational and vibrational modes of many molecules, especially organic ones, are distributed across the THz band. These modes can be observed as absorption peaks in the THz spectra. The specific location and amplitude of these absorption peaks can be used to identify the molecules.
- *See through*: Many dry and nonmetallic materials, such as plastic, paper, cardboard, and textiles, are transparent to THz waves. This property allows THz waves to inspect samples that are under cover or inside nonoptically transparent containers. Microwaves have this see-through capability as well, but their larger wavelength compared to THz waves does not allow achieving high-resolution images. IR can provide much better resolution than THz, but IR waves cannot see through covers.
- *Coherence*: THz TDS measures electric field, and phase and amplitude are readily available. The phase and amplitude can be used to measure the dielectric constant of a material in a more straightforward manner than with FTIR or Raman.

The basic application for THz TDS is the characterization of materials in either reflection or transmission geometries. The goal of this characterization is to extract the "optical" constants of the material, which can be related with the complex dielectric constant. Because the THz waveform is a measurement of the electric field of the THz wave, the absorption coefficient refers to the attenuation in electric field, which is half of the absorption coefficient in energy. The factor multiplying the reference amplitude in the denominator for the absorption coefficient considers the Fresnel losses due to refractive index mismatch between the material and the reference measurement, which is typically done in air. If the sample has a low refractive index, the Fresnel loss factor can be omitted.

Typical values for the refractive index of plastics in the THz range are in the range 1.3–2.0. For ceramics, typical values are between 2 and 3. For semiconductors, refractive index values are often above 3. Absorption

**TABLE 8.1**

Refractive Index and Absorption Coefficient of
Some Commonly Used Materials

| Material | Refractive Index | Absorption Coefficient |
|---|---|---|
| Sapphire | 3.096 | 1.9 |
| LaAlO$_3$ | 4.980 | 5 |
| HDPE | 1.534 | 0.45 |
| LiNbO$_3$ | 5.156 | 16 |
| PTFE | 1.431 | 0.2 |
| SiO$_2$ | 1.957 | 3.3 |
| Quartz | 2.109 | 0.15 |
| BK7 | 2.593 | 50 |
| GaAs | 3.550 | <1 |
| Si (regular) | 3.416 | 0.3 |
| Si (high resistivity) | 3.413 | 0.3 |
| FZ Si (high resistivity) | 3.418 | 0.3 |
| InP:Ir | 3.509 | <5 |
| ZnSe | 3.021 | 3 |
| GeBi | 4.042 | 5.9 |

*Source:* Courtesy of Zomega, http://www.z-thz.com/
index.php?option=com_content&view=article
&id=48&Itemid=55.

coefficient typically increases with frequency in plastic and ceramic materials due to scattering as the grain size and trapped air pockets have a typical size (~100 µm) that matches the wavelength of the frequencies in the higher part of the THz band. Table 8.1 shows the refractive index and absorption coefficient of commonly used materials in the THz range at 1 THz.

## Applications and Trends

### Biomedical Imaging

In vivo molecular imaging is considered as the next frontier in medical diagnostics. In the ideal situation, this imaging would be performed noninvasively. However, it is difficult to conceive of such a system; instead, we will consider an endoscopic THz imaging probe capable of near-field micrometer resolution and spectroscopic analysis. The development of such a tool would be an entirely new technology that would provide potential for earlier detection and characterization of disease, understanding of biology, and evaluation of treatment. Physicians currently rely on relatively gross parameters of disease. In the case of cancer, this typically includes tumor burden, anatomic location, and similar parameters. Disease characterization may be improved by using more specific parameters, such as the detection of premalignant molecular abnormalities, growth kinetics, angiogenesis growth factors, tumor cell markers, or genetic alterations.

This technology would also allow for the assessment of therapeutic effectiveness at a molecular level, long before phenotypic changes occur. In vivo molecular imaging may allow for the study of pathogenesis within intact microenvironments of living systems. From the current clinical perspective, the THz microscope has tremendous potential. A tool featuring a sampling probe is envisioned that would be placed within an organ through the skin. Once the probe was in place, an in vivo image would be obtained consisting of a tomographic image and a multispectral analysis of a 3D volume of tissue. A picture of a tumor, for example, would be reconstructed using multiple THz images at different angles. This method creates a biologically accurate 3D picture that gives a researcher a better understanding of the disease and how it has spread (Zhang, 2002).

One of the breakthrough elements of sampling molecular information is the development of an imaging system capable of high spatial resolution and sensitivity. The most commonly used imaging techniques for extraction of molecular information are positron emission tomography (PET), magnetic resonance (MR), and optical techniques. PET is frequently used when a substrate to a given target exists that can only be labeled to a positron emitter. PET has high sensitivity; however, its spatial resolution is poor. MR imaging has two potential advantages over PET. It has a higher spatial resolution (micrometer compared to several millimeters) and the fact that physiologic and anatomic detail can be extracted simultaneously. In comparison with PET imaging, MR imaging is several magnitudes less sensitive (millimolar rather than picomolar), which is why reliable signal amplification strategies must be developed. Recently, cell labeling techniques have been developed that will allow efficient in vivo tracking of cell lines expressing transgenes, potentially at the single-cell level.

## Explosive Identification

THz TDS enhances noncontact and real-time detection of explosive and related compounds (ERCs) at standoff distances. Early spectroscopic measurements showed that most explosives had unique fingerprints in the THz band that could be used to identify them. Furthermore, those fingerprints were measurable even with covers such as plastic, cloths, paper, and cardboard. Therefore, THz TDS had the potential to be used in applications such as mail inspection and suicide bomber identification (see Figure 8.7), reported by several well-known media as realization of a science fiction: "A new optical system can identify explosives uses terahertz wave technology" (BBC), "Bomb-detecting sensors advance at 'terahertz' scale" (USA Today),

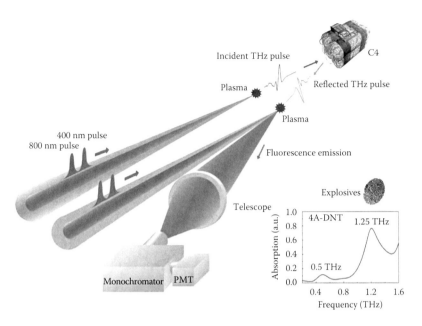

**FIGURE 8.7**
Application of the THz REEF in detection of explosives' fingerprints.

"Physics: Detection from a distance" (Nature), and "Unique THz 'finger-prints' will identify hidden explosives from a distance" (News Blaze). The THz TDS offers several advantages compared to other identification techniques, for instance,

- Penetrability into cloth, plastic, paper, and other nonmetallic and dry materials, thus providing the possibility to identify the explosives behind nonoptically transparent covers
- Capability of identification either through transmission or reflection geometry (for a real application, reflection geometry is the most realistic approach)

Explosives such as TNT, RDX, HMX, PETN, tetryl (2,4,6-trinitrophenyl-n-methylnitramine), 2-amino-4,6-DNT (2-amino-4,6-dinitrotoluene), 4-amino-2,6-DNT, 4-nitrotoluene, 1,3,5-TNB (1,3,5-trinitrobenzene), 1,3-DNB (1,3-dinitrobenzene), 1,4-DNB, 2,4-DNT, 2,6-DNT, 3,5-dinitroaniline, and 2-nitrodiphenylanine can be identified with THz wave spectroscopy. Figure 8.8 shows the spectra of some of the explosives. As observed, each ERC has a distinct spectral profile. However, not all ERCs show absorption features in the THz band. The position and relative amplitude of the peaks can be used to identify a particular ERC.

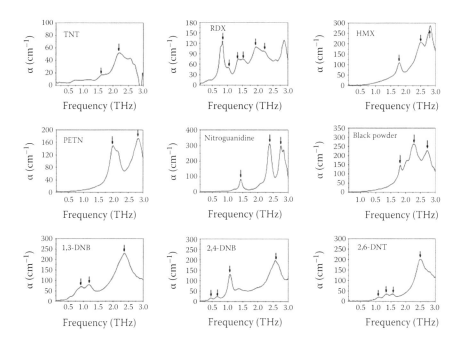

**FIGURE 8.8**
ERC fingerprints in the THz range up to 3 THz. The arrows show the resonance features.

## Pharmaceuticals

THz TDS has been applied in the pharmaceutical industry for drug analysis and discovery and coating thickness measurements. In particular, THz TDS has been used to study polymorphisms and chiral symmetry in drug compounds. Different polymorphs and chiral symmetry have an impact on the effectivity and toxicity of a drug. THz TDS has also been used to study phase transitions between different molecular configurations as a function of temperature and humidity (hydration). For example, sulfamethoxazole (SMZ) is a compound used to treat bacterial infections that interacts with caffeine and may cause side effects in certain patients. The bonding of SMZ with caffeine can be studied with THz TDS by analyzing the evolution of the absorption peaks in time. Figure 8.9 shows an example of studying the hydration and dehydration of D-glucose with THz wave spectroscopy.

Depending on the temperature, we can see that the speed of the hydration changes. It is also possible to do spectral imaging, in which each pixel contains the spectroscopic information of that location. Figure 8.10 shows an example identifying three pellets corresponding to different compounds. From an industrial perspective, THz TDS could be used to identify and classify pills based on the drug compound they contain. This capability could be used to discard those pills that do not meet certain quality standards

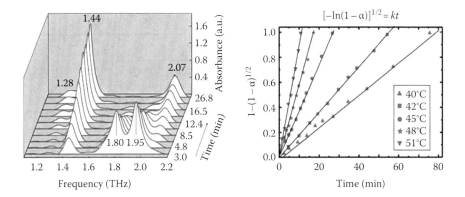

**FIGURE 8.9**
Study the chemical reaction dynamics with THz TDS.

because they do not have the proper compounds or do not have the compound in the right amount or concentration.

## Food Safety

Food safety is another sector that will benefit from the progress of THz TDS technology. Because THz waves can penetrate plastic and paper, THz TDS has the potential to analyze food that is already packaged in search of contaminants and hazardous compounds such, antibiotics or pesticides. For instance, a recent application has been studied for the detection and identification of antibiotics in highly scattering dry food matrices. Several antibiotics commonly used in the veterinary field were found to have absorption features in the THz range. These features could be used to detect the presence of such antibiotics in dry food for chicken stock. Eventually, the identification of antibiotic residues may be performed in chicken-derived dry products to ensure that the concentration is below the levels established by governmental entities, such as the FDA. Figure 8.11 shows the main results of the described research work.

## Handheld THz Instrumentation

Technological progress in the area of THz instrumentation and systems has been driven mostly by THz wave spectroscopy applications, in recent years. Commercial THz spectrometers have become more compact and are easier to use for nonexpert operators. They also offer improved performance compared to home-made systems built in the laboratory. The trend of accelerated technological progress in THz instrumentation indicates that the gap between application needs and performance and functionality is getting narrower as time advances. The micro-Z, a handheld, battery-operated THz

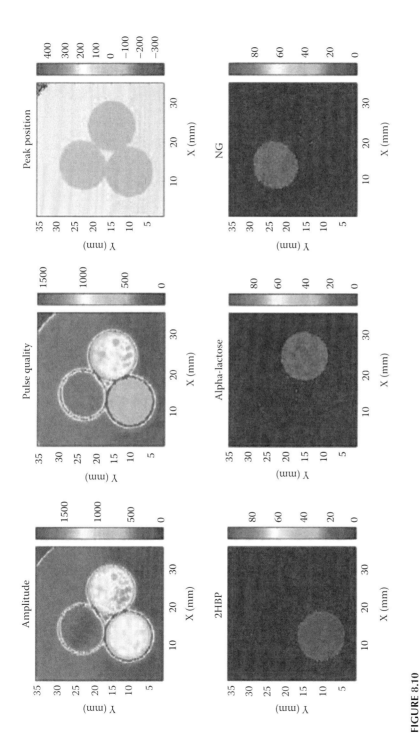

**FIGURE 8.10**

**(See color insert.)** Spectral imaging. Each pixel contains the spectral information of the sample, which allows identification of the position of a particular compound.

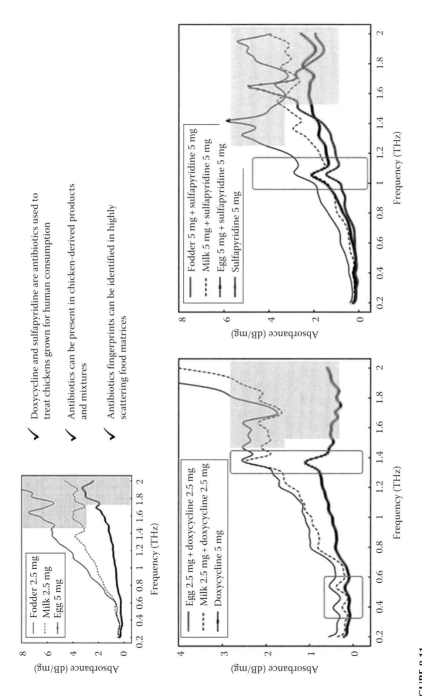

**FIGURE 8.11**

The presence of antibiotics could be determined upon looking at the characteristic absorption peaks. (Courtesy of University of Barcelona, Barcelona, Spain.)

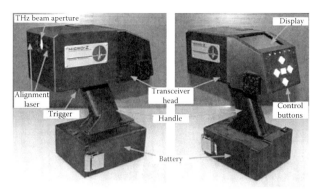

**FIGURE 8.12**

Micro-Z handheld THz TDS. The micro-Z is a compact, handheld, battery-powered THz TDS that has the total freedom of operation previously unattainable with stationary instruments. Featuring an easy-to-navigate interface, the device can be targeted for a variety of on-site inspection tasks using THz waves, including real-time chemical identification. (Courtesy of Zomega, http://www.z-thz.com/index.php?option=com_content&view=article&id=48&Itemid=55)

wave broadband spectrometer, is an example of recent progress in the THz field that realizes devices that were almost unthinkable a few years ago. The trend toward size reduction, higher component integration, and performance improvement is expected to have an impact on high-demanding THz applications such as THz wave sensing, especially in THz pulsed systems for homeland security and nondestructive evaluation (NDE).

In 2011, a Rensselaer–Zomega team designed and built a fully integrated, turnkey, handheld, and battery-operated THz wave spectrometer, the micro-Z (Figure 8.12). The 5 lb device is the size of a cordless drill and was a finalist in the 2011 Prism Awards for Photonics Innovation. Designed to be a lightweight tool for detection and identification of chemicals such as ERCs in the open field, this compact system is configured to work in normal-incidence reflection in a point-and-shoot fashion. It produces and measures pulsed THz waves from 0.1 to 3.0 THz using time-domain techniques in both transmission and reflection geometries, with a waveform acquisition rate of 500 Hz and a time delay >100 ps. The high waveform acquisition rate allows compensating for the operator hand tremor frequencies, measured to be between 1 and 100 Hz. Although some progress is still necessary for a fully deployable system in a battlefield environment, the headway achieved with micro-Z is just one example of the efforts to address the needs of users. Once thought very difficult or impossible to achieve, we expect that this trend of miniaturization, compact design, and ease of use of THz systems will continue in the future.

## Applications in Defense

THz wave spectroscopy covers the part of EM spectrum between 0.1 and 10 THz. Funding agencies have driven much of the recent technological evolution in THz instrumentation and systems, and most of the progress has been toward the development of THz TDS systems. THz TDS has been demonstrated to be an excellent technique for characterizing and identifying many organic compounds, especially explosives and their related compounds (Shen et al., 2005; Baker et al., 2007; Chen et al., 2007). A TDS system generates and detects a THz pulse in the time domain and computes the spectrum by performing a Fourier transform on the waveform. Most applications of THz TDS and NDE have been explored intensively in a laboratory environment (Tonouchi, 2007).

### Industrial Applications

In any field, there is always a gap between the demonstration of a new technology in an academic lab and the ability of a wide user base to put the technology into practice for their own particular applications. Usually, it is the role of industry to close that gap by providing user-friendly tools, instruments, and systems to meet practical needs. Companies such as Picometrix (United States), TeraView (UK), and Zomega Terahertz Corporation (United States) have been pushing the boundaries of THz technology to realize promising applications with THz waves. For example, Picometrix was one of the first companies to develop a system targeted to NDE applications. TeraView has focused its efforts on the development of THz spectroscopic systems for the pharmaceutical industry, and Zomega has focused on developing compact and portable THz spectroscopic systems for research and NDE applications.

A major driver for the technological development is anticipated to be imaging applications. THz wave imaging has shown its potential in NDE and security applications (Kawase et al., 2003; Dobroiu et al., 2004; Karpowicz et al., 2005, 2006; Becker et al., 2007; Dougherty et al., 2007; Adam et al., 2009), because THz waves can penetrate dry, nonmetallic, and nonpolar materials such as cloth, paper, cardboard, and ceramic, which enable the interrogation of optically nontransparent targets. Because the wavelength is on the order of millimeters or less, the resolution of the images is also a few millimeters, which can provide great detail in images of macroscopic objects. The use of THz for imaging has also been driven by security applications, such as full-body scanners now being tested at many airports. An example of the potential is illustrated in Figure 8.13a, a racquetball bat in an optically opaque case clearly imaged with a THz probe. Due to the potentials in finding other obscured objects, THz imaging is already being tested in Australia for secondary scanning of suspicious individuals at airports in order to enhance security for all passengers. In this application, companies such as Thruvision (UK) and Brijot (United States), acquired by Microsemi in 2011, have been leading the technical effort. Figure 8.13b shows a different application of THz wave spectroscopy, imaging a moth's wings.

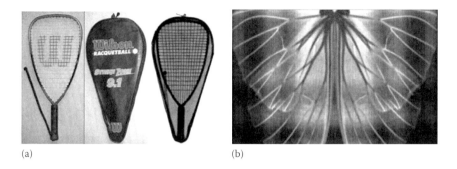

(a)                                                (b)

**FIGURE 8.13**
(a) Left, image of a racquetball bat; middle, image of the racquetball bat in case; right, THz image of the racquetball bat in case (Zhang 2009). (b) THz image of moth wings from a Zomega system. (Courtesy of Zomega, http://www.z-thz.com/index.php?option=com_content&view=article&id=48&Itemid=55)

## Conclusions

THz wave air photonics is proven as an appealing technique in dealing with the limitations concerning (1) the ambient moisture absorption in the THz range and (2) the bandwidth of semiconducting PCAs. THz wave air photonics technique has made the broadband standoff THz spectroscopy, imaging, remote sensing, and detection/identification feasible. The "all-optical" standoff THz wave detection circumvents the mentioned limitation in the ambient moisture absorption in the THz range. Moreover, the broad spectral information of a target material can be obtained at lower false-alarm rates, which enhances application of the THz wave air photonics techniques in homeland security and environmental scopes. The introduced THz REEF technique, with its omnidirectional emission pattern and minimal ambient water vapor absorption, together with the long-distance THz wave generation, makes broadband standoff THz spectroscopy feasible (Dai and Zhang, 2009). The advantages of THz wave photonics technique over previous techniques can be summarized as

1. High THz field: (a) >100 kV cm$^{-1}$ (~100 fs, 1 mJ laser pulses) and (b) >1.5 MV cm$^{-1}$ (~28 fs, 2.7 mJ laser pulses) with ultrabroadband smooth spectrum
2. Good THz beam profile and directionality
3. No emitter damage issue
4. Potential applications in remote THz wave sensing and nonlinear THz spectroscopy

Besides its unique features and characteristics, there are some drawbacks concerning the THz wave air photonics technique to be addressed in the future, for example, amplified laser requirements (about mJ pulse$^{-1}$), laser safety issues, and third-order nonlinearity effects (compared to solid-state THz sources that are liner or quadratic to the optical excitation). A few years ago, THz imaging was almost impractical, as it could take tens of minutes to acquire an image. With recent advancements in fast waveform acquisition rates and smaller footprint sizes, pulsed systems have suddenly become more feasible for imaging applications because system operation is no longer limited by the speed of waveform acquisition but by the scanning speed. Technological development is expected to target at increasing the imaging speed and performance of these THz imaging systems for a variety of applications, including in biomedical testing, clinical trials, homeland security, defense, and environmental fields.

## Acknowledgment

The author wishes to thank doctoral student and PRISM Center researcher Mohsen Moghaddam for help in preparing the chapter manuscript.

## Bibliography

Adam, A.J.L., Planken, P.C.M., Meloni, S., Dik, J., 2009. Terahertz imaging of hidden paint layers on canvas. *Optics Express*, 17, 3407–3416.

Azarm, A., Xu, H.L., Kamali, Y., Bernhardt, J., Song, D., Xia, A., Teranishi, Y., Lin, S.H., Kong, F., Chin, S.L., 2008. Direct observation of super-excited states in methane created by a femtosecond intense laser field. *Journal of Physics B*, 41, 225601.

Baker, C., Lo, T., Tribe, W., Cole, B., Hogbin, M., Kemp, M.C., 2007. Detection of concealed explosives at a distance using terahertz technology. *Proceedings of IEEE*, 95, 1559–1565.

Chen, J., Chen, Y., Zhao, H., Bastiaans, G.J., Zhang, X.-C., 2007. Absorption coefficients of selected explosives and related compounds in the range of 0.1–2.8 THz. *Optics Express*, 15, 12060–12067.

Chen, Q., Zhang, X.-C., 1999. Polarization modulation in optoelectronic generation and detection of terahertz beams. *Applied Physics Letters* 74(23), 3435–3437.

Corkum, P.B., Burnett, N.H., Brunel, F., 1989. Above-threshold ionization in the long-wavelength limit. *Physical Review Letters*, 62, 1259–1262.

Dai, J., Karpowicz, N., Zhang, X.-C., 2009. Coherent polarization control of terahertz waves generated from two-color laser-induced gas plasma. *Physical Review Letters*, 103, 023001.

Dai, J., Liu, J., Zhang, X.-C. 2011. Terahertz wave air photonics: Terahertz wave generation and detection with laser-induced gas plasma. *IEEE Journal of Selected Topics in Quantum Electronics*, 17, 183–190.

Dai, J., Xie, X., Zhang, X.-C., 2006. Detection of broadband terahertz waves with a laser-induced plasma in gases. *Physical Review Letters*, 97, 103903.

Dai, J., Zhang, X.-C., 2009. Terahertz wave generation from gas plasma using a phase compensator with attosecond phase-control accuracy. *Applied Physics Letters* 94, 021117.

Dobroiu, A., Yamashita, M., Ohshima, Y.N., Morita, Y., Otani, C., Kawase, K., 2004. Terahertz imaging system based on a backward-wave oscillator. *Applied Optics* 43, 5637–5646.

Dougherty, J.P., Jubic, G.D., Kiser, W.L., 2007. Terahertz imaging of burned tissue. *Proceedings of SPIE*, 6472, 64720N.

Exter, M.V., Fattinger, Ch., Grischkowsky, D., 1989a. High-brightness terahertz beams characterized with an ultrafast detector. *Applied Physics Letters*, 55, 337–339.

Exter, M.V., Fattinger, Ch., Grischkowsky, D., 1989b. Terahertz time-domain spectroscopy of water vapor. *Optics Letters*, 14, 1128–1130.

Federici, J.F., Schulkin, B., Huang, F., Gary, D., Barat, R., Oliveira, F., Zimdars, D., 2005. THz imaging and sensing for security applications—Explosives, weapons and drugs. *Semiconductor Science and Technology*, 20, 266–280.

Ferguson, B., Zhang, X.-C., 2002. Materials for terahertz science and technology. *Nature Materials*, 1, 26–33.

Filin, A., Compton, R., Romanov, D.A., Levis, R.J., 2009. Impact-ionization cooling in laser-induced plasma filaments. *Physical Review Letters*, 102, 155004.

Karpowicz, N., Dawes, D., Perry, M.J., Zhang, X.-C., 2006. Fire damage on carbon fiber materials characterized by THz waves. *Proceedings of SPIE*, 6212, 62120G.

Karpowicz, N., Redo-Sanchez, A., Zhong, H., Li, X., Xu, J., Zhang, X.-C., 2005. Continuous-wave terahertz imaging for non-destructive testing applications. *IRMMW-THz*, 1, 329–330.

Karpowicz, N., Zhang, X.-C., 2009. Coherent terahertz echo of tunnel ionization in gases. *Physical Review Letters*, 102, 093001.

Kawase, K., Ogawa, Y., Watanabe, Y., Inoue, H., 2003. Non-destructive terahertz imaging of illicit drugs using spectral fingerprints. *Optics Express*, 11, 2549–2554.

Kim, K.Y., Taylor, A.J., Glownia, J.H., Rordriguez, G., 2008. Coherent control of terahertz supercontinuum generation in ultrafast laser–gas interactions. *Nature Photon*, 2, 605–609.

Kresβ, M., Löffler, T., Thomson, M.D., Dörner, R., Gimpel, H., Zrost, K., Ergler, T. et al., 2006. Determination of the carrier-envelope phase of few-cycle laser pulses with terahertz-emission spectroscopy. *Nature Physics*, 2, 327–331.

Kulander, K.C., Schafer, K.J., Krause, J.L., 1991. Dynamic stabilization of hydrogen in an intense, high-frequency, pulsed laser field. *Physical Review Letters*, 66, 2601–2604.

Liu, J., Dai, J., Chin, S.L., Zhang, X.-C., 2010. Broadband terahertz wave remote sensing using coherent manipulation of fluorescence from asymmetrically ionized gases. *Nature Photonics*, 4, 627–631.

Liu, J., Zhang, X.-C., 2009. Terahertz radiation enhanced emission of fluorescence from gas plasma. *Physical Review Letters*, 103, 235002.

Mittleman, D., 2003. *Sensing with Terahertz Radiation*. Springer series in optical sciences. Vol. 85. Springer, New York.

Mlejnek, M., Wright, E.M., Moleney, J.V., 1999. Moving-focus versus selfwaveguiding model for long-distance propagation of femtosecond pulses in air. *IEEE Journal of Quantum Electronics*, 35, 1771–1776.

National Research Council, 2004. *Existing and Potential Standoff Explosives Detection Techniques*. National Research Council of the National Academies Press, Washington, DC.

Phelps, A.V., 1968. Rotational and vibrational excitation of molecules by low-energy electrons. *Reviews of Modern Physics*, 40, 399–410.

Schumacher, D.W., Weihe, F., Muller, H.G., Bucksbaum, P.H., 1994. Phase dependence of intense field ionization: A study using two colors. *Physical Review Letters*, 73, 1344–1347.

Shen, Y., Lo, T., Taday, P.F., Cole, B.E., Tribe, W., Kemp, M., 2005. Detection and identification of explosives using terahertz pulsed spectroscopic imaging. *Applied Physics Letters*, 86, 241116.

Sun, F.G., Jiang, Z., Zhang, X.-C., 1998. Analysis of terahertz pulse measurement with a chirped probe beam. *Applied Physics Letters*, 73(16), 2233–2235.

Talebpour, A., Chien, C.Y., Chin, S.L., 1996a. Population trapping in rare gases. *Journal of Physics B*, 29, 5725–5733.

Talebpour, A., Liang, Y., Chin, S.L., 1996b. Population trapping in the CO molecule. *Journal of Physics B*, 29, 3435–3442.

Tonouchi, M., 2007. Cutting-edge terahertz technology. *Nature Photon*, 1, 97–105.

Wang, S., Ferguson, B., Abbott, D., Zhang, X.-C., 2003. T-ray imaging and tomography. *Journal of Biological Physics*, 29(2–3), 247–256.

Wen, H., Lindenberg, A.M., 2009. Coherent terahertz polarization control through manipulation of electron trajectories. *Physical Review Letters*, 103, 023902.

Wu, Q., Zhang, X.-C., 1995. Free-space electro-optic sampling of terahertz beams. *Applied Physics Letters*, 67, 3523–3525.

Xu, H.L., Azarm, A., Bernhardt, J., Kamali, Y., Chin, S.L. 1996. The mechanism of nitrogen fluorescence inside a femtosecond laser filament in air. *Journal of Chemical Physics*, 360, 171–175.

Yeh, K.-L., Hoffmann, M.C., Hebling, J., Nelson, K.A., 2007. Generation of 10 mJ ultrashort terahertz pulses by optical rectification. *Applied Physics Letters*, 90, 171121.

Zhang, X.-C., 2002. Terahertz wave imaging: Horizons and hurdles. *Physics in Medicine and Biology*, 47, 3667–3677.

Zomega Terahertz Corporation, 2012. *The Terahertz Wave eBook: Technical Review*. Ed. Zomega Terahertz Corporation, http://www.z-thz.com/index.php?option= com_content&view=article&id=48&Itemid=55.

# 9

## Precision Collaboration and Advanced Integration Using Laser Systems and Techniques

Avital Bechar, Shimon Y. Nof, and Juan P. Wachs

**CONTENTS**

Lasers were introduced five decades ago, and since then, they have impacted almost every aspect of our life, from healthcare to defense systems, by enabling a new ultraprecise, multipurpose technology. Laser technology has been implemented as an end product and as part of the production chain. This dual functionality is possible due to unique characteristics, such as high-rate energy transmission, high irradiance, and spatial and temporal coherence and precision. While the most common applications are found in medical and communication technologies, other areas such as manufacturing, agriculture, construction, and defense also benefit from this groundbreaking scientific discovery. In spite of the rapid dissemination of laser technologies to diverse and varied application fields, its role in support of collaboration and discovery is still in its infancy. Research activities brining laser-based technology to these areas have been relatively limited. Nevertheless, the translation to this domain has been recognized as vital for activities that demand increasingly more coordinated effort among interacting agents, including humans, machines, and digital, possibly photonic agents.

In the following chapter, recent advances in laser technology and a number of application domains are discussed and reviewed, focusing primarily on lasers' applications in precision interactions and collaboration. To this end, a number of laser characteristics and roles are observed; a framework with four collaboration functions and five collaboration dimensions is defined. Such taxonomy framework is believed to enable a better understanding of the existing opportunities that laser-based technology offers, its advantages and opportunities.

## Introduction

Laser technology was introduced five decades ago, and since then, it has affected most aspects of our life, from health and manufacturing to defense and energy systems. Laser technology has been implemented not only as an end product but also as a means of production through a variety of industries, exploiting its unique characteristics, such as energy transmission, high irradiance, and spatial and temporal coherence and precision. The main applications are found in medical, information, and communication technologies and materials processing, where it is playing an increasingly important economic role.[1]

The implementation of lasers beyond communication and processing to collaboration systems, however, is still in its infancy. The research activity found in these areas so far is relatively small, even though effective and precise interaction and collaboration have been recognized as vital for the increasingly complex, networked engineering and service activities. Recent advances in laser technology increase the potential feasibility of harnessing certain laser solutions for new and improved collaboration systems. These advances include using new wavelength bands and reducing pulse duration, together with increasing precision and power efficiency, while decreasing costs.

The volume of research conducted about lasers is vast: over 490,000 refereed articles have been found, according to Web of Science,[2] including (1) physics, (2) optics, (3) engineering, (4) chemistry, (5) material sciences, (6) spectroscopy, and (7) instrumentation, which comprise about 42%, 20%,

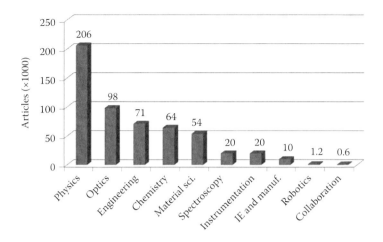

**FIGURE 9.1**

Relative research efforts on lasers in various subject topics, measured by the number of published, refereed articles. Each article may contain one or several topics. (Source: Web of Science, 2012.)

14.5%, 13%, 11%, 4%, and 4% of the total volume, respectively. Only 2% of the research work on lasers has been in the area of industrial engineering and manufacturing, 0.25% on robotics, and half than that on collaboration (Figure 9.1).

The potential value and promise of laser systems have already been proven in various areas. One application of laser technology is in precision collaboration. While most collaboration activities can benefit from precision, both in signals and measurements, in inputs and outputs, precision collaboration implies processes and situations that require particular attention to precision in terms such as exact position, identification, location, dimensions, proximity, and other attributes. Such precision collaboration activities integrate communication, synchronization, control, geometrical environmental data, and information collection. Laser technology can be applied in operating each of these functions effectively, because of its unique ability to accurately and rapidly transfer and transmit data, information, and energy with no spatial or temporal residuals. This multiple functionality enables lasers to be used as a tool, information medium, energy source, marker, and/or sensor in precision collaboration, for instance, in metal welding, tissue joining, chemical detection,[3] and in certain types of target recognition.[4]

The objectives of this chapter are to (1) review and analyze the opportunities in applying advanced laser techniques for precision interactions and collaboration; (2) evaluate the impact of these techniques on the quality, performance, and effectiveness of the collaboration activities; and (3) reveal how the improvement by the laser precision can transform and influence the collaboration processes and interactions' ability to achieve superior outcomes.

### Laser Characteristics Relevant for Precision Interactions and Collaboration

Precision collaboration consists of several functions requiring mutual and situational precision, such as communication, synchronization, geometrical environmental data, and information collection, and control features, including commands, interactions, interfaces, performance, and status data evaluation (Figure 9.2).

Following the model of impact of collaboration on decision and action quality,[5] each decision and its implied action involves three components: (1) problem features (policy, action, and choice), (2) knowledge (information, logic, and experience), and (3) decision process (tools, models, and methods). Lasers have a useful role in all three decision and action components. For instance, laser sensors help understand the problem's current features and constraints and help identify potential actions; lasers can gain information about ranges and target features; and lasers can participate in

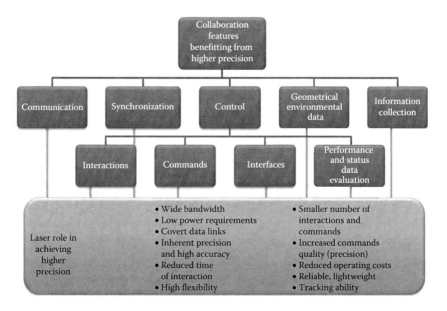

**FIGURE 9.2**
Collaboration features, their precision requirements, and relevant laser advantages to satisfy them.

methods and tools for the collaborative decision and action process. In all these roles, which depend on collaboration among systems' components, lasers can provide precision, fast response, and other advantages, as discussed later. Each of the collaboration features in Figure 9.2 has its own specific requirements and its own scale of accuracy/precision. Laser systems can be used to operate each of these features and their combinations in an accurate and precise way by virtue of their unique abilities to optically transfer and transmit data, information, and energy with exact spatial and temporal specifications.

Together with computational power, the abilities and the precision implications of laser systems are even more advantageous. Laser systems are characterized by flexibility and accuracy. They require relatively low maintenance. Specifically, laser characteristics that are relevant for communication and information transfer in collaborative systems are bandwidth, noise resistance, low power requirements, narrow broadcasting beam, and secure data transmission. For geometrical and environmental data, laser accuracy and reliability are especially useful. For control and interfacing, the unique targeting and tracking abilities and ability to issue timely commands using structured light and laser spot detecting are also advantageous.

## Overview: Laser Advantages and Opportunities

The relative advantages of lasers, in comparison to the common, non-laser-based collaboration and interaction methods and techniques, are numerous. Lasers incur relatively lower cost and higher functional flexibility[6]; are lightweight, point to targets accurately, and are highly reliable[7]; and can have wider bandwidth, lower power requirements, and increased potential for covert data links in comparison with radio-based systems.[8] Lasers can enable better tracking[9] and have inherent precision, which results in better functionalities for precise collaboration.

An important advantage of such precision is the reduction in collaboration errors and overall time of interaction, less interactions, and less commands required to achieve the collaboration goal, compared with less precise means. Furthermore, such precision in collaboration typically results in overall increased quality (precision) of the commands issued by the collaborating parties. In addition, more accurate information is conveyed, resulting in more effective control and more successful decisions. Due to the added accuracy with laser-based interactions and collaboration, the total operating costs are reduced, as well as savings of any additional costs required to achieve a certain level of accuracy while interacting with the aid of non-laser-based techniques.

So far, relatively few researchers exploited these laser characteristics for precision collaboration and interaction. Some illustrative examples are as follows:

- Ultraprecision motion control of a magnetic suspension actuation system, obtained by employing a laser measuring system to calculate control parameters.[10]
- Accurate laser projection of surgical plans onto patients enables reduction in risk and operating time, by following the accurately projected plans.[11]
- Laser technology enables measurement of the optimal location of surgery entry points.[11]
- Precise laser pointing can improve collaboration of multiple users in graphical environments.[12]
- In the development of automated vine cutting transplanter, the process and interaction between different systems that require precision synchronization could be automated by using laser sensor and a tracking system called DGNSS-RTK.[13]
- Collaboration and tracking ability of mobile robots were improved by applying laser scanners for collaborative localization.[14]
- Structured light and laser beam projector can improve human perception and simplify human action, thus improving the efficiency and precision of teleoperation.[15]

Significant relative advantages have been demonstrated, as in the following illustrative examples:

- In certain machining processes and material processing, using lasers is more economical than traditional measurement methods and requires less maintenance due to less wear and less need for replacing scuffed or broken instruments. It is also faster and has higher energy efficiency, enabling superior surface properties.[1]
- Transition from welded to hydroformed beams in the frames of vehicles in the United States was enabled mainly by using robots and laser cutting, significantly less costly and more flexible compared with Computer Numerical Control (CNC) machines.[6]
- In surgery and dental operations, lasers can minimize cellular destruction, tissue swelling, and hemostasis and reduce the need for anesthesia, resulting with reduced postoperative pain.[16,17]
- Lasers can also circumvent the need for suturing,[3] which could have additional benefits, for example, in battlefields, by reducing risk of infection and minimizing the weight of supplies.

- In the medical field and in nanotechnology industries, lasers are often the most suitable technique for micro- and nanofabrication, precision microcutting, and surface microtexturing and microdeposition.[1]

The following section introduces a new precision collaboration framework. This framework is then applied in the review and evaluation of lasers' systematic application for precision collaboration and interaction in various areas.

## Framework of Precision Collaboration with Lasers

Lasers can be used in a collaborative mission as a tool to be monitored by the collaboration and interaction architecture, or as part of the collaboration methodology, protocol, and process. In medical, manufacturing, exploration, and military tasks that require teleoperation or telerobotics of a processing tool, the remote control of lasers (processing tools) would typically be relatively more accurate, flexible, and safe compared with a mechanical tool, as realized in the cases of tissue welding, chemical detection,[3] and laser metal welding.[18-20]

Although collaboration and interaction are inherent subactivities in many areas, such as dental operations and surgery, manufacturing, robotics, transportation, defense, and agriculture, the use of lasers for enabling and improving them in an integrated way has been relatively limited. When they have been applied, it was usually designed for isolated functions, including its use as a tool to map the environment and detect specific objects, and for targeting, localization, marking, and other interactions. The emerging collaboration functions and systems that have been implemented effectively with lasers and the emerging role and methods in which lasers are being used to execute these functions are summarized in a precision collaboration framework (Figure 9.3). The first precision collaboration element in this framework (left column) specifies who the collaborating participants are. The second element indicates the collaboration functions involved, and the third element is the role (or roles) of laser (or lasers) in this collaboration. The four main precision collaboration functions are explained next, including the role of lasers applied and illustrated in Figure 9.4 for these four functions. Then, Table 9.1 follows this framework in illustrating research studies where precision collaboration has been introduced experimentally with laser components.

## Precision Collaboration Function 1: Targeting

Targeting, as a precision collaboration function, implies the precise definition of an object and its exact location. Example: Defining the position and

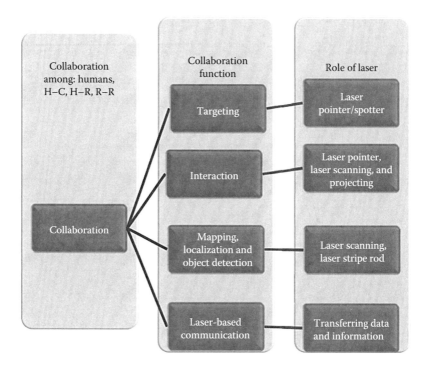

**FIGURE 9.3**
Precision collaboration framework: emerging laser roles in precision collaboration (H, human; C, computer; R, robot).

orientation of a pepper to be harvested right now by a robotic device. A laser pointer provides a powerful alternative as a user interface for robot control. It is cheap and lightweight, points accurately, and is reliably recognized.[7] Previous research showed that a laser pointer is useful for providing target designation for a picking-up robot[25] and developed a button-pushing robot.[26] A wearable interface for remote collaboration, enabling in addition remote human collaborators to point to real objects in the task space with a laser spotter, was developed.[27] A body-worn steerable laser pointer and a fixed camera to enable remote collaboration were used[28] and applied for hands-free telepointing with completely self-contained, wearable, visual augmented reality (AR).

For interactive virtual environments, adaptive sensor fusion-based object tracking and human action recognition were developed.[29] This system tracks people in a virtual environment robustly and can recognize their action view invariantly, based on adaptive sensor fusion approach, which interacts between a laser scanner and a video camera. It also enables a person in an interactive virtual environment to simultaneously and conveniently interact with virtual agents. Cruz–Ramírez et al. investigated targeting ceiling objects, such as light panels, with a laser pointer for dismantling interior

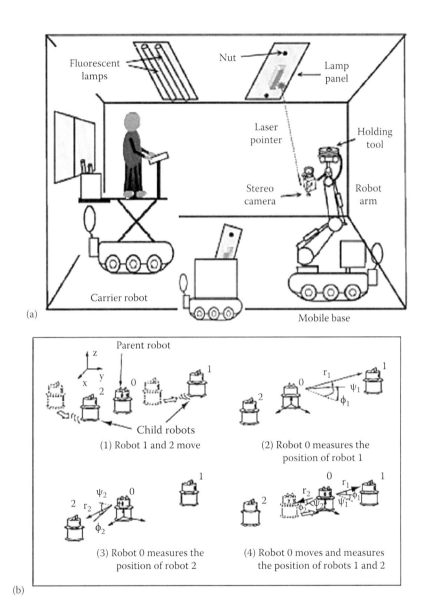

**FIGURE 9.4**
Illustration of precision collaboration functions. (a) Targeting[21]; (b) interaction.[22]

(*continued*)

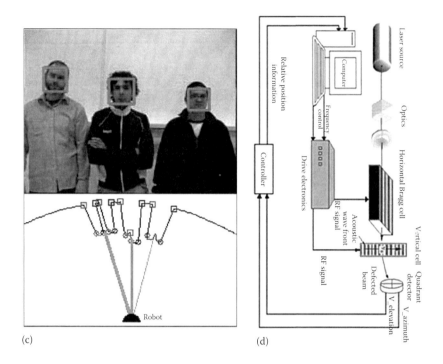

(c)             (d)

**FIGURE 9.4 (continued)**
Illustration of precision collaboration functions. (c) mapping, localization, and object detection[23]; and (d) laser-based communication.[24]

facilities in buildings, using human–robot collaboration.[21] A laser pointer fixed at the robot's tip was applied both as a guide in the teleoperation process and as a first indicator of the current holding position. Positioning tasks for industrial robots and objects mapping were performed by projecting laser spots over a workpiece, in order to gather information about the workpiece geometry with a multicamera system.[18] Young Sang et al. developed a mobile robotic manipulator that grasps objects identified with a laser pointer by a human operator.[30] In this task, the grasping was done autonomously after a human operator targeted the object with a laser pointer.

A human–robot interaction (HRI) system for robot tool path teaching in AR environment was developed and tested on a laser welding industrial robot.[19,20] The HRI system used information from a camera and a laser range finder (LRF). In addition, the system detected and tracked a laser spot projected by the LRF. As indicated earlier, Glossop et al. combined laser with AR for projecting surgical plans, entry points for probes directly onto patients.[11]

In the earlier targeting examples, precision is essential for cost-effective and high-quality fulfillment of tasks. The collaborating parties, humans, robots, machines, and instruments, could also be damaged if measurements or communicated information are imprecise.

**TABLE 9.1**

Research and Applications Survey of Laser Technology for Collaboration and Controlled by a Collaborative System

| Domain | Application | Collaboration Activities | Collaboration Objectives | References | Role of Lasers | Laser Advantages | Collaboration Function[a] |
|---|---|---|---|---|---|---|---|
| Dental operations and surgery | Soft tissue: gingivectomies, gingivoplasties, operculectomies, biopsies, incising and draining procedures, frenectomies, and treatment of aphthous ulcers | No interaction, integration, or collaboration | Dental drill, dental decay prevention, and decay detection | [16,17,59] | Tool | Minimizes cellular destruction, tissue swelling, and hemostasis; reduces need for anesthesia; reduces postoperative pain | 0 |
| Dental operations and surgery | Hard tissue: vaporizing decay, etching enamel and dentin, desensitizing exposed root structure, and creating temporary analgesia | No interaction, integration, or collaboration | Development of new dental adhesives and composite systems, new methods for managing caries, and new endodontic treatments | [16,17,59] | Tool | Minimizes cellular destruction, tissue swelling, and hemostasis; reduces need for anesthesia; reduces postoperative pain | 0 |
| Dental operations and surgery | Tissue welding | Telerobotics | Tissue welding | [3] | Tool | Circumvents the need for suturing | 0 |
| Manufacturing | Mechanical machining, cutting | No interaction, integration, or collaboration | Laser-based processing | [6] | Tool | Lower cost and higher flexibility | 0 |

*(continued)*

**TABLE 9.1 (continued)**
Research and Applications Survey of Laser Technology for Collaboration and Controlled by a Collaborative System

| Domain | Application | Collaboration Activities | Collaboration Objectives | References | Role of Lasers | Laser Advantages | Collaboration Function[a] |
|---|---|---|---|---|---|---|---|
| Manufacturing | Automating computer-aided design (CAD)/computer-aided manufacturing (CAM) production | CAD/CAM functions only | Measurement data acquisition | [60] | Laser scanning, 3D modeling | Sensor advantages | 0 |
| HRI | Intelligent chair tool | Low-level control | Obtain better position and orientation performance | [61] | Laser stripe rod to measure actuator movement | Accurate positioning | 0 |
| Robots | Casualty extraction and care | Process control only | None | [3] | Tissue welding, chemical detection | Processing and data acquisition | 0 |
| Manufacturing | Laser processing | No collaboration or interaction | Processing only | [1] | Laser processing | Laser-based processes | 0 |
| Motion controller | Ultraprecision motion control | Laser sensor | Measurement accuracy only | [10] | Laser measurements | More precise motion control | 0 |
| Agriculture | Automated vine cutting transplanter | Interaction of systems | None | [13] | Laser measurement | Precise measurement | 0 |
| Mobile robots | People detection and tracking | Responsive collaboration/interaction | Improved HRI, human activity understanding, and intelligent cars | [9] | Laser scanning, mapping, people detection | Improved human tracking | 1 |

| HRI | Dismantling interior facilities in buildings | Collaboration and interface functions | Smoother, better quality execution of tasks | [21] | Point to a target in the task space with the laser pointer | Accurate and fast marking | 1 |
|---|---|---|---|---|---|---|---|
| Robots | Assisting in cases of chronic stroke | Direct, intuitive interaction | Enable intuitive control | [30] | Selection of objects with laser pointer | Accurate and timely data acquisition and transmission | 1 |
| HRI | Wearable collaborative systems | Remote collaboration/ interfaces | Sensor-based collaboration | [27,28] | Point to real objects in the task space with laser spot, laser pointer | Tracking moving objects accurately | 1 |
| Service/ manufacturing | Control for room cleaning | Collaboration of techniques | Improved process control | [38] | Mapping | Control advantages | 2 |
| Telerobots | Remote guidance | HRI, teleoperation | Intuitive interaction | [46] | Mapping | Enabling precise guidance | 2 |
| Mobile robots | Pose/gestures detection | HRI, corporal gestures | Precise cognition | [48] | Laser range scanner, detection | Accurate and fast data acquisition | 2 |
| Mobile robots | Pose/gestures detection navigation | HRI | Precise cognition | [52] | Laser range scanner, detection | Accurate and fast data acquisition | 2 |
| Social robots | Feature scanning | Human–computer interaction | Precise recognition | [47] | Initial face scanning | Accurate response | 2 |
| User interface | SDG | Collaborative interaction | Computer-controlled laser pointers | [31] | Input device for collaboration, laser pointers | Interface advantages | 1,2 |

*(continued)*

**TABLE 9.1 (continued)**

Research and Applications Survey of Laser Technology for Collaboration and Controlled by a Collaborative System

| Domain | Application | Collaboration Activities | Collaboration Objectives | References | Role of Lasers | Laser Advantages | Collaboration Function[a] |
|---|---|---|---|---|---|---|---|
| HRI | Telerobotics | Telerobotic manipulation | Teleoperation guidance | [15] | Laser beam projector | Ability to project targets | 1,2 |
| Human–robot interface | Robot control | User interface | Specifying command and target location | [7] | User interface, laser pointers | Lower cost, lightweight, points accurately, and reliable | 1,2 |
| Robots | Positioning tasks | Mapping objects, positions | Joint team operations | [18] | Point to a workpiece in the task space with the laser pointer | Accurate and timely sensing | 1,2 |
| User interface | Man–machine interface for intelligent presenter–audience collaborative environment | Collaboration, interaction | Computer-controlled laser pointers | [62] | Input device for collaboration, laser pointers | User interface | 1,2 |
| User interface | SDG | Collaboration, interaction | Interaction with multiple pointers | [12] | Input device for collaboration, laser pointers | Accurate and timely sensing | 1,2 |
| Biofied rooms | Acquire environmental information | HRI | Multisensor integration | [49] | Laser scanning | Accurate and timely data acquisition | 3 |
| Mobile robot | Network surveillance | Multisensor integration | Sensor data integration | [41] | Laser scanning, distance measurements | Accurate and timely data acquisition and transmission | 3 |

| | | | | | | | |
|---|---|---|---|---|---|---|---|
| HRI | Localization | Interactive communication | Collaborative positioning | [39] | Laser range scanner, mapping | Accuracy for better responsiveness | 3 |
| Mobile robots | Navigation in urban areas | Laser-sensed data transfer | Better, safer team movements | [54] | Laser range scanner, mapping | Accurate and fast scanning | 3 |
| Mobile robots | Human detection and tracking | HRI | Joint team performance | [23] | Laser scanning, detect people legs | Accurate and timely multisensor data acquisition | 3 |
| Social robots | Calculate people trajectories | No interaction, integration, or collaboration | Detecting people, people behavior | [63] | Laser scanning, mapping, detect people | Accurate and timely multisensor data acquisition | 3 |
| Mobile robots | Collaborative navigation | Collaborative control | Better performance | [64] | Laser scanning, mapping, obstacle detection | Accurate and timely data acquisition and transmission | 3 |
| Mobile robots | Navigation | Teleoperation | Multisensor-based mobility | [43] | Laser scanning, mapping, obstacle detection | Accurate and timely data acquisition and transmission | 3 |
| Face recognition | Multimedia | Human–computer interaction | Precise face recognition | [51] | Face scanning | Accurate and timely data acquisition and transmission | 3 |

(*continued*)

**TABLE 9.1 (continued)**

Research and Applications Survey of Laser Technology for Collaboration and Controlled by a Collaborative System

| Domain | Application | Collaboration Activities | Collaboration Objectives | References | Role of Lasers | Laser Advantages | Collaboration Function[a] |
|---|---|---|---|---|---|---|---|
| Robots | Mapping | Collaborative robots | Integration of data from distributed sources for battery navigation | [44] | Laser scanning, mapping, obstacle detection | Accurate and timely data acquisition | 3 |
| Mobile robots | Navigation and tracking | Collision avoidance | Multisensor integration | [65] | Laser range scanner, mapping | Accurate object detection and mapping | 3 |
| Face recognition | Multimedia | Human–computer interaction | Multisensor integration | [50] | Face scanning | Accurate and timely sensing | 3 |
| Mobile robots | Cooperative positioning | Intelligent spatial interactions | Sensor-based collaboration | [39] | Laser scanning, mapping, people detection | Tracking moving objects accurately | 3 |
| Mobile robots | AR | Teleguide, HRI | Enable intuitive control | [46] | Laser scanning, mapping, obstacle detection, and representation by color | Accurate and timely data acquisition and transmission for stereoscopic visualization | 3 |
| Mobile robots | Multirobot collaboration | Mapping and object detection | More precise performance | [66] | Structured light, mapping, detection | More accurate and timely maps | 3 |

| | | | | | | | |
|---|---|---|---|---|---|---|---|
| Mobile robots | Navigation and tracking | Object recognition and obstacle avoidance | Observe partially available objects | [45] | Laser range scanner, mapping | Tracking moving objects accurately | 1,3 |
| HRI | Laser-based tracking | HRI | Obtain better position and orientation details | [56] | Laser scanning, detect human position and orientation | Accurate tracking | 1,3 |
| Manufacturing | Human detection and tracking | Laser sensors input | Accurate data | [55] | Laser scanning, detect people legs | Accurate and timely data acquisition | 1,3 |
| Dental operations and surgery | AR apparatus, projecting surgical plans onto patient | AR | More effective and accurate surgery | [11] | Laser scanning and projecting and marking, distance measurements | Accurate and timely data acquisition and transmission | 1,3 |
| AR | AR apparatus | AR | Multisensor integration | [29] | Laser scanning and projecting and marking, distance measurements | Objects tracking and human action recognition | 1,3 |
| Robots | Laser welding, path teaching | HR collaboration, AR | Better welding quality | [20] | Laser scanning, laser spot for marking and laser welding | Accurate and timely data acquisition and transmission | 1,3 |
| Mobile robots | Localization | Robot collaboration | Better performance of robot team | [40] | Mapping, motion state estimation | Accuracy | 2,3 |

*(continued)*

**TABLE 9.1 (continued)**

Research and Applications Survey of Laser Technology for Collaboration and Controlled by a Collaborative System

| Domain | Application | Collaboration Activities | Collaboration Objectives | References | Role of Lasers | Laser Advantages | Collaboration Function[a] |
|---|---|---|---|---|---|---|---|
| Mobile robots | Navigation | HRI | Better performance of human robot team | [67] | Laser range scanner, mapping | Accuracy for better responsiveness | 2,3 |
| Robots | Mapping, localization, navigation | Mapping for collaboration | Responsive collaboration/ interaction | [53] | Laser range scanner, mapping | More accurate and timely maps | 2,3 |
| Mobile robots | Search and rescue | Collaboration (tracking) | Better performance | [22] | Laser scanning, mapping, determine relative position | Accurate and timely data acquisition and transmission | 2,3 |
| Robots | AR | HR collaboration | Collaborative classification | [68] | Laser scanning, mapping, objects detection | Accurate and reliable detection and marking | 2,3 |
| Mobile robots | Navigation | HRI, corporal posture detection | Better performance of human robot team | [69] | Structured light, mapping, detection | Accuracy for better responsiveness | 2,3 |
| HRI | Pose/gestures detection | HRI, gestures detection | Precise control | [70] | Structured light, object detection | Accurate processing by industrial robots | 2,3 |
| HRI | HRI, telerobotics | HR collaboration, user interface, gesture detection | Teleoperation guidance | [71] | Structured light, object detection | Accuracy for better control | 2,3 |

| Robots | Laser arc welding, path teaching | HR collaboration, AR | Collaboration among laser sensors, target marking, and laser processor | [19] | Laser scanning, laser spot for marking and laser welding | Accurate processing by industrial robots | 1,2,3 |
|---|---|---|---|---|---|---|---|
| Mobile robots | Development of pointing, acquisition, and tracking for laser communication | Development of control for pointing, acquisition, and tracking (PAT) system | Joint team performance | [8] | Laser communication | Wider bandwidth, lower power requirements, and increased potential for covert data links in comparison to radio-based systems | 4 |
| Mobile robots | Development of pointing, acquisition, and tracking for laser communication | Multisensory interactions | Joint team performance | [57] | Laser communication | Accurate and timely data acquisition and transmission | 4 |
| Mobile robots | Development of pointing, acquisition, and tracking for laser communication | Multisensory interactions | More precise performance | [58] | Laser communication | Accurate and timely data acquisition and transmission | 4 |
| Mobile robots | Development of pointing, acquisition, and tracking for laser communication | Multisensory interactions and communication | Better performance | [24] | Laser communication | Accurate and timely data acquisition and transmission | 4 |

[a] The collaboration functions as defined earlier as the four precision collaboration functions are the following: 0, lasers are used with no collaboration so far; collaboration function 1, targeting; collaboration function 2, interaction; collaboration function 3, mapping, localization, and object detection; and collaboration function 4, laser-based communication.

## Precision Collaboration Function 2: Interactions

In the past, laser pointers used in the field of robotics have been limited to simple target designation, without exploring their potential as versatile input devices. As indicated by Ishii et al., target designation on its own is far from sufficient for a human–robot interface.[7] They suggested that laser pointers can be effective also for specifying commands and target locations for a robot to follow and execute. They used laser pointer-based user interface for issuing various instructions to a robot by applying stroke gesture recognition over the laser's trajectory. Through this interface, the operator can draw stroke gestures using a laser pointer to specify target objects and commands for the robot to execute accordingly.

In another interesting role combination, it was found that a laser pointer can be used as a robot-controlling interface. Such a laser gesture interface was designed and implemented to issue instructions to a mobile robot. Oh and Stuerzlinger describe a system that uses multiple computer-controlled laser pointers as interaction devices for one or more displays and present several alternatives for distinguishing between different laser pointers' arcs.[31] They demonstrated the use of laser pointers as input devices that can provide concurrent input streams to a single-display groupware (SDG) environment and its groupware applications. A framework that supports multiple pointing devices to explore the collaboration utility of multiple laser pointers interaction in graphical environments has also been developed.[12]

Laser technology for tracking fingers has been introduced for gesture recognition in human–computer interfaces.[32,33] Active tracking has been achieved by using a laser diode (with invisible or visible light), steering mirrors, and a photodetector. Successful tracking has been achieved through the analysis of the backscattered light, measured by the photoreceptor during a millisecond-long circular laser saccade around the target position. This procedure has enabled the acquisition of real time, 3D coordinates in a relatively narrow window (concerning the target size). By implementing such a sterile finger tracker, this procedure can be applied to extend current efforts in the area of interactive medical imaging for healthcare operations.[34–37]

In the illustrated cases of interaction, besides operational cost-effectiveness and quality based on the precision of communicated information, there is also reduction in total process time. When errors due to imprecise or less precise commands are eliminated, in some cases, for example, medical applications, significant damage can be prevented; in other cases, repeated commands to correct interaction errors would not be necessary.

## Precision Collaboration Function 3: Mapping, Localization, and Object Detection

In certain interaction and collaboration systems, mapping, localization, and object detection are essential for maintaining the purpose of the collaboration.

In all reported cases of using laser techniques for mapping, localization, and object detection, the devices commonly used have been laser range scanners. However, these laser techniques and devices were not part of the collaboration or interaction, but merely a tool enabling decisions to be made based on the scanning results. Harb et al. demonstrated the significant role played by a mobile robot in a clean-room medical facility. But this role was impeded when a certain control function became difficult to complete by applying a single control technique; the integration of several different collaborating techniques was necessary for successful performance of certain complicated missions.[38] In that case, mapping data extracted from a laser range scanner was fed to several control techniques.

In other cases, where cooperative localization or mapping is the objective, it is accomplished by interactive communication between mobile robots and a network of laser range scanners.[39] Zhuang et al. studied the problem of cooperative localization based on environmental information interaction in an unknown, complex environment. They developed an algorithm for the adaptive selection of cooperative localization based on multirobot collaboration using extended Kalman filter in the cluttered environment.[40] Li et al.[41] developed a network surveillance system based on data fusion algorithm of mobile robot team for collaborative target localization.

Laser scanning was used in a number of mapping applications, for example, for (1) mapping and determination of relative positions for a team of robots collaborating in a search and rescue task,[22] (2) mapping and obstacle detection in a collaborative navigation of autonomous unmanned vehicles,[42] and (3) force-feedback-enhanced teleoperation navigating of outdoor robotic vehicles.[43] Collaborative mapping was developed for a team of robots where each was equipped with its own LRF.[44] In this application, a distributed data fusion algorithm was used to combine the local maps produced by each robot into a shared global map. Collaborative localization of mobile robots' tracking ability or structured movement using data fusion from laser scanners mounted on robots was investigated in a number of studies.[14,45]

In telerobotics systems, mapping the environment has an important purpose for the success of the interaction and the situation awareness by human operators. The use of graphical elements extracted from a 3D mapping produced by a laser range scanner can improve telerobotics performance. Livatino et al. conducted an experiment to examine the value of a graphical elements approach for an actual telerobotic system, where an operator manipulates a robot located at a distance of approximately 3000 km away. The results validated the simplicity and effectiveness of the experimental approach and provided a base for further investigations.[46]

Related to mapping and localization, several HRI systems apply laser range scanners to detect people and identify faces and gestures. For instance, Marcos et al. used a laser scanning technique to create virtual head orientations for human–computer interaction with social robots. Interpretation of human attitudes was conducted by detection and tracking of human movements throughout interior environments using a pan–tilt–zoom (PTZ) camera and an laser range finder (LRF) mounted on a mobile robot platform.[47] After detection of a human, the person's legs were detected using LRF and then the walking gestures were analyzed.[48] Sakurai and Mita[49] developed a robot that interacts with people and collects information in biofied rooms. Ghys et al.[50] and Rama et al.[51] developed 3D face recognition systems for multimedia applications. Svenstrup et al.[52] developed a method to determine a person's pose based on laser range measurements for robot navigation tasks in human environments.

Several studies using laser range scanners were conducted for environments where people and mobile robots collaborate and while collaborating may share the same space. In these studies, people are detected and tracked with laser range scanners, which are also applied for constructing 3D mapping of the environment and for analyzing the constrains for people movements based on the current map,[9] based on an annotated data set,[53,54] based on leg detection,[23,48,55] or based on parametric shape modeling.[56] In all these examples, precision collaboration is essential both for the effective operation and for safety reasons.

## Precision Collaboration Function 4: Laser Communication

In collaboration and interaction tasks among mobile robots, laser-based communications can be the solution for the increasing demand for communications bandwidth. Researches show that using laser technology will enable use of relatively higher bandwidth, lower power requirements, and increased potential for covert data links, in comparison to radio-based systems.[8] So far, the communication requires a visual contact between the transmitter and the receiver in both interacting robots. This requirement may be complicated by the inherent strict requirements of agility and accurate steering of the laser beam over wide angular ranges. Such systems will have to cope with the high-frequency vibrations of the optical platform and atmospheric disturbances. The solutions proposed are based on gimbals and gyroscopes,[57] omnidirectional sensor mount,[58] high-speed nonmechanical laser beam steering devices,[24] and a feed forward vibration rejection systems.[8]

A summary of the research and application of lasers in systems that comprise collaboration aspects is given in Table 9.1. In all these cases, laser is applied as one or a combination of the following: a tool, information media, energy source, pointing device, or sensor to collect data, all as part of the collaboration process/collaborative system or as a tool to be controlled by the collaborative system.

## Opportunities and Challenges

The opportunities in applying advanced laser techniques for precision collaboration are derived from the framework in Figure 9.2 and the review in Table 9.1. They are a combination of five collaboration dimensions, as defined in the following, according to four attributes of collaboration:

1. Type of interaction
2. Type of tasks
3. Specific application
4. The relevant environment

Each unique combination of collaboration, interaction, tasks, application, and environment will require specific communication; synchronization; geometrical environmental data and information collection; control characteristics, including commands, interactions, and interfaces; performance and status data evaluation; and, in turn, specific laser characteristics and technology. Five dimensions of collaboration emphasizing precision can be defined, as described next.

### Collaboration Dimensions

Based on the earlier survey and the abilities and possibilities of lasers for precision collaboration, we propose to classify precision collaboration and integration along five main dimensions:

D1. Virtual shared surfaces among users: conceptual design, sketching, and planning—When physical design and development are required, teamwork is often necessary. Certain types of work can be carried out remotely. It means that the designers can be distributed and use a shared virtual model for their conceptual design, reflecting the characteristics of the true physical locations and surface features. In such contexts, accurate registration between the virtual spaces and the physical ones is crucial to assure that all the participants are working under the same coordinate system.[72] This level of accuracy can be achieved only by projecting the regions of interest with lasers. Those regions of interests are the same surface regions on which one or more of the collaborating participants are sketching.[73]

D2. Shared spaces with robots (cowork): pointing robots to targets—Assistive, service, and surveillance robots interact with a user usually through gestures and speech commands.[74] In terms of gestures, the most common type is the deictic class, such as pointing to a location. When such a gesture is evoked, the robots move and pick an object, or drop it.[75] These systems suffer from ambiguity of target locations, especially when used outdoors. It means that when a user intends to point to one location, by following its line of sight, the robot perceives a different location, typically due to the parallax effect.[76] In such situations, a laser is mounted on a user's wrist to pinpoint an unambiguous target location, easily recognizable by the robot through a computer vision system.

D3. Shared gestures: pointing gestures—In the first dimension, laser projected points were used for common registration among different virtual views; nevertheless, the problem of representing accurately the sketches made by designers is not considered. These sketches are commonly designed on a projected surface,[77] such as the one provided by Microsoft Surface. A problem of this type of interactive displays is that the trajectories drawn over them are as thick as the fingers creating them. Still, some designs require thin links (trajectories) between the nodes representing instances of their model. An alternative approach is to sketch in the air, using a special glove that projects a single point coming from a laser installed on the user's palm.[78] By adopting such hardware/software combination, the trajectories represented by the movement of a finger or hand are translated into thin lines, resulting from the precise trajectory by the laser's dot projection.[79]

D4. AR: AR is the virtual environment generated by integration of the captured views from cameras with objects superimposed on those images.[80] Thus, users interact with virtual objects as if they were interacting with real ones. This technique has also been used to project textured patterns on real objects lacking any patterns (e.g., uniform colors) in order to generate a realistic illusion of how the object would appear if it had that pattern.[81] In such scenarios, accurate registration between the object and the pattern is needed to assure a realistic view, rather than using only a lighting artifact. This registration is achieved with a user of laser rendering on the edges/corners of the object, to set the boundaries of the projection.[82]

D5. Shared patterns: projected structured light from individual sources to assess depth—Modern depth cameras use structured infrared (IR) light projected on the environment to assess the distance of the objects within the environment to the camera.[83] This solution is cheap and precise and requires low energy consumption.[84] An example of this technology is the Microsoft Kinect.[85] In spite of the revolutionary solution for depth imaging, this technology does not work outdoors due to the diffusion of the IR light and the high interference with day light. To overcome this problem, a solution is to use lasers to generate the beam of light for projection on the environment. While this light is visible to the human eye, it can be filtered from the imaging system using the coefficients of signal in the frequency domain.[86]

The five dimensions earlier are illustrated by collaboration applications based on visual information and can be generalized to other, nonvisual types of information along the same five dimensions.

## Research Needs and Opportunities

Initial progress in laser-based precision collaboration has been made and also points to further opportunities and challenges. Fifteen key areas are shown in Table 9.2 along the earlier five collaboration dimensions, taken from application areas of industrial engineering, medicine, manufacturing, robotics, HRI, and human interfaces. In these research areas and applications, lasers are used as a tool, information medium, energy source, pointing device, and sensor to collect data as part of the collaboration process/ collaborative system or as a tool to be controlled by the collaborative system.

While initial progress has been reported along these dimensions, further research is needed. Specifically, six areas of need and opportunity are observed:

1. Display groupware interfaces and synchronization. This area focuses on research of collaborative computer environment for multiple users that are physically in relative proximity,[31] and development of a framework that supports multiple pointing devices, such as mouse and laser pointer in graphical environments.[12]
2. Graphical representation 46,60 and rendering using lasers.[20,39]
3. Gesture recognition using lasers.[7,48,52]

Lasers in combination with servo mirrors and photoreceptors have been used to track users' fingers. Their spatial and temporal information are used to recognize specific gestures.

4. Human factors of integrated laser systems for collaboration[27,28]
5. AR using lasers

Laser projectors have been previously used to create AR,[11,20] instead of the standard head-mounted displays (HMD). This technology overcomes problems such as small field of view, limited resolution, multiple focus planes, and eye fatigue.

**TABLE 9.2**

Emerging Contribution and Challenge Areas for Systems Interaction, Collaboration, and Integration with Lasers

| Collaboration Dimension | Area/Application | Function/Device | Sample References |
|---|---|---|---|
| D1 | Mobile robots/ communication | Laser communication, development of pointing, acquisition, and tracking for laser communication | [8] |
| D1–D2 | Robot control | User interface, specifying command and target location; laser pointers | [7] |
| D2 | User interface and display groupware | Computer-controlled laser pointers; input device for collaboration | [31] |
| D2 | User interface and display groupware | Input device for collaboration, laser pointers | [12] |
| D2 | Mobile robots/ intelligent space | Cooperative positioning, laser scanning, mapping, people detection | [39] |
| D2 | HRI | Dismantling interior facilities in buildings, pointing to a target in the task space with the laser pointer | [21] |
| D2 | Assistive robot (chronic stroke) | Selection of objects with laser pointer | [30] |
| D2–D3 | Mobile/service robots, HRI | Pose/gestures detection, laser range scanner, detection | [48] |
| D2–D3 | HRI, wearable collaborative systems | Collaboration/interface, point to real objects in the task space with laser spot, laser pointer | [27,28] |
| D2–D3 | Industrial robots, positioning tasks | Mapping objects, point to a work piece in the task space with a laser pointer | [18] |
| D4 | Industrial robots, laser welding, path teaching | HR collaboration, AR, laser scanning, laser spotting for marking, and laser welding | [20] |
| D4 | Dental operations and surgery/AR | Projecting surgical plans onto patient, laser scanning, projecting and marking | [11] |
| D5 | HRI, manipulator control, robot guidance | HR collaboration, user interface, gesture and posture detection | [69,70] |
| D5 | HRI, teleoperation guidance | HR collaboration, user interface, gesture detection | [71] |
| D5 | Mobile robots | Multirobot collaboration, mapping, and object detection | [66] |

6. Enhanced depth sensor

Technology is needed for indoor/outdoor use[87]: Monocular cameras using structured light are relatively cheap and reliable and do not require calibration. While some cameras deliver satisfactory performance indoors using IR structured light, other devices use stereoscopic[88] or rely on laser-structured projections to deliver depth information.[89]

In collaborative control theory (CCT) that is central to collaborative e-work, e-production, and e-service systems,[90,91] the use of laser technology can be considered as an enabler of better, more precise interaction for the purpose of collaborative activities: laser-based interactions can improve the efficiency and effectiveness of the collaboration processes and the support protocols designed for optimized collaboration. They can also reduce the time required for interactions needed for collaboration, including setup and trial interactions, and eliminate or reduce interaction-related errors and conflicts.

There are also key contributions to cyber–physical systems interaction with lasers: laser advantages and roles in interactive collaboration can reduce errors and conflicts, thus reducing the time of interaction and the number of interactions and commands required to achieve the collaboration goal, compared with less precise means of interaction. In addition, more precise collaboration based on lasers, as discussed earlier and illustrated in Tables 9.1 and 9.2, can increase the quality (precision) of the commands, hence reducing the number of resulting actions and improving the overall performance.

The following two examples illustrate the six opportunity areas:

1. *Welding robots path teaching example*: In a path teaching task of welding laser robots in an industrial environment, for example,[20] the human–robot collaboration is conducted through AR. In this case, laser technology is used as
   a. Processing tool—laser welding
   b. Sensor—laser range scanner
   c. Marker—laser pointer/spot

An HRI system has been designed by fusing information from a camera and an LRF. The LRF captures the Cartesian information given by a user and the robot working paths and trajectories. An AR environment has been designed where the virtual tool is superimposed onto live video. The operator simply needs to point and click on the image of the workpiece, in order to generate the required welding tool path. A laser spot is marking the workpiece and then the workpiece is tracked by the system throughout the process.[20]

In this example, the collaboration dimensions are D1, D2, and D4.

2. *Selective harvesting example*: Human–robot collaborative task sharing system for selective harvesting[92] can apply laser technology for the following:
   a. Cutting tool and energy transfer—laser cutter
   b. Sensing to detect fruits and obstacles—laser range scanner
   c. Targeting—laser pointer
   d. Collaboration and interaction by detecting and tracking human operator/supervisor positions; tracking human operator's hands and following human operator's commands with laser pointers, scanners, and projected light

In this example, the collaboration dimensions are D1, D2, D3, and D5.

## Conclusions and Recommendations

Lasers play an important role in modern life, society, and industry. Although laser technology has evolved for over 50 years, there are scientific and engineering challenges and areas that need to be explored for gaining further benefits with the advantages of laser technology.[1]

The assimilation of lasers beyond communication, processing, and discovery to interaction and collaboration systems is still in its infancy. Only 0.12% of the reviewed papers published on lasers are devoted to these areas. Recent advances in laser technology raise the potential feasibility of harnessing certain laser solutions for new and improved integration through more precise interaction and collaboration systems. Precision and efficiency advantages of lasers include relatively more accurate and more timely sensory data and signals; relatively more precise, more responsive, and more timely laser-based control and communication of information; and relatively more accurate and more timely laser-based processing by accurate, fast, and efficient energy transfer and transmission.

Lasers can be applied in a collaborative mission as a tool to be monitored by the collaborative and interaction architecture, or as part of the collaboration methodology and techniques. Laser-based collaboration and interaction are used in many areas, such as dental and medical operations and surgery, manufacturing, production, robotics, military, transportation, security, construction, and agriculture. So far, the use of lasers for precision collaboration and interaction is relatively limited; lasers are mainly applied as a tool to map the environment, detect objects, for targeting, localization, and other interactions.

Table 9.3 presents areas of direct impacts of laser-based interaction and collaboration, which are anticipated to expand in the near future, and their

**TABLE 9.3**

Direct Impacts of Laser-Based Interactions Anticipated in Eight Areas of Emerging Collaborative/Interactive Systems and Networks

| Impacts on Safety, Security, Productivity, and Quality | Examples | Dimensions of Sharing among Interacting Agents/Operators/Participants | | | | |
|---|---|---|---|---|---|---|
| | | D1. Virtual Surfaces | D2. Space with Robots | D3. Gestures | D4. AR | D5. Patterns |
| 1. Manufacturing | Collaborative CAD/CAM production[60]; collaborative processing[20] | ■ | ■ | | ■ | |
| 2. Production | Facility monitoring for safety and quality[93]; sensor automation for safety and productivity[94] | ■ ■ | ■ ■ | ■ | | ■ ■ |
| 3. Agriculture | Precision human–robot tasks[92] | ■ | ■ | ■ | ■ | ■ |
| 4. Transportation | Airport security and safety[95] | ■ | | ■ | | ■ |
| 5. Supply chains | Supply chain security[96] | ■ | ■ | | ■ | ■ |
| 6. Training | Security training[97] | ■ | | | ■ | ■ |
| 7. Healthcare | Operating room image browser, robotic scrub nurse[98] | | ■ | ■ | | |
| 8. Environment | Environmental monitoring by laser-based wireless sensor networks[99] | | ■ | ■ | | ■ |

required collaboration dimensions. Harnessing lasers for precision collaboration is a significant and promising research area: It is expected to reduce costs, increase flexibility, and decrease system weight. Lasers will record, sense, mark, and inform about points on surfaces and in space relatively more accurately and with inherent precision. They are reliable, can have wide bandwidth and low power requirements, and can transmit both secure data and deliver efficiently high, focused energy. All these advantages and improvements will enable the creation and implementation of precision collaborative systems.

## Acknowledgment

Research reported in this chapter was partially supported by the PRISM Center at the Production, Robotics, and Integration Software for Manufacturing and Management at Purdue University and by the Agricultural Robotics Lab at the Institute of Agricultural Engineering, Volcani Center, Israel.

## References

1. Li, L., The challenges ahead for laser macro, micro and nano manufacturing, in *Advances in Laser Materials Processing: Technology, Research and Applications*, Lawerence, J., Pou, J., Low, D. K. Y., and Toyserkani, E. (eds.), Woodhead Publishing Limited, Cambridge, U.K., 2010, pp. 20–39.
2. WOS , Web of Science, 2012. http://guides.lib.purdue.edu/WebOfScience.
3. Yoo, A. C., Gilbert, G. R., and Broderick, T. J., *Military Robotic Combat Casualty Extraction and Care*, Springer, New York, 2010.
4. Bechar, A., Meyer, J., and Edan, Y., An objective function to evaluate performance of human-robot collaboration in target recognition tasks, *IEEE Transactions on Systems Man and Cybernetics Part C-Applications and Reviews* 39(6), 611–620, 2009.
5. Nof, S. Y., Cultural factors: Their impact in collaboration support systems and on decision quality, in *Cultural Factors in Systems Design, Decision Making and Action*, Proctor, R. W., Nof, S. Y., and Yih, Y. (eds.), CRC Press, Boca Raton, FL, 2011.
6. Brogardh, T., Robot control overview: An industrial perspective, *Modeling Identification and Control* 30(3), 167–180, 2009.
7. Ishii, K., Zhao, S., Inami, M., Igarashi, T., and Imai, M., Designing laser gesture interface for robot control, in *Proceedings of the 12th IFIP International Conference on Human-Computer Interaction*, Uppsala, Sweden, 2009, pp. 479–492.
8. Sofka, J., Nikulin, V. V., Skormin, V. A., Hughes, D. H., and Legare, D. J., Laser communication between mobile platforms, *IEEE Transactions on Aerospace and Electronic Systems* 45(1), 336–346, 2009.
9. Luber, M., Tipaldi, G. D., and Arras, K. O., Place-dependent people tracking, *International Journal of Robotics Research* 30(3), 280–293, 2011.
10. Shan, X. M., Kuo, S. K., Zhang, J. H., and Menq, C. H., Ultra precision motion control of a multiple degrees of freedom magnetic suspension stage, *IEEE-ASME Transactions on Mechatronics* 7(1), 67–78, 2002.
11. Glossop, N., Wedlake, C., Moore, J., Peters, T., and Wang, Z. H., Laser projection augmented reality system for computer assisted surgery, in *Proceedings of the Medical Image Computing and Computer-Assisted Intervention—Miccai 2003, Pt 2*, Montréal, Quebec, Canada, 2003, pp. 239–246.
12. Vogt, F., Wong, J., Po, B. A., Argue, R., Fels, S. S., and Booth, K. S., Exploring collaboration with group pointer interaction, in *Proceedings of the Computer Graphics International*, Crete, Greece, 2004, pp. 636–639.

13. Mazzetto, F. and Calcante, A., Highly automated vine cutting transplanter based on DGNSS-RTK technology integrated with hydraulic devices, *Computers and Electronics in Agriculture* 79(1), 20–29, 2011.

14. Jiang, R. and Chen, Y., Multi-robot precise localization based on multi-sensor fusion, in *Proceedings of the 2008 IEEE Conference on Robotics, Automation, and Mechatronics*, Chengdu, China, 2008, pp. 967–972.

15. Park, Y. S., Structured beam projection for semi-automatic teleoperation, in *Proceedings of the Conference on Optomechatronic Systems*, Boston, MA, 2000, pp. 192–201.

16. Myers, T. D. and McDaniel, J. D., The pulsed Nd:YAG dental laser: Review of clinical applications, *Journal of the California Dental Association* 19(11), 25–30, 1991.

17. Wigdor, H. A., Walsh, J. T., Featherstone, J. D. B., Visuri, S. R., Fried, D., and Waldvogel, J. L., Lasers in dentistry, *Lasers in Surgery and Medicine* 16(2), 103–133, 1995.

18. Gonzalez-Galvan, E. J., Cruz-Ramirez, S. R., Seelinger, M. J., and Cervantes-Sanchez, J. J., An efficient multi-camera, multi-target scheme for the three-dimensional control of robots using uncalibrated vision, *Robotics and Computer-Integrated Manufacturing* 19(5), 387–400, 2003.

19. Li, H. C., Gao, H. M., and Wu, L., Teleteaching approach for sensor-based arc welding telerobotic system, *Industrial Robot: An International Journal* 34(5), 423–429, 2007.

20. Chuen Leong, N., Teck Chew, N., Thi Anh Ngoc, N., Yang, G., and Chen, W., Intuitive robot tool path teaching using laser and camera in augmented reality environment, in *Proceedings of the 11th International Conference on Control, Automation, Robotics and Vision*, Singapore, 2010, pp. 114–119.

21. Cruz-Ramirez, S. R., Ishizuka, Y., Mae, Y., Takubo, T., and Arai, T., Dismantling interior facilities in buildings by human robot collaboration, in *Proceedings of the 2008 IEEE International Conference on Robotics and Automation*, Vol. 1–9, Pasadena, CA, 2008, pp. 2583–2590.

22. Guarnieri, M., Kurazume, R., Masuda, H., Inoh, T., Takita, K., Debenest, P., Hodoshima, R., Fukushima, E., and Hirose, S., HELIOS system: A team of tracked robots for special urban search and rescue operations, in *Proceedings of the 2009 IEEE-RSJ International Conference on Intelligent Robots and Systems*, St. Louis, MO, 2009, pp. 2795–2800.

23. Bellotto, N. and Hu, H., Multisensor-based human detection and tracking for mobile service robots, *IEEE Transactions on Systems Man and Cybernetics Part B: Cybernetics* 39(1), 167–181, 2009.

24. Nikulin, V., Khandekar, R., Sofka, J., and Mecherle, G. S., Performance of a laser communication system with acousto-optic tracking: An experimental study, in *Proceedings of SPIE, the International Society for Optical Engineering SPIE*, San Diego, California, USA, 2006, pp. 61050C.1–61050C.10.

25. Kemp, C. C., Anderson, C. D., Nguyen, A. J., and Xu, Z., A point and click interface for the real world: Laser designation of objects for mobile manipulation, in *Proceedings of the 3rd ACM/IEEE International conference on Human-Robot Interaction*, Amsterdam, Netherlands, 2008, pp. 241–248.

26. Suzuki, T., Ohya, A., and Yuta, S., Operation direction to a mobile robot by projection lights, in *Proceedings of the 2005 IEEE Workshop on Advanced Robotics and its Social Impacts*, Nagoya, Japan, 2005, pp. 160–165.

27. Kurata, T., Sakata, N., Kourogi, M., Kuzuoka, H., and Billinghurst, M., Remote collaboration using a shoulder-worn active camera/laser, in *Proceedings of the 8th International Symposium on Wearable Computers*, Arlington, VA, 2004, pp. 62–69.

28. Mann, S., Telepointer: Hands-free completely self contained wearable visual augmented reality without headwear and without any infrastructural reliance, in *Proceedings of the fourth International Symposium on Wearable Computers*, Atlanta, GA, 2000, pp. 177–178.

29. Yeong Nam, C., Young-Ho, K., Jin, C., Kyusung, C., and Hyun, S. Y., An adaptive sensor fusion based objects tracking and human action recognition for interactive virtual environments, in *Proceedings of the eigth International Conference on Virtual Reality Continuum and its Applications in Industry*, ACM, Yokohama, Japan, 2009.

30. Young Sang, C., Cressel, D. A., Jonathan, D. G., and Charles, C. K., Laser pointers and a touch screen: Intuitive interfaces for autonomous mobile manipulation for the motor impaired, in *Proceedings of the 10th International ACM SIGACCESS Conference on Computers and Accessibility*, ACM, Halifax, Nova Scotia, Canada, 2008.

31. Oh, J. Y. and Stuerzlinger, W., Laser pointers as collaborative pointing devices, in *Proceedings of the Graphics Interface 2002 Conference*, Calgary, Alberta, Canada, 2002, pp. 141–149.

32. Perrin, S., Cassinelli, A., and Ishikawa, M., Gesture recognition using laser-based tracking system, in *Proceedings of the 6th IEEE International Conference on Automatic Face and Gesture Recognition*, Seoul, South Korea, 2004, pp. 541–546.

33. Cassinelli, A., Perrin, S., and Ishikawa, M., Smart laser-scanner for 3d human-machine interface, in *Proceedings of the International Conference on Human Factors in Computing Systems*, Portland, OR, 2005, pp. 1138–1130.

34. Wachs, J., Stern, H., Edan, Y., Gillam, M., Feied, C., Smith, M., and Handler, J., Gestix: A doctor-computer sterile gesture interface for dynamic environments, *Advances in Soft Computing* 39, 30–39, 2007.

35. Wachs, J. P., Stern, H. I., Edan, Y., Gillam, M., Handler, J., Feied, C., and Smith, M., A Gesture-based tool for sterile browsing of radiology images, *Journal of the American Medical Informatics Association* 15(3), 321–323, 2008.

36. Mithun, J., Cange, C., Packer, R., and Wachs, J., Intention, context and gesture recognition for sterile mri navigation in the operating room, in *Proceedings of the Progress in Pattern Recognition, Image Analysis, Computer Vision, and Applications*, Buenos Aires, Argentina 2012, pp. 220–227.

37. Mithun , J., Wachs, J., and Packer, R., Hand-gesture-based sterile interface for the operating room using contextual cues for the navigation of radiological images, *Journal of the American Medical Informatics Association* 20, e183–e186, 2012.

38. Harb, M., Abielmona, R., and Petriu, E., Speed control of a mobile robot using neural networks and fuzzy logic, in *Proceedings of the 2009 International Joint Conference on Neural Networks*, Vols. 1–6, IEEE, New York, 2009, pp. 1346–1352.

39. Morioka, K., Oinaga, Y., and Nakamura, Y., Control of human-following robot based on cooperative positioning with an intelligent space, *Electronics and Communications in Japan* 95(1), 20–30, 2012.

40. Zhuang, Y., Gu, M. W., Wang, W., and Yu, H. Y., Multi-robot cooperative localization based on autonomous motion state estimation and laser data interaction, *Science China—Information Sciences* 53(11), 2240–2250, 2010.

41. Li, G., Du, J., Zhu, C., and Sheng, W., A cost-effective and open mobile sensor platform for networked surveillance, in *Proceedings of the Conference on Signal and Data Processing of Small Targets*, SPIE, San Diego, CA, 2011.

42. Ying, W., Sun, F., Liu, H., and Song, Y., A collaborative navigation system for autonomous vehicle in flat terrain, in *Proceedings of the IEEE International Conference on Control and Automation*, Christchurch, New Zealand, 2009, pp. 1615–1620.

43. Jarvis, R., Terrain-aware path guided mobile robot teleoperation in virtual and real space, in *Proceedings of the Third International Conference on Advances in Computer–Human Interactions: ACHI 2010*, Los Alamitos, CA, 2010, pp. 56–65.

44. Rogers, J., Paluri, M., Cunningham, A., Christensen, H. I., Michael, N., Kumar, V., Ma, J., and Matthies, L., Distributed autonomous mapping of indoor environments, in *Proceedings of the Conference on Micro- and Nanotechnology Sensors, Systems, and Applications III*, SPIE-Int Soc Optical Engineering, Orlando, FL, 2011.

45. Tseng, K.-S. and Tang, A. C.-W., Self-localization and stream field based partially observable moving object tracking, *Eurasip Journal on Advances in Signal Processing* 2009, 12, 2009.

46. Livatino, S., Muscato, G., De Tommaso, D., and Macaluso, M., Augmented reality stereoscopic visualization for intuitive robot teleguide, in *Proceedings of the IEEE International Symposium on Industrial Electronics (ISIE 2010)*, IEEE, Bari, Italy, 2010, pp. 2828–2833.

47. Marcos, S., Gomez-Garcia-Bermejo, J., and Zalama, E., A realistic, virtual head for human-computer interaction, *Interacting with Computers* 22(3), 176–192, 2010.

48. Ramirez-Hernandez, A. C., Rivera-Bautista, J. A., Marin-Hernandez, A., and Garcia-Vega, V. A., *Detection and Interpretation of Human Walking Gestures for Human–Robot Interaction*, IEEE, New York, 2009.

49. Sakurai, F. and Mita, A., Biofied room integrated with sensor agent robots to interact with residents and acquire environmental information, in *Proceedings of the Conference on Sensors and Smart Structures Technologies for Civil, Mechanical, and Aerospace Systems 2011*, SPIE; Amer Soc Mech Engineers; KAIST, San Diego, CA, 2011.

50. Ghys, C., Paragios, N., and Bascle, B., Graph-based multi-resolution temporal-based face reconstruction, in *Proceedings of the Advances in Visual Computing*, Pt 1, Lake Tahoe, NV, 2006, pp. 803–812.

51. Rama, A., Tarres, F., and Rurainsky, J., 2D-3D pose invariant face recognition system for multimedia applications, in *Recent Advances in Multimedia Signal Processing and Communications*, Grgic, M., Delac, K., and Ghanbari, M. (eds.), Springer, Berlin, Germany, 2009, pp. 121–144.

52. Svenstrup, M., Tranberg, S., Andersen, H. J., and Bak, T., Pose estimation and adaptive robot behaviour for human-robot interaction, in *Proceedings of ICRA: 2009 IEEE International Conference on Robotics and Automation*, Vol. 1–7, IEEE, New York, 2009, pp. 3222–3227.

53. Zivkovic, Z., Booij, O., Krose, B., Topp, E. A., and Christensen, H. I., From sensors to human spatial concepts: An annotated data set, *IEEE Transactions on Robotics* 24(2), 501–505, 2008.

54. Yang, S.-W., Wang, C.-C., and Thorpe, C., The annotated laser data set for navigation in urban areas, *International Journal of Robotics Research* 30(9), 1095–1099, 2011.

55. Martinez-Otzeta, J. M., Ibarguren, A., Ansuategi, A., Tubio, C., and Aristondo, J., People following behaviour in an industrial environment using laser and stereo camera, in *Proceedings of the 23rd International Conference on Industrial, Engineering and Other Applications of Applied Intelligent Systems*, Cordoba, Spain, 2010, pp. 508–517.

56. Glas, D. F., Miyashita, T., Ishiguro, H., and Hagita, N., Laser-based tracking of human position and orientation using parametric shape modeling, *Advanced Robotics* 23(4), 405–428, 2009.

57. Marins, J. L., Xiaoping, Y., Bachmann, E. R., McGhee, R. B., and Zyda, M. J., An extended Kalman filter for quaternion-based orientation estimation using MARG sensors, in *Proceedings of the IEEE Intelligent Robots and Systems*, Vol. 4, New York, 2001, pp. 2003–2011.

58. Rosheim, M. E. and Sauter, G. F., New high-angulation omni-directional sensor mount, in *Proceedings of the Free-Space Laser Communication and Laser Imaging II*, SPIE, Seattle, WA, 2002, pp. 163–174.

59. Boulnois, J. L., Photophysical processes in recent medical laser developments: A review, *Lasers in Medical Sciences* 1(1), 47–66, 1986.

60. Sickel, K., Baloch, S., Melkisetoglu, R., Bubnik, V., Azernikov, S., and Fang, T., Toward automation in hearing aid design, *Computer-Aided Design* 43(12), 1793–1802, 2011.

61. Faudzi, A. A. M., Suzumori, K., and Wakimoto, S., Development of an intelligent chair tool system applying new intelligent pneumatic actuators, *Advanced Robotics* 24(10), 1503–1528, 2010.

62. Naren, V. and Shaleen, V., Towards development of a non-touch man-machine interface for intelligent presenter-audience collaborative environment, *ASME Conference Proceedings* 2006(42578), 841–849, 2006.

63. Kanda, T., Glas, D. F., Shiomi, M., and Hagita, N., Abstracting people's trajectories for social robots to proactively approach customers, *IEEE Transactions on Robotics* 25(6), 1382–1396, 2009.

64. Wenjian, Y., Fuchun, S., Huaping, L., and Yu, S., A collaborative navigation system for autonomous vehicle in flat terrain, in *Proceedings of the IEEE International Conference on Control and Automation*, IEEE, Christchurch, New Zealand, 2009, pp. 1615–1620.

65. Rongxin, J., Xiang, T., Li, X., and Yaowu, C., A robot collision avoidance scheme based on the moving obstacle motion prediction, in *Proceedings of the International Colloquium on Computing, Communication, Control, and Management*, Los Alamitos, CA, 2008, pp. 341–345.

66. Tavakoli, M., Cabrita, G., Faria, R., Marques, L., and de Almeida, A. T., Cooperative multi-agent mapping of three-dimensional structures for pipeline inspection applications, *International Journal of Robotics Research* 31(12), 1489–1503, 2012.

67. Mehdi, S. A. and Berns, K., Ordering of robotic navigational tasks in home environment, in *Proceedings of the 13th Federation-of-International-Robot-soccer-Association Robot World Congress*, Springer-Verlag Berlin, Germany/Bangalore, India, 2010, pp. 242–249.

68. Giesler, B., P. S., M. W., and Dillmann, R., Sharing skills: Using augmented reality for human-robot collaboration, in *Proceedings of the SPIE*, Bellingham, WA, 2004, pp. 446–453.

69. Cho, K. B. and Lee, B. H., Intelligent lead: A novel hri sensor for guide robots, *Sensors* 12(6), 8301–8318, 2012.

70. Wang, B. C., Li, Z. J., Ye, W. J., and Xie, Q., Development of human-machine interface for teleoperation of a mobile manipulator, *International Journal of Control Automation and Systems* 10(6), 1225–1231, 2012.

71. Du, G. L., Zhang, P., Mai, J. H., and Li, Z. L., Markerless kinect-based hand tracking for robot teleoperation, *International Journal of Advanced Robotic Systems* 9, 10, 2012.

72. Craig, D. L. and Zimring, C., Support for collaborative design reasoning in shared virtual spaces, *Automation in Construction* 11(2), 249–259, 2002.

73. Sareika, M. and Schmalstieg, D., Urban sketcher: Mixed reality on site for urban planning and architecture, in *Proceedings of the 2007 sixth IEEE and ACM International Symposium on Mixed and Augmented Reality*, ISMAR, Washington, DC, 2007, pp. 27–30.

74. Breazeal, C., Hoffman, G., and Lockerd, A., Teaching and working with robots as a collaboration, in *Proceedings of the Third International Joint Conference on Autonomous Agents and Multiagent Systems*, Washington, DC, 2004, pp. 1030–1037.

75. Sato, E., Yamaguchi, T., and Harashima, F., Natural interface using pointing behavior for human-robot gestural interaction, *IEEE Transactions on Industrial Electronics* 54(2), 1105–1112, 2007.

76. Cipolla, R. and Hollinghurst, N. J., Human-robot interface by pointing with uncalibrated stereo vision, *Image and Vision Computing* 14(3), 171–178, 1996.

77. Ishii, H. and Kobayashi, M., Clearboard: A seamless medium for shared drawing and conversation with eye contact, in *Proceedings of the Conference on Human Factors in Computing Systems*, New York, 1992, pp. 525–532.

78. Fruchter, R., Biswas, P., and Yin, Z., USA, DiVAS-a cross-media system for ubiquitous gesture-discourse-sketch knowledge capture and reuse, U.S. Patent Application 11/132,171, May 17 2005.

79. Jackson , B. and Keefe, D. F., Sketching over props: Understanding and interpreting 3d sketch input relative to rapid prototype props, in *Proceedings of the IUI 2011 Sketch Recognition Workshop*, New York, 2011, 75 pp.

80. Azuma, R. T., A survey of augmented reality, *Presence: Teleoperators and Virtual Environments* 6(4), 355–385, 1997.

81. Leykin, A. and Tuceryan, M., Automatic determination of text readability over textured backgrounds for augmented reality systems, in *Proceedings of the Third IEEE and ACM International Symposium on Mixed and Augmented Reality*, Los Alamitos, CA, 2004, pp. 224–230.

82. Lee, J. C., Hudson, S. E., and Tse, E., Foldable interactive displays, in *Proceedings of the 21st Annual ACM Symposium on User Interface Software and Technology*, New York, 2008, pp. 287–290.

83. Salvi, J., Pagès, J., and Batlle, J., Pattern codification strategies in structured light systems, *Pattern Recognition* 37(4), 827–849, 2004.

84. Tsalakanidou, F., Malassiotis, S., and Strintzis, M. G., Face localization and authentication using color and depth images, *IEEE Transactions on Image Processing* 14(2), 152–168, 2005.

85. Silberman, N. and Fergus, R., Indoor scene segmentation using a structured light sensor, in *Proceedings of the IEEE International Conference on Computer Vision*, Los Alamitos, CA, 2011, pp. 601–608.

86. Maimone, A. and Fuchs, H., Reducing interference between multiple structured light depth sensors using motion, in *Proceedings of the IEEE Virtual Reality*, Costa Mesa, California, USA 2012, pp. 51–54.

87. Miller, A., White, B., Charbonneau, E., Kanzler, Z., and LaViola, J. J., Interactive 3D model acquisition and tracking of building block structures, *IEEE Transactions on Visualization and Computer Graphics* 18(4), 651–659, 2012.

88. Kim, H., Lee, G. A., Yang, U., Kwak, T., and Kim, K. H., Dual autostereoscopic display platform for multi-user collaboration with natural interaction, *Etri Journal* 34(3), 466–469, 2012.

89. Chavez, A. and Karstoft, H., Improvement of kinect (tm) sensor capabilities by fusion with laser sensing data using octree, *Sensors* 12(4), 3868–3878, 2012.

90. Nof, S. Y., Design of effective e-Work: Review of models, tools, and emerging challenges, *Production Planning and Control* 14(8), 681–703, 2003.

91. Nof, S. Y., Collaborative control theory for e-Work, e-Production, and e-Service, *Annual Reviews in Control* 31(2), 281–292, 2007.

92. Bechar, A., Wachs, J. P., and Nof, S. Y., Developing a human-robot collaborative system for precision agricultural tasks, in *Proceedings of the 11th ICPA, International Conference on Precision Agriculture,* Indianapolis, IN, 2012.

93. Ko, H. S., Wachs, J. P., and Nof, S. Y., Web-based facility monitoring system using wireless sensor network, in *Proceedings of the Industrial Engineering Research Conference,* Reno, NV, 2011.

94. Jeong, W. and Nof, S. Y., A collaborative sensor network middleware for automated production systems, *Computers and Industrial Engineering* 57(1), 106–113, 2009.

95. Chen, X. W., Landry, S. J., and Nof, S. Y., A framework of enroute air traffic conflict detection and resolution through complex network analysis, *Computers in Industry* 62(8–9), 787–794, 2011.

96. Tkach, I., Edan, Y., and Nof, S. Y., Security of supply chains by automatic multi-agents collaboration, in *Proceedings of the INCOM*, Bucharest, Romania, 2012.

97. Velasquez, J. D. and Nof, S. Y., Best-matching protocols for assembly in e-work networks, *International Journal of Production Economics* 122(1), 508–516, 2009.

98. Jacob, M. G., Li, Y. T., and Wachs, J. P., A gesture driven robotic scrub nurse, in *Proceedings of the IEEE International Conference on Systems, Man and Cybernetics*, Piscataway, NJ, 2011, pp. 2039–2044.

99. Liu, S. C., Gao, L., Yin, Z. W., Shi, Y. C., Zhang, L., Chen, X. F., and Cheng, J. C., Simple hybrid wire-wireless fiber laser sensor by direct photonic generation of beat signal, *Applied Optics* 50(12), 1792–1797, 2011.

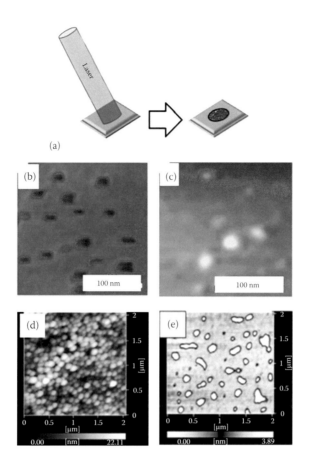

(a)

(b)                          (c)

100 nm                       100 nm

(d)                          (e)

**FIGURE 2.3**
(a) Schematic diagram for laser nanostructuring of silicon surface; laser nanostructuring at laser fluence below ablation threshold at (b) 0.10 J/cm² and (c) 0.25 J/cm² and above ablation threshold at 2 J/cm² in (d) air medium and (e) in water medium. (From Kumar, P., *Appl. Phys. A*, 99, 245, 2010; Kumar, P. et al., *J. Nanosci. Nanotechnol.*, 9, 3224, 2009; Kumar, P., *Adv. Sci. Lett.*, 3, 67, 2010.)

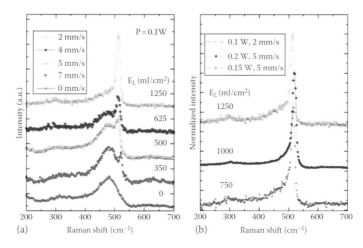

**FIGURE 2.8**
Raman spectra of P-doped *a*-Si:H films with two-layer structures (P-doped films (II)) on glass substrates (a) before and after laser annealing at laser power of 0.1 W with different scan speed and (b) after laser annealing at laser fluence from 0.75 to 1.25 J/cm². (From Jin, J., *Appl. Surf. Sci.*, 256, 3453, 2010.)

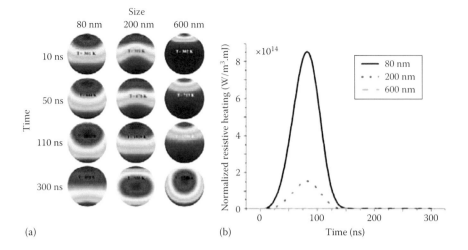

(a)                          (b)

**FIGURE 2.13**
(a) Temperature distribution for nanoparticles of varying diameter with time and (b) comparison of normalized resistive heating for nanoparticles with varying diameter. (From Zhang, M.Y. and Cheng, G.J., *J. Appl. Phys.*, 108, 113112, 2010.)

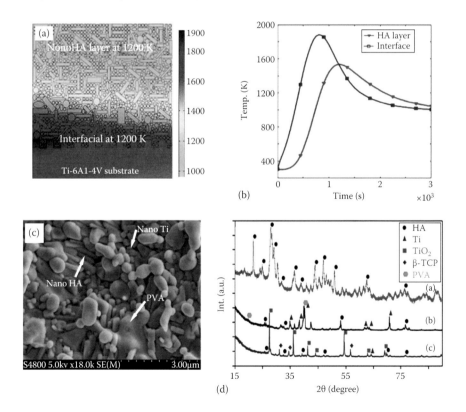

**FIGURE 2.15**
(a) A model showing heating of HAp and Ti nanoparticles, (b) rise and fall of temperature, (c) SEM image of the nanocomposite of HAp and Ti nanoparticles, and (d) x-ray diffraction pattern showing various materials components in the nanocomposite. (From Zhang, M.Y. et al., *ACS Appl. Mater. Interfaces*, 3, 339, 2011.)

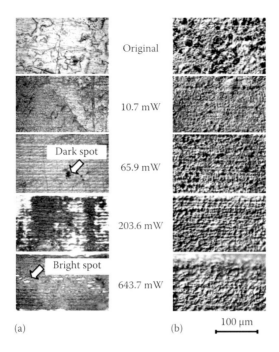

(a)                    (b)

Original

10.7 mW

65.9 mW

203.6 mW

643.7 mW

Dark spot

Bright spot

100 μm

**FIGURE 2.21**
Top surface of shock-peened specimens for select fluencies for (a) galvanized and (b) galvan-nealed steel. The term "Original" refers to the top surfaces before LSP process. (From Lee, D. and Asibu, E.K., Jr., *J. Laser Appl.*, 23, 022004-1, 2011.)

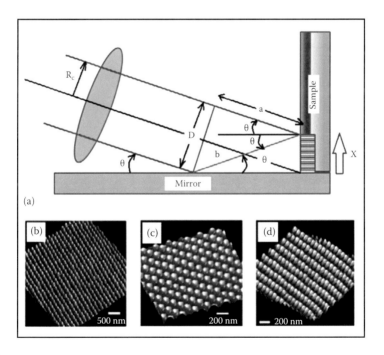

**FIGURE 2.23**
2D nanopatterns on PMMA produced by EUV laser interference lithography using Lloyd's mirror interferometer (schematic shown in (a)) with two exposures at different angles, (b) dots with 60 nm FWHM feature size and a period of 150 nm, (c) regular-shaped dots, and (d) elongated dots. (From Marconi, M.C. and Wachulak, P.C., *Prog. Quant. Elect.*, 34, 173, 2010.)

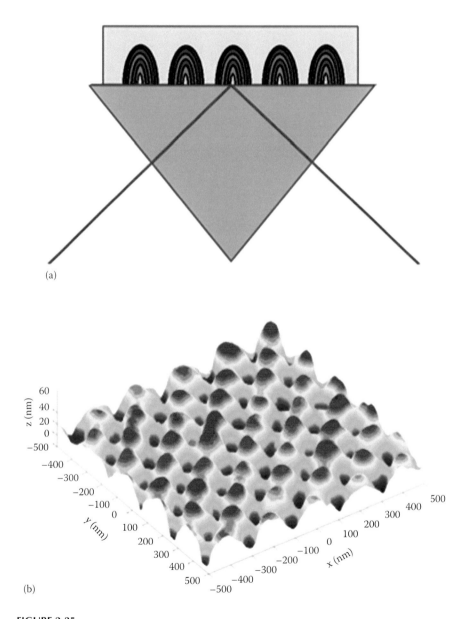

(a)

(b)

**FIGURE 2.25**
(a) Schematic diagram for EIL for TIR of four p-polarized incident beams and (b) AFM image of the nanostructures so obtained. (From Chua, J.K. and Murukeshan, J.K., *Micro Nano Lett.*, 4, 210, 2009.)

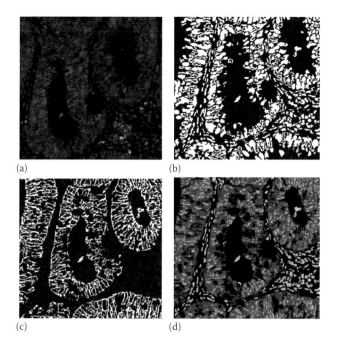

**FIGURE 4.5**
Colon cancer tissue images stained to show distribution of MET protein and segmentation.[4] (a) Combined image with red (membrane), blue (nuclei), and green (MET protein); (b) probability map for nuclei; (c) probability map for membrane; and (d) compartments with red (membrane), blue (epithelial nuclei), gray (stromal nuclei), and green (cytoplasm). Pink regions are excluded from quantification and black is background and extracellular matrix.

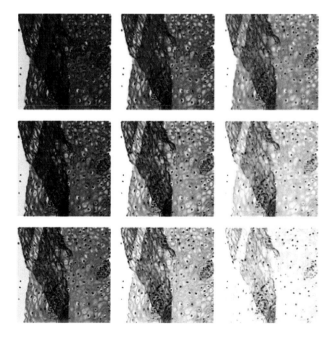

**FIGURE 4.9**
Adjusted H&E images of cervix tissue with folded region.[12] Left to right corresponds to increasing and decreasing hematoxylin levels and top to bottom to decreasing eosin levels. A pathologist may digitally adjust stain levels to better understand a tissue without processing a new sample.

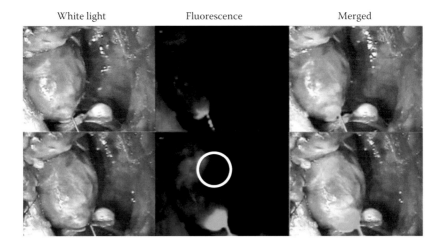

White light          Fluorescence          Merged

**FIGURE 4.11**
Preclinical FIGS[17] for vasculature mapping in rat heart at time of injection (top) and 60 s after injection (bottom). Fluorescent regions show circulating methylene blue and vasculature, while ischemic regions (circle) remain dark.

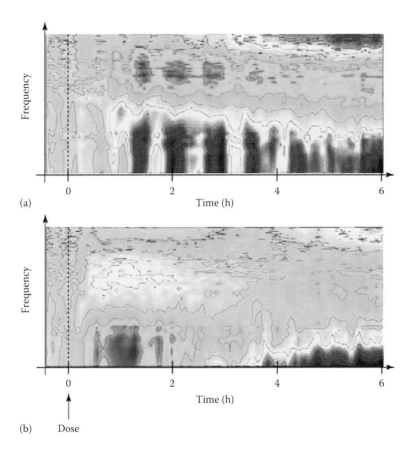

(a)

(b)     Dose

**FIGURE 5.5**
Drug-response spectrograms for phenotypic profiling of (a) iodoacetate and (b) cytochalasin D; Time–frequency analysis is performed on the time-dependent signals received from fluctuating speckles. The difference between the drug responses is highlighted according to their *relatively unique* voice prints and signatures in the spectrograms.

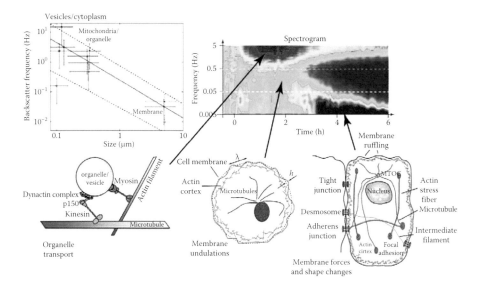

**FIGURE 5.6**
Light-scattering mechanisms showing backscatter frequencies as a function of size for vesicle transport, organelle motions, membrane undulations, and membrane forces and shape changes.

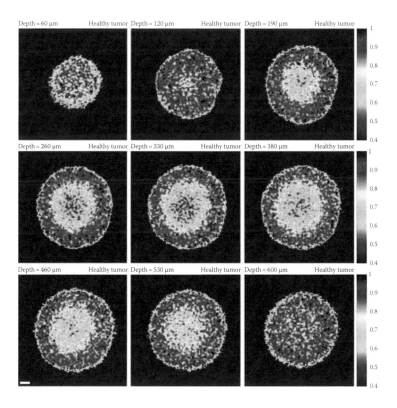

**FIGURE 5.8**
MCI of a 0.8 mm diameter tumor; cross sections of tissue activity are color-coded red, as active, and blue, as quiet.

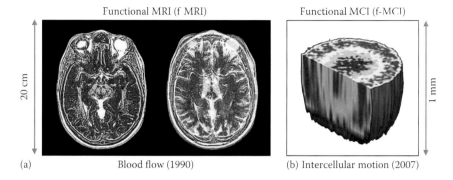

**FIGURE 5.9**
Functional imaging: (a) f-MRI the blood flow in the brain and (b) f-MCI volumetric imaging of cellular motion in a tumor spheroid; reds and oranges represent high motility in the outer shell surrounding the low-motility region of the necrotic core.

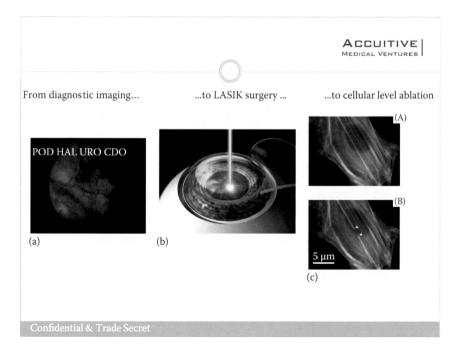

**FIGURE 6.1**
Medical application examples. (a) Compounds find the tumor and a laser is used to activate fluorescence. (From Peng, Q. et al., *Rep. Prog. Phys.*, 71, 056701, 2008.) (b) Correct vision by vaporizing tissue to permanently change shape of the cornea. (From www.advancedvisi onnetwork.com.) (c) Ablation of actin filaments inside a cell with femtosecond pulsed laser, (A) before laser and (B) after laser. (From Peng, Q. et al., *Rep. Prog. Phys.*, 71, 056701, 2008.)

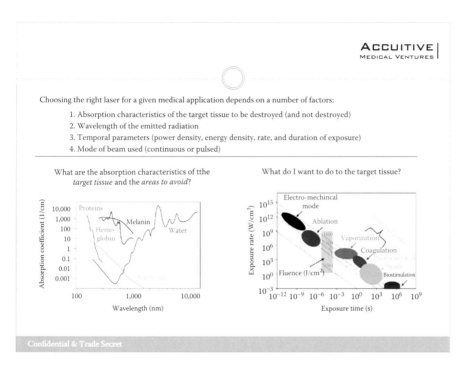

Choosing the right laser for a given medical application depends on a number of factors:

1. Absorption characteristics of the target tissue to be destroyed (and not destroyed)
2. Wavelength of the emitted radiation
3. Temporal parameters (power density, energy density, rate, and duration of exposure)
4. Mode of beam used (continuous or pulsed)

What are the absorption characteristics of tthe *target tissue* and the *areas to avoid*?

What do I want to do to the target tissue?

**FIGURE 6.2**
Selecting a medical laser. (From Peng, Q. et al., *Rep. Prog. Phys.*, 71, 056701, 2008.)

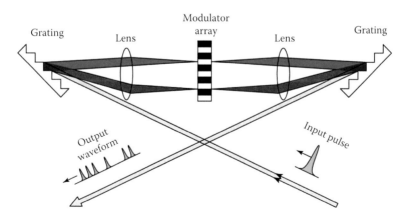

**FIGURE 7.1**
Basic setup for Fourier transform optical pulse shaping.

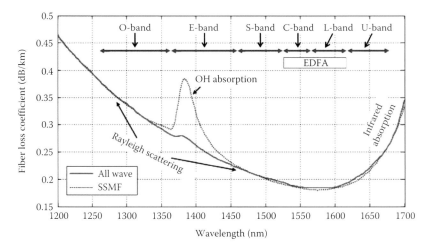

**FIGURE 7.8**
Transmission loss of modern glass fibers (dB/km vs. wavelength). Different frequency bands considered for fiber-optic transmission are noted, as is the operating band for the dominant optical amplifier technology (EDFAs). The principal physical mechanisms defining the fiber loss are also indicated. (From Essiambre, R. J. et al., *J. Lightwave Technol.*, **28**(4), 662, 2010.)

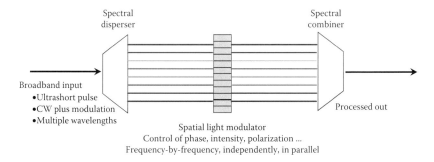

**FIGURE 7.10**
Generalized view of pulse shaping, including not only ultrashort pulses but also continuous-wave lasers that are modulated with data and multiple wavelength sources. In the optical communications community, signal manipulation based on generalized pulse-shaping geometries is often termed dynamic wavelength processing.

**FIGURE 8.3**
Standoff THz generation with ~30 fs laser pulses.

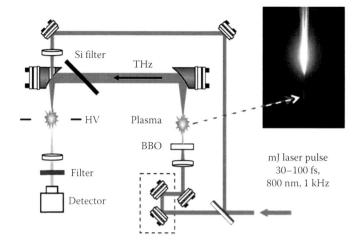

**FIGURE 8.4**
The THz wave ABCD setup schema.

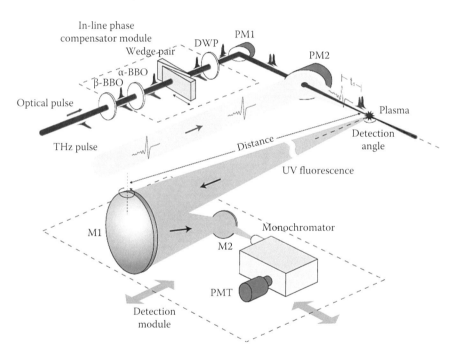

**FIGURE 8.5**
Schematic of the THz wave remote sensing technique. (From Liu, J. et al., *Nat. Photonic.*, 4, 627, 2010.) The $2\omega$ pulse is generated by passing the fundamental beam through a type-I β-BBO crystal. Both the fundamental and second-harmonic optical pulses are linearly polarized along a vertical direction. The relative phase change between the $\omega$ and $2\omega$ pulses is tuned by the lateral translation of fused silica wedges in the optical beam path after the α-BBO. The two optical pulses are focused by a parabolic mirror (effective focal length, 150 mm) into air to generate plasma. The time delay $t_d$ is defined as the delay between the optical pulse peak and THz pulse peak. The fluorescence detection system consists of a UV concave mirror (M1; diameter of 200 mm and focal length of 500 mm), a UV plane mirror (M2), a monochromator, and a PMT. The distance of remote sensing is varied by moving the fluorescence detection system with respect to the plasma (DWP).

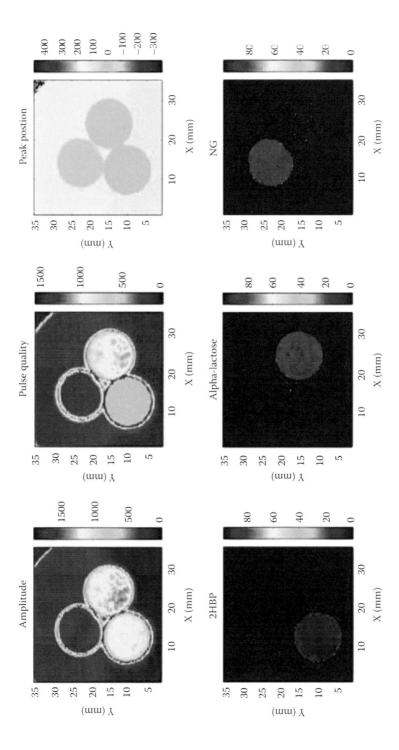

**FIGURE 8.10**
Spectral imaging. Each pixel contains the spectral information of the sample, which allows identification of the position of a particular compound.

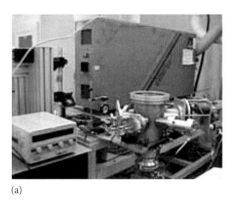
(a)

(b)

**FIGURE 10.5**
(a) Excimer laser PLD setup, (b) plasma-assisted MW CVD setup. (From Raghavan, D., Synthesis of multi-walled carbon nanotubes by plasma enhanced microwave CVD using colloidal form of iron oxide as a catalyst, Oklahoma State University, Stillwater, OK, 2005; Ramakrishnan, M.P., Experimental study on microwave assisted CVD growth of carbon nanotubes on silicon wafer using cobalt as a catalyst, Oklahoma State University, Stillwater, OK, 2005; Nidadavolu, A.G.R., Synthesis of carbon nanotubes by microwave plasma enhanced CVD on silicon using iron as catalyst, Oklahoma State University, Stillwater, OK, 2005.)

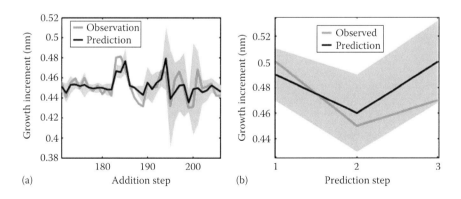

(a)

(b)

**FIGURE 10.7**
Comparison of growth increment from simulation and LGP prediction model; the shaded region represents the 95% confidence interval: (a) one-step-ahead prediction; (b) three-step-ahead prediction. (From Cheng, C. et al., *Proc. NAMRI/SME*, 40, 371, 2012.)

**FIGURE 14.5**
Laser-based intrusion alert system. (From FLM—Applications.)

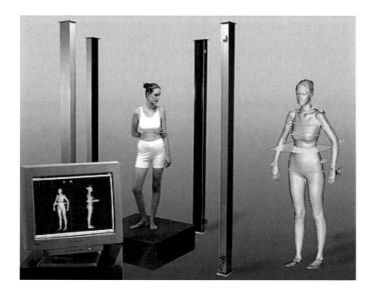

**FIGURE 14.8**
3D body measurements. (From 3D body scan laser measurements, Retrieved February 11, 2013, from http://www.assystbullmer.co.uk/3d_body_measurements.shtml, n.d.)

**FIGURE 14.14**
Advertisement for Tobii eye trackers showing a possible application for their product (scanned from a Tobii flyer).

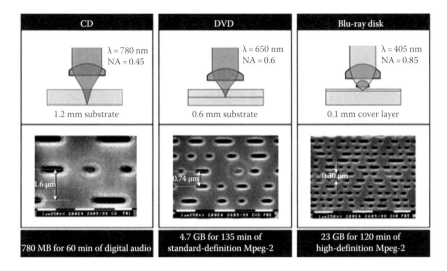

**FIGURE 15.1**
Laser beams of the CD, DVD, and Blue-ray. (From Bergh, A.A., Commercial applications of optoelectronics, February 2006, Retrieved April 1, 2013, from Photonics Spectra: http://www.photonics.com/Article.aspx?AID=24317.)

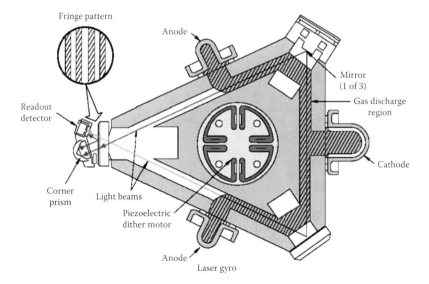

**FIGURE 15.6**
Diagram of an RLG. (From Makris, K., Ring laser gyro, December 22, 2012, Retrieved March 4, 2013, from Kostas Makris website: http://www.k-makris.gr/AircraftComponents/Laser_Gyro/laser_gyro.htm.)

**FIGURE 15.8**
Comparison of the live route and the database route. (From Westcott, R., Self-driving car given test run at Oxford University, February 14, 2013, Retrieved February 28, 2013, from *BBC News*: http://www.bbc.co.uk/news/technology-21465042.)

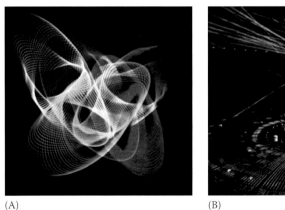

(A)                                        (B)

**FIGURE 15.15**
(A) Illustration of a laser beam art. (Cross, L., Inventing the laser light show, 2005, Retrieved March 4, 2013, from Lowell Cross: http://www.lowellcross.com/home/.). (B) Laser show at the February 2010 Super Bowl LVIV halftime show in Miami, FL. (From Daukantas, P., *Opt. Photon. News*, 21(5), 42, 2010.)

# 10

## Sensing and Informatics in Laser-Based Nanomanufacturing Processes

Changqing Chang, Satish T.S. Bukkapatnam, and Ranga Komanduri

### CONTENTS

### Introduction

As foretold by Professor Taniguchi some 20 years ago [1], lasers have emerged as a primary energy source for material/phase transformation, removal, or addition in manufacturing processes to realize atomic-scale precisions and nanometric scale features, such as films, tubes, wires, honeycomb, as well as various macromolecular structures [2]. Nanomanufacturing is a process of using precision machines that can generate precision tool motions to fabricate designed surface forms/dimensions with nanometric tolerance [3].

Nanomanufacturing has attracted considerable attention recently, due to the requirement of increasingly sophisticated devices and structures with outstanding properties and the trend of decreasing component sizes, material usages, and energy consumption of products. The development of nanomanufacturing technologies are highly desired to achieve nanoprecision and resolution to realize the novel functionality and properties of miniaturization, nanomaterial, and nanostructures [4]. Lasers have provided important opportunities in the realization of nanomanufacturing, and laser-based manufacturing of nanomaterials has shown advantages in narrow size distribution, excellent material property, yield, and purity. For example, laser deposition techniques can allow the generation and growth of nanomaterials at a particular location of a substrate, making them very flexible for device manufacturing. Laser-based materials processing has been successfully applied in industry for several decades such as laser-based additive manufacturing, surface modification, and micromachining.

As a widely used laser technology, pulse laser deposition (PLD) is critical for the growth of dense, vertically aligned carbon nanotubes (CNTs). Apart from being the energy source, lasers have also been employed for in situ monitoring of nanomanufacturing processes. Dimension measurement of the workpiece and the machine is always an essential process for the purpose of quality control in all fields of manufacturing. Because accuracy is the most important requirement for nanomanufacturing, the dimensional measurement is a much crucial process for nanomanufacturing than traditional manufacturing. For instance, optical systems typically based on splitting the primary laser beam have been attempted for in situ measurement of nanostructure geometries in laser-based nanofabrication processes [5].

Many advanced nanomanufacturing processes, which use lasers as energy sources for material transformation, are data intensive, and incorporation of informatics is anticipated to improve process and system control, increase productivity, and speed up innovation. Unlike conventional manufacturing systems, nanomanufacturing systems exhibit complex features and, in most cases, cannot be described by conventional scale model. Hence, signal-based informatics, often referred to as *time-series informatics*, can be a good option to study laser-based nanoprocesses, especially to provide a new way to monitor those processes and improve the scale and production rates of nanomaterials and structures. The remainder of the chapter is organized as shown in Figure 10.1. A brief review of different laser-based nanomanufacturing processes categorized from quality monitoring and informatics standpoint into fabrication processes for surface nanostructures and nanomaterial structures, nanoscale machining, and nanomanufacturing of 3D structures is provided in the "Laser-based nanomanufacturing processes: an in situ process monitoring and informatics perspective" section; the signals from both chemical and physical sources, of laser-based nanomanufacturing processes, as well as their sensing methods are reviewed in the "Sensors and sensing of laser-based nanomanufacturing processes" section; alternative laser-based

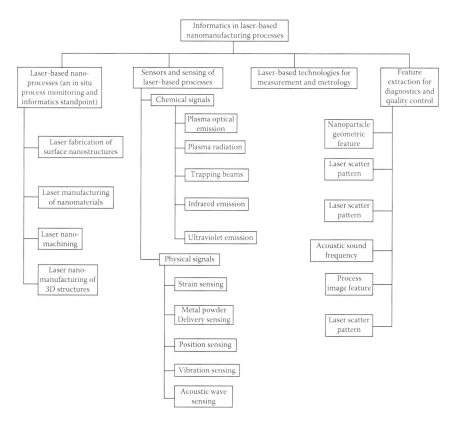

**FIGURE 10.1**
Flowchart for the study of informatics in laser-based nanomanufacturing processes.

technologies for the emerging measurement and metrology applications are summarized in the "Laser-based nanomanufacturing processes: an in situ process monitoring and informatics perspective" section; feature extraction methods reported in the literature for diagnostics and quality control of various laser processes are delineated in the "Feature extraction for diagnostics and quality control in laser processes" section; and a case study on the application of time-series informatics methods for laser-based CNT synthesis process modeling and control is provided in the "Case study" section.

## Laser-Based Nanomanufacturing Processes: An In Situ Process Monitoring and Informatics Perspective

This section will review the various widely used laser-based nanomanufacturing process from an in situ process monitoring standpoint.

## Laser Fabrication of Surface Nanostructures

*Scanning near-field photolithography* (SNP): Near-field optical methods offer a unique way for optical measurement and manipulation of materials [6]. In near-field techniques, the laser source is placed in the near field of the substrate, where exponentially decaying evanescent waves dominate. Often, this nanometric distance ensures the evanescent wave arrives at the substrate with sufficient energy. With an optical fiber-based near-field scanning optical microscope coupled with an ultraviolet (UV), the photochemical reaction can be initiated. Chemical etching, plasma etching, or UV light radiation can be further used to create nanoscale patterns on the surface [7]. *Nanoridge aperture beam* transmission enhanced nanofabrication: nanoscale ridge aperture beam generally has high transmission efficiency and confined nanoscale radiation in the near-field region compared with other regularly shaped aperture beams. Dubay et al. [8] found that high-energy laser beams, passing through an aperture with much less attenuation than a circular aperture, have sufficient energy to produce nanoscale patterns on a surface through contact lithography.

*Laser interference lithography*: Laser interference lithography is a large-area, noncontact, and maskless lithographic fabrication approach using the interference pattern of two or more coherent laser waves. Nanosurface structures and repeatable structures can be created over large areas using this technique. Some earlier efforts have been reported on far-field laser interference lithography methods [9,10]. Near-field interference lithography is based on evanescent wave interferences to defeat the diffraction limit of the lasers and to fabricate a variety of nanostructures [11,12]. Chua and Murukeshan [13] demonstrated complicated 2D nanostructures fabrication using multiple beam interference through polarization tuning, based on internal light reflection evanescence wave near-field lithography.

*Surface modification using femtosecond and nanosecond (ns) lasers*: Single and multiple coatings on cutting tools were introduced in the 1970s to increase the wear resistance and reduce friction of cemented carbide and high-speed steel (HSS) cutting tools [14]. The coating materials used include TiC, TiN, $Al_2O_3$, and diamond, and coating processes was done through chemical vapor deposition (CVD) and physical vapor deposition (PVD). The growth from CVD process is typically columnar in nature. Carbide and HSS tools coated with different materials can be used as the work material and may be subjected to shorter (>1 fs) and longer (>1 ns) pulses of laser ablation under various conditions to modify the microstructure of the coatings from columnar to amorphous/nanocrystalline structures. Such a PLD technique provides for an effective means for surface modification. PLD has emerged as an attractive process for microelectronics, coatings, and micro electro-mechanical systems (MEMS)/nanoelectromechanical systems (NEMS) fabrication. It is a simple but powerful technique to grow thin films from the vapor phase. In PLD, an intense, short-pulsed laser beam is focused onto the target material, resulting in emission of a cloud of vaporized material and subsequent deposition on a substrate to grow thin films. These thin films can be utilized in many ways. One such field demands exploration of new super hard tool coatings, which can be implemented in various processes, such as high-speed machining, precision machining, dry machining, and environmentally conscious machining. This can be further extended to other types of coatings such as corrosion-resistant coatings for composites.

*Laser microwelding of similar/dissimilar materials*: In many MEMS applications, dissimilar materials (e.g., glass and Cu), similar materials (glass and glass), and optically transparent *materials* have to be bonded. High energy, chemical incompatibility, and poor thermomechanical properties can limit weld quality. Femtosecond laser with a pulse width of <120 fs, repetition rate of 1–5 kHz and pulse energy of 1–6 mJ at 1 kHz peak power, and low average energy (6 W) localizes the heat energy and can produce good weld [15] (see Figure 10.2). Additionally, femtosecond lasers can offer fast processing times, noncontact processing, and precise positioning. A femtosecond laser can also be advantageous for joining dissimilar materials. A transparent and an opaque material can be joined together without much difficulty.

## Laser Manufacturing of Nanoscale Materials and Structures

Nanomaterials have unique physical and chemical properties. They are building blocks of nanodevices in the bottom-up fabrication approach.

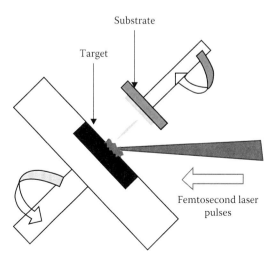

**FIGURE 10.2**
Schematic of welding of two materials using a femtosecond laser.

### Laser Manufacturing of Nanoparticles

Nanoparticles have been used in many products and have a broad spectrum of applications. Nanoparticles can be manufactured by various conventional processes, such as chemical reduction, phase decomposition, electrical discharge, and mechanical grinding [4]. While those processes have been used for large-scale nanoparticle fabrications, they are restricted by the expensive and tedious postprocessing. For example, purification is required to remove by-products [16] after the manufacturing process, which may contaminate the produced particles. Therefore, the quality and purity of the nanoparticles produced by these traditional approaches are usually not sufficient. Recently, laser-based manufacturing approaches for nanoparticle, such as laser ablation of solid targets, are gaining notable attention [17].

Cheng et al. [18] studied the formation of carbon nanohorns via laser ablation and investigated the in situ particle quality control with the use of a scanning differential mobility analyzer (DMA). This measurement technique provided real-time size distributions of the nanoparticles every minute during the course of a production run.

Henglein et al. [19] used a pulsed ruby laser with a wavelength of $\lambda = 694$ nm to ablate thin gold films to produce gold nanoparticles. They focused the laser into an aqueous solution with suspended microscale particles, leading to the formation of colloids, which are nanoparticles dispersed in liquid media. Tsuji et al. [20] studied nanoparticle fabrication using a femtosecond laser. This technique produced nanoparticles at a lower yield rate compared to nanosecond laser pulses but had a much narrower particle size variation. Barcikowski et al.'s study [21] further showed that the size of nanoparticles generated with femtosecond lasers was mostly determined by the pulse overlap and fluence for different materials. Pulsed laser ablation (PLA) for the nanomaterial synthesis has the advantages of synthesizing pure nanomaterials.

*Laser-induced self-organization*: Favazza et al. [22] studied the synthesis of ultrathin metal films using laser-induced self-organization. They concluded that with different levels of laser irradiation, different nanostructures can be realized. For instance, nonuniform laser irradiation initiates a *tunable* thermocapillary effect in the film giving rise to nanowires, and continued laser irradiation leads to a Rayleigh-like breakup of the nanowires producing nanoparticles with spatial long-range and short-range order. They further suggested that laser-induced self-organization in thin films could be an attractive route to produce well-defined nanoparticle arrangements for applications in optical information processing, sensing, and solar energy harvesting.

### Laser Manufacturing of Nanotubes

Nanotubes have many outstanding properties. For example, CNTs have very high Young's modulus (greater than 1 TPa), high electrical capacity

(1000 times of copper [23]), high thermal conductivity, low density, and super hydrophobic. As a result, nanotubes, particularly CNTs, have been used widely for stain-resistant textiles, transparent conductors in flat display panels, solar cells, optical limiting, hydrogen storage, composites, gas sensors, conducting plastics, antibacterial coatings, $CO_2$ filer, nanolasers, ultrafast oscillators, field effect emitters, and conducting adhesives [5].

Much of the prior works in laser-based nanotube synthesis mainly focuses on CNT production by laser ablation. Guo et al. [24] first reported laser synthesis of CNTs in 1995 [43]. They used a frequency doubled Nd:YAG laser of a 6–7 ns pulse width to ablate a metal/graphite composite. The role of the bimetals (e.g., Co/Cu) is to act as a catalyst to enable carbon diffusion at high furnace temperatures (around 1200°C) after the graphite is vaporized by the laser. Saturation of diffused carbon in the nanocatalyst particles (also by laser vaporization) allows the carbon to be extruded in a tubular form. They found both single-walled and multiple-walled CNTs in their experiments. Later, Maser et al. [25] used $CO_2$ pulsed lasers to improve beam absorption of the graphite composite. Under similar materials and environmental conditions as in [43], they found around 80% beam absorption and 0.2 g/h CNT production rate, which is comparable to other nonlaser techniques for CNT manufacturing, indicating the possibility of large-scale CNT production via laser ablation.

Klanwan et al. [26] studied single-walled CNT synthesis using laser ablation under atmospheric conditions and produced CNTs with diameter smaller than 2 nm and length of over 500 nm. In their study, a graphite target with nickel–cobalt catalyst was ablated by Nd:YAG laser in an electrical furnace under atmospheric pressure of continuous $N_2$ gas flow. They measured the size distribution of CNTs using size classification by a DMA. They found that the diameters of the CNTs are controlled by the catalyst nanoparticle sizes and the fiber length is controlled by the carbon vapor deposition time.

### Laser Manufacturing of Nanowires

Nanowires are nanostructures with the diameter of the order of a nanometer and have specific anisotropic physical and chemical properties that are significantly different from other bulk materials. Laser-manufactured nanowires are increasingly used in the microelectronics (e.g., transistors), photonics (e.g., waveguides), instrumentation (e.g., gas sensors), and energy industries (e.g., solar cells). Various different types of nanowires exist, including metal oxides (e.g., ZnO), metallic (e.g., Ni), semiconducting (e.g., Si), and insulating (e.g., $SiO_2$). They can be manufactured by PLA, PLD, laser interference lithography, and laser-assisted molecular beam epitaxy (LAMBE) techniques. ZnO is one of the most important nanowires, thanks to the great potentials for photonic industry applications [4].

PLA is a well-established method for nanowire manufacturing. Morales and Lieber [27] first used the laser ablation technique to produce Si and Ge nanowires in 1998. It was based on vapor–liquid–solid (VLS) mechanism, that is, a vapor phase material diffuses into the small catalyst particles to form liquid catalysts, which allow nanowires to grow from them until they are cooled down to a solid state. However, the catalyst particle diameter was limited to >0.1 µm by conventional VLS method. The PLA approach overcame the limitations in nanowire diameter, and 6–20 nm diameter Si and 3–9 nm diameter Ge nanowires were produced [27]. Wang et al. [28] reported the use of a similar technique but avoided the use of a metallic nanoparticle catalyst. They used mixed $SiO_2$ and Si powders ablated by a pulsed laser to significantly improve the yield rate. Jia et al. [29] produced uniform Zn–Se nanowires on the surface of Zn–Se crystal irradiated by femtosecond lasers. They claimed that the nanowire length and diameter can be controlled by varying laser pulse energy and pulse number, and the synthesis mechanism as self-catalyzed VLS process.

PLD is also widely applied for the synthesis of nanowires on a substrate. It is based on the laser ablation of a target material and depositing the laser-generated target material vapor onto a substrate that has a layer of catalyst nanoparticles (e.g., Au). The laser-ablated target material vapor/plasma is deposited on the heated target material coated with catalyst nanoparticles that enable the growth of nanowires from them. The nanowire size and distribution are controlled by the laser ablation processing parameters, such as the catalyst nanoparticle size and the gas pressure [30].

Laser interference lithography [31] has been used to combine with metal-assisted etching for nanowire manufacturing on a substrate. The method uses standard laser interference lithography, followed by a thin metal (e.g., Au) film coating. Then the substrate in direct contact with the metal is dissolved. Consequently, the metal film sinks into the substrate, leaving nanowires at the position of the holes in the metal film. To avoid the use of a catalyst, a number of variations of the PLD were developed for nanowire growth on a substrate. These include laser-induced forward transfer (LIFT), LAMBE, and the use of self-organized nanopolystyrene spheres as a template in PLD. In the LIFT technique, PLD is used to deposit a nanostructured film of the target material on a glass substrate at relatively higher gas pressure compared to standard PLD. This nanostructured film is then transferred to a substrate (same material as the target material) by LIFT. Nanowire growth can be realized without an additional catalyst. For silicon nanowire growth by LIFT technique, $SiO_2$ nanoparticles formed in PLD were considered as the nuclei for the nanowire growth [32]. In LAMBE  technique, a nanostructured template such as anodized aluminum oxide (AAO) is used and LAMBE is used to deposit target materials (e.g., a metal) onto the AAO surface.

## Laser Nanomachining

Lasers are widely used for macro- and nanomachining applications in numerous industries such as automotive, electronics, and medical manufacturing. However, there are many challenges encountered in the utilization of lasers, particularly for nanomachining. The most critical requirement is that the diffraction limit of laser light must be overcome. With recent developments in laser technology in terms of short-wavelength and ultrashort pulse width, there are a lot of opportunities to beat the diffraction limit for nanomachining of structures, devices, and materials [33].

Joglekar et al. [34] indicated that a remarkable feature of material damage induced by short-pulsed lasers is that the energy threshold becomes deterministic for sub-picosecond pulses. This effect, coupled with the advent of kHz and higher repetition rate chirped pulse amplification systems, has opened the field of femtosecond machining. They found that they can consistently machine features as small as 20 nm, demonstrating great promise for applications ranging from MEMS construction and microelectronics to targeted disruption of cellular structures and genetic materials. Tan et al. [35] utilized thin film laser micromachining for repairing semiconductor masks, creating solar cells, and fabricating MEMS devices. A unique high repetition rate femtosecond fiber laser system capable of variable repetition rates from 200 kHz to 25 MHz along with helium gas assist was used to study the effect of pulse repetition rate and pulse energy on femtosecond laser machining of gold-coated silicon wafer. Chimmalgi et al. [36] investigated ultrashort-pulsed laser radiation for precision materials processing and surface nanomachining, and their study demonstrated that controllable surface nanomachining can be achieved by femtosecond laser pulses through local field enhancement in the near field of a sharp probe tip. Nanomachining of thin gold films was accomplished by coupling 800 nm femtosecond laser radiation with a silicon tip in ambient air. Finite-difference time-domain numerical predictions of the spatial distribution of the laser field intensity beneath the tip confirmed that the observed high spatial resolution was due to the enhancement of the local electric field.

Micromachining of bulk materials using longer pulse (>1 ns) and higher energy lasers has been a challenge due to melting and vaporization around the machining zone, recast layer, and general edge quality and geometry obtainable. Higher energy and longer pulse length are attributed for this problem. In contrast, these phenomena are minimized or almost eliminated when shorter wavelengths (femtosecond) are used. Superior quality, thus, can be achieved using ultrashort laser pulses at low fluencies close to the ablation threshold. The location of the laser can be precisely controlled using a motion stage (Figure 10.3), thus producing structures that can have a wide variety of applications in various areas, such as optical and biomedical. With high repetition rate of the laser, it can be used in producing structures with high aspect ratio that can be utilized for microfluidic applications [37–39]. Because it is a noncontact process, one potential application is in dicing of silicon wafers that is usually done with the help of a blade saw resulting in serious surface damages.

A femtosecond laser system provides a unique opportunity to investigate pulsed laser micropolishing of surfaces of Spinel ($MgAl_2O_4$) and other such gem-quality materials. These materials are increasingly been sought

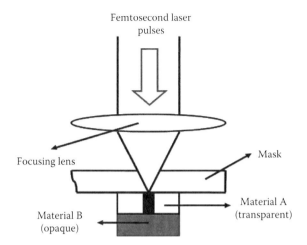

**FIGURE 10.3**
Schematic of micromachining/drilling.

after for the emerging defense (e.g., body armor), lightweight aerospace, and high-temperature microelectronic applications. These materials present significant challenges for polishing. The current practice involves chemical–mechanical polishing at elevated temperatures (>100°C) using concentrated acidic (pH < 1) or alkaline (pH > 10) slurries. The current polishing practices are considered hazard-prone and not amenable for effective control of local topographies. The femtosecond laser pulses are advantageous over traditional polishing approaches in that local melting coupled with 3+ degree-of-freedom table movement allows polishing of multimaterial surfaces with speeds, focus depths, and laser intensities adapted to the local material properties and geometry [40]. For many emerging applications of microelectronic power devices, MEMS sensors and actuators, the surface roughness of the films needs to be controlled to Ra < 5 Å. Such smooth finishes are needed for applications such as silicon direct bonding [41]. Popular microelectronic processes, such as chemical mechanical planarization (CMP), only provide a partial alleviation of the issues. The use of nanosecond and femtosecond laser pulses offers an attractive alternative.

The pulsed laser micropolishing process (see Figure 10.4) uses carefully controlled pulses of laser just for rapid melting, redistribution (flattening), and resolidification of a microscopic layer of material on the surface. The laser fluence (i.e., the energy density) $F$ should be such that the resulting depth of the melt pool $x_d$ has to be high enough to melt away the asperity crests, but must not be deeper than the valleys. The process depends mainly on two factors: the surface materials and topography and the fluence $F$ of the laser beam. Fluence $F$ of the beam depends on the diameter $D$ of the melt pool, feed rate $V_r$ of the scanner, and the laser intensity $I$ as $F = 6000\, I/(D\, V_r)$.

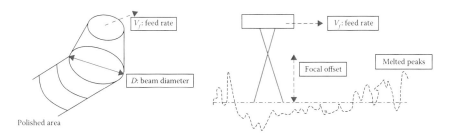

**FIGURE 10.4**
Schematic of pulsed laser micropolishing. (From Lamikiz, A. et al., *Int. J. Mach. Tool Manufact.*, 47(12–13), 2040, 2007.)

The melt-pool diameter $D$ is given by $D = r\,[2 \ln (I/I_{th})]^{0.5}$, where $r$ is the radius of the focus laser beam. If the incident laser intensity $I$ exceeds a material-dependent threshold $(I_{th})$, the high fluence can cause the material to rapidly melt or vaporize and may even ionize causing plasma. The plasma temperatures can be as high as 50,000°C and the induced pressures can be adequately high to affect certain phase and geometric transformations [42]. The melt-pool depth is given by $x_d = a^{-1} \ln\left(F/F_a^{th}\right)$, where $F_a^{th}$ and $a$ are material-dependent thresholds for fluence and optical energy absorptivity, respectively. Thus, a key to laser micropolishing is to match the laser wavelengths and pulse durations to material's molecular properties, specifically the characteristics of the electronic energy absorption bands [43,44].

The current state of laser-based micropolishing processes can be advanced to subnanometric polishing by the use of shorter pulses to minimize the extent of melt pool [44]. The use of a femtosecond laser can improve the surface quality by effectively melting surface without causing significant ablations, in addition to providing a more effective control of pulse train profiles as well as reflected radiation in complex geometries and multimaterial settings. Here, the melt duration $t_m$ has to be sufficiently long to allow fusion and settling down of the molten surface to smooth out rough ridges of frequency $f_{min}$ and above. While this may well be suited for smoothed and/or uniformly patterned surfaces with asperities in µm range, significant issues may arise whenever the surface consists of multiple materials (as in a modern microelectronic element) or highly intricate geometric details in sub-µm ranges. Longer pulses may lead to the introduction of capillary waves on the surface that may produce additional undesirable ridges on the surface [40]. Therefore, a shorter femtosecond laser pulsing may be advantageous in these applications. Through appropriate design of a series of steps of pulsed laser polishing with appropriate selection of fluence (0.05–10 J/cm²), frequency (10–10³ Hz), and pulse duration (>25 fs), one can provide effective means to smooth complex multimaterial surfaces. It may be noted that a similar setup can be investigated for micromachining by operating the system at 1000 J/cm² fluence ranges.

While significant efforts have been made toward applying pulsed lasers (at nano-, pico-, and femtosecond ranges) for polishing of metals, including Ni, Ti, steel, and W, ceramics, and organic polymers [40,41,43,46,47], not much research has been conducted toward optimizing the various process parameters for effective polishing to achieve both planarity and finish. As stated in the foregoing, polishing of such materials is currently effected through chemical–mechanical means at elevated temperatures. Effective local control of topographies and geometry can be highly challenging in such scenarios. The use of femtosecond laser pulses can alleviate many of the current challenges.

## Laser Nanomanufacture of 3D Structures and Devices

To build nanoscale 3D devices and structures, a number of laser-based techniques have been developed including two-photon polymerization (TPP), nanowelding, and nanoforming. Among these techniques, the TPP technique has been the most widely used. TPP is different from UV photopolymerization in stereolithography where the absorption of the laser or UV light is on the surface of the liquid (although there is a penetration depth); thus, to build up 3D structures, a new layer of liquid needs to be placed for each layer for standard stereolithography. Therefore, the layer thickness in UV light stereolithography is limited by the fluid layer thickness, which is difficult to reach nanoscale resolution due to surface tension [48].

Assembly of nanomaterials for micro-/nanoproduct manufacture can be carried out in a number of ways including joining, welding, deposition, and laser holographic radiation. Optical tweezers can be applied for the manipulation and positioning of micro- and nanocomponents in a liquid environment [49]. Unfortunately, a simple scaling down of macro and micro laser beam joining techniques is not always successful, mainly, because the achievable resolution is limited by the optical wavelength of the applied laser radiation. Furthermore, nanosystems in many cases behave quite differently compared to their macroscopic counterparts. The optical penetration depth may be on the scale of the microsystem itself; near-field effects can lead to field enhancement and interference in the proximity of small nanoparticles [50].

In micro- and nano-assembly, highest accuracy and resolution in terms of positioning of the functional components among each other are necessary. However, due to the increasing complexity of the microsystems, these accuracy requirements can hardly be fulfilled during the whole production chain by means of conventional methods. The manufacturing strategy involves the assembly of microcomponents with relatively larger tolerances using conventional methods. In a second step, the critical parts of the system are aligned to each other with the required accuracy using a laser beam. Laser-based adjustment techniques have been used in numerous micro-/nano-assembly applications as they allow an easily automatable and contact free positioning. In order to overcome these restrictions, a nano-adjustment approach, using ultrashort lasers and allowing high alignment accuracies without undesired thermal effects, called micro shockwave bending, has been developed [51]. When focusing an ultrashort laser pulse on a material surface, the energy is absorbed by the free and valence electrons and the material is transformed in a state of plasma. The rapidly expanding plasma induces shockwaves leading to a deformation of the substrate if the plasma pressure exceeds the bulk material's yield stress.

## Sensors and Sensing of Laser-Based Nanomanufacturing Processes

The development of nanomanufacturing has an impact on the types of sensors and sensing systems required—they must be cheap, designed to do the minimum required, and fast. But these requirements run counter to those of the types of sensors and signals needed for the fundamental understanding of nanomanufacturing processes, which call for flexibility and precision and do not need to be fast or cheap [52]. The sensor signals used in nanoscale sensing techniques for laser-based manufacturing process monitoring may be classified into chemical or physical signals.

### Chemical Signals

Chemical signals mainly include the emissions from the chemical processes involved in laser processing. Some of the methods based on chemical signals are summarized in the following paragraphs:

Material wear detection by *plasma spot sensing*: Klinger et al. [53] developed a new sensor to measure online the material wear in laser material processing, during which it is hardly possible to measure the material wear through the plasma occurring in the working zone at temperatures above 3000 K. The authors utilized the plasma spot as a signal and evaluated the distance between sensor and workpiece. The sensor is insensitive against the spot size and shape, and hence accurate, even with low aperture and within the turbulent atmosphere. The time for one measurement is one millisecond.

*Plasma optical emission sensing*: In order to control the laser welding process, Sibillano et al. [54] utilized optical fibers to capture the strong plasma optical emission, which was observable during the process. They showed that plasma plume optical spectroscopy can be very promising for realizing a reliable online monitoring of the quality of welded joints and in general for keeping under control the welding process. Plasma optical spectra were characterized by the presence of emission lines coming both from the excited atoms and from the ions produced during the laser–surface interaction and can determine the chemical composition and the dynamics of interaction of the different chemical species inside the plume.

*Optical emission of plasma plumes sensing*: Ancona et al. [55] developed an optical sensor for real-time monitoring of laser welding based on a spectroscopic study of the optical emission of plasma plumes. The welding plasma's electron temperature was contemporarily monitored for three of the chemical species that constitute the plasma plume by the use of related emission lines. The evolution of electron temperature was recorded and analyzed during several welding procedures carried out under various operating conditions. A clear correlation between the mean value and the standard deviation of the plasma's electron temperature and the quality of the welded joint was found. This information was used to find optimal welding parameters and for real-time detection of weld defects such as crater formation, lack of penetration, weld disruptions, and seam oxidation.

*Plasma radiation*: A popular technique to monitor laser welding processes is to record laser-*induced* plasma radiation with a high-speed camera [56]. The recorded image sequences can be analyzed using pattern recognition systems.

*Trapping beams*: Optical tweezers are a promising tool for nanomanufacturing, but the efficiency of optical tweezers manufacturing depends on the number of trapping beams available. Micro optics technology offers the opportunity to significantly increase the number of trapping beams without a significant increase of the cost or size of the optics. Dagalakis et al. [57] reported their research on optical-tweezers-based nanomanufacturing using an array of laser beams generated by a single laser diode. Their array of laser beams is generated by an array of servo-controlled scanning dual-axis micromirrors. Capacitor electrodes underneath the micromirror plates provide electrostatic actuation, which allows control of the micromirror position. With proper reflecting surfaces, it is possible to control the impact angle of the individual laser beams onto the micro-nano-particles, thus generating an optical beam gripper effect. Other chemical signals extracted include weld pool infrared emissions, visible (Vis) light [55,58,59], and UV emissions from laser-induced plasma [60].

## Physical Signals

*Physical signals* refer to the parameter measurements of a laser process. The methods based on physical signals are summarized in the following paragraphs:

*Strain sensing*: Anwander et al. [61] developed laser-based noncontacting strain sensor to record large mechanical and thermal strains for high temperatures up to 1200°C, based on tracking laser speckles through a digital correlation technique.

*Metal powder delivery sensing*: Hu et al. [62] studied real-time control in laser-based additive manufacturing process. Metal powder delivery real-time sensing and control is studied to achieve a controllable powder delivery for fabrication of functionally graded material.

*Position sensing*: Chen et al. [63] studied automatic laser beam position sensing in the laser material processing, which can be used to relocate the laser beam. Atomic force microscopy (AFM) is widely used in the sensing of nanomanufacturing process. The sample to be imaged is placed on the scanner tube and brought close to a very sharp tip mounted at the free end of a thin compliant beam. The beam bends in proportion to the attractive or repulsive force acting on the atom(s) at the apex of the tip to atom(s) on the substrate. Among the subsequently developed techniques for detecting the cantilever displacement, the optical beam deflection technique [64] measures the displacement by detecting the deflection of a laser beam reflected from the backside of the microcantilever, which is directed onto a position-sensitive detector (photodiode sensor). In noncontact or tapping mode operation, the cantilever tip is made to vibrate near the sample surface with spacing on the order of a few nm or intermittently touching the surface at the lowest deflection [65].

*Displacement sensing*: Ryabov et al. [66] developed an in-process tool geometry measurement system for milling using a laser displacement sensor. The system reconstructs and displays the 3D image of a milling tool and evaluates the tool geometric failures via a hybrid laser sensor measurement method that uses displacement and intensity techniques simultaneously. The system is designed to compensate for interruptions in the laser path brought about by chips and coolant drops present under actual cutting conditions. In experimental tests, the reconstructed 3D geometrical tool image and the intensity measurements enabled us to detect the size and the location of the chipped part and to determine the length of the flank wear.

*Vibration sensing*: Golnabi [67] studied continuous measurements of parameters (e.g., temperature) using optical laser sensor technology in automation and flexible manufacturing. Gozen et al. [68] studied the 3D dynamic characterization of nanotool motions (piezoelectric actuator) in the nanomilling process using a custom-designed laser Doppler vibrometer-based measurement system. The measurement system is also used to correlate the orientation of the nanotool motions and nanopositioning stage axes with the orientation of the sample surface.

*Acoustic wave measurements*: Luo et al. [69] studied acoustic sounds in keyhole and conduction laser welding process. The characteristic signals representing good welding quality were from 10 to 20 kHz. The more the welded metal vaporizes, the higher the plasma temperature and the stronger the acoustic signals. Furthermore, keyhole shape also affected the acoustic signal intensities.

*Distance sensing*: The probe-to-sample distance is a critical parameter to control both the nanofeature size and shape in near-field scanning optical microscopy (NSOM) nanofabrication technique. At a small probe diameter and probe-to-substrate distance, the NSOM overcomes the traditional far-field diffraction limit and can be used to obtain subwavelength-size patterns. In addition, higher writing speed leads to shorter exposure time and thus lower exposure dose, resulting in a narrower line width and shallower depth [70].

*Surface-enhanced Raman scattering*:  Ma et al. [71] introduced a novel fabrication method for surface-enhanced Raman scattering (SERS) sensors that used a fast femtosecond laser scanning process to etch uniform patterns and structures on the end face of a fused silica optical fiber, which is then coated with a thin layer of silver through thermal evaporation that is presented. The uniform SERS sensor built on the tip of the optical fiber tip was small and lightweight and could be especially useful in remote sensing applications. Other physical signals extracted to study laser-based manufacturing process include acoustic emission (AE) signal [72] and voltage difference between workpiece and nozzle [73,74].

## Laser-Based Technologies for Measurement and Metrology

Ever since the audiences were captivated by Edison's invention of the phonograph, scientists have labored to device means to extend the principle idea that grooves on a substrate can store information. Over the last century, these efforts have led to the phonograph being supplanted by vinyl records, compact discs (CD), laser discs (LD), digital video discs (DVD), and, as of present, Blu-ray. The phonograph needle has now been replaced by a noncontacting beam of collimated light—the laser. It is interesting to view these developments in storing information in grooves from the perspective of the science metrology; after all, the fidelity of the reproduced information can only be as good as the means of reading the information—be it a finely crafted vinyl player or a high-end CD system (indeed, we might have been billed for a DVD at checkout using a bar-code scanner [an application of measuring the breadth of patterns using a laser beam]). In a similar analogy, as one gains in terms of reliability, longevity, and repeatability of the record by eschewing contact altogether, it is plausible to foresee the metrological advantages of using a laser beam instead of a contacting stylus.

It has been shown that the morphology of a manufactured surface not only determines the functional integrity of a component but also stores information regarding the process. In this context, the function of surface metrology ranges from the domain of quality assurance to process control. For instance, consider the case of a part being turned on the lathe; the spiral pattern made by the tool on the workpiece, besides determining the surface finish, is also indicative of a host of process conditions prevalent during manufacture. For example, the spacing between the spiral grooves correlates with the feed, the depth of the grooves with the depth of the cut taken, burn marks with thermal aspects, irregular-shaped grooves with tool wear, and sudden changes in the groove spacing with stability of the machine. Thus, the measurement of surfaces not only is critical for quality assurance purposes but can serve as a means for process diagnosis. In this section, we will explore the use of laser informatics in measurement ranging from the extremely small scale (tens of nanometers) to cartography (several hundred meters). Given the breadth of the discussion, we will eschew a nuanced discussion of each of these aspects. We now motivate two brief examples of laser informatics before moving on to detailed review.

Consider, for instance, an ultraprecision manufacturing process such as diamond turning, also called ultraprecision machining (UPM). In UPM, a natural (single crystal) diamond is typically used to produce parts with extremely tight geometric tolerances often specified to within a few nanometers, for example, very large (>50 ft) mirrors used in radiometry applications often have aspheric geometry that is mandated to within hundredth of a micrometer (waviness, curvature error, etc.) with surface roughness specified to single digits on the nanometer scale [75]. These characteristics are critical to ensure the functional integrity of the component and are therefore tightly controlled during the manufacturing process. To accomplish these tight specifications, UPMs are specially constructed to minimize extraneous errors from sources such as extraneous vibrations, thermal fluctuations, and slide and workpiece positioning errors that can have a deleterious effect on the surface morphology. In such UPMs, accurate slide movement is achieved using a network of laser interferometers, and the workpiece is often positioned by laser alignment [76]. Subsequent to manufacturing, the part itself is often measured using laser metrology.

It is easy to recognize that without laser-based techniques, it would become difficult (if not impossible) to accomplish any of the foregoing tasks. For example, if the errors from the tool slides are not controlled and negated at an early stage, the workpiece geometry may be compromised. Given that geometry tolerances are within a few hundredth of a micrometer, the slide velocity is usually tens of mm/min. There are utmost a few milliseconds available to recognize the onset of anomalies and make corrective adjustments. Conventional tachometers and position measurement implements are ineffective under such circumstances as they suffer from slower response, and more importantly, their inferior kinematic characteristics due to striction, creep, inertia, and backlash may lead to errors that are greater than the specified tolerance.

Finally, consider an ultraprecision machined part measuring a few feet in diameter, both the geometric and surface morphology specifications of which are important. However, these characteristics occur at scales several magnitudes apart. Also, there is the added concern that any contact with the surface may irrevocably (e.g., with the stylus of a profilometer) hamper the functional performance of the part. We are indeed faced with an engineering dilemma; we cannot measure the part because doing so would necessitate contacting it, and contacting the part in turn could damage the part and void the objective for measurement. Under such circumstances, there is a need for metrological approaches that are eventually noncontact, capable of measuring large surfaces for both geometry and rugosity and doing both within a few minutes. Tools such as scanning electron microscopy (SEM) and AFM [77] can surmount at most one of these constraints. Laser-based implements, in contrast, can be tailored to meet these demands and have consequentially emerged as the favored option.

As another example, Burge et al. [78] describe the construction of the Giant Magellan Telescope (GMT) that consists of seven mirrors each having an 8.4 m aspheric geometry. The radius of curvature of these sections is specified at 36 m ± 1 mm, that is, within three orders of measurement, consequently, the manufacturing tolerance for this feature is even more tighter at ±0.3 mm. Abrasive polishing is used to finish the mirror on a machine equipped with a specially designed laser tracker in order to remove specific amounts of material at predetermined positions. These positions are themselves mapped a priori using a laser interferometric system. The final optical testing metrology is also laser-based and uses a helium–neon (He–Ne) holographic system as a comparator gage.

Wang et al. [79] designed an optical probe for in situ measurement of surface roughness. Based on light scattering, a probe that consists of a laser diode, a measuring lens, and a linear photodiode array was built to measure surface roughness. A coaxial design that incorporates a dual-laser probe and compressed air makes the proposed system insensitive to the position of the test surface and to surface conditions such as the presence of debris, vibration, and lubricants that result from machining. The in situ measurement on a diamond-turning lathe showed that the technique is stable and accurate and applicable to online measurement of surface roughness of an engineering surface.

Wang et al. [80] developed a technique to measure surface roughness in the submicrometer range, based on laser scattering from a rough surface. A telecentric optical setup using a laser diode as a light source was used to record the light field scattered from the surface. The light intensity distribution of the scattered band, which was correlated to the surface roughness, was recorded by a linear photodiode array and analyzed using a single-chip microcomputer.

## Feature Extraction for Diagnostics and Quality Control in Laser Processes

This section presents how signals, such as full penetration based on image information, geometric features from laser point cloud, and appearance features for the plasma radiation, and/or features from the laser-based processes can be used for quality monitoring and diagnosis purposes.

For instance, signals from a plasma can be used to generate information on the possible presence of defects during the welding process [81]. Effective optical sensors based on the measurement of the spatially integrated optical intensity as well as the spectroscopic analysis of the UV/Vis emission [82] can also be applied for defect detection. Such kind of optical sensors based on the plasma plume optical emission has been used successfully in industrial environments for the detection of several types of welding defects [54].

Through their study of the in situ nanoparticle quality control using measurement technique, Cheng et al. [18] found that the geometric mean diameter of the produced particles tended to increase with increases in pulse width. For example, for a given laser peak power and repetition rate, carbon nanoparticles of mobility diameter close to 100 nm were produced in a large abundance using longer laser pulse lengths (e.g., 10 ms) as compared to the shorter pulse lengths (e.g., 1 ms). This technique has been used for continuous in situ control of the nanoparticle synthesis.

Traditionally, tool condition monitoring techniques employ force, spindle motor torque, current or power, and AE signals, or a combination of these signals, during machining. Those systems are mostly used for machining in the roughing range. Wong et al. [83] investigated the scatter pattern of reflected laser light to monitor tool conditions in the roughing to near-finishing range. The scatter pattern was created by a low-power laser beam that is reflected from the surface of the workpiece, and it was captured with a digital camera and processed using image processing approaches. The standard deviation of the intensity distribution in the region of interacted state captured by the digital camera was extracted. They found good correlation between the standard deviation parameter and tool wear, which was used for tool condition monitoring.

Laser scatter tested was performed in [84] to test the surface roughness in the sheet metal parts. Sárosi et al. [84] employed a one-shot deflectometry method in the infrared range for the test. At sufficiently long wavelengths, the sheet metal's surface becomes specular-reflecting, thus enabling the use of the deflectometry method.

Liu et al. [85] presented in-time motion adjustment strategy to remedy and eliminate defects occurring during laser cladding to improve the dimensional accuracy and surface finish. Based on the relationship between the motion of laser head relative to the growing part and other parameters in effects on clad profile, the laser traverse speed, stand-off distance, and laser approach orientation to the existing clad layer were adjusted by instructions from a close-loop control system in real time to remedy and eliminate defects.

Solid freeform fabrication (SFF) methods for metal part building, such as 3D laser cladding, are generally less stable and less repeatable than other rapid prototyping methods. A large number of parameters govern the 3D laser cladding process. These parameters are sensitive to the environmental variations, and they also influence each other. Hu et al. [86] introduced the research work to improve the performance of its developed 3D laser cladding process.

Laser-based ultrasonic (LBU) measurement shows great promise for online monitoring of weld quality in tailor-welded blanks. Tailor-welded blanks are steel blanks made from plates of differing thickness and/or properties butt-welded together; they are used in automobile manufacturing to produce body, frame, and closure panels. LBU uses a pulsed laser to generate the ultrasound and a continuous wave laser interferometer to detect the ultrasound at the point of interrogation to perform ultrasonic inspection. LBU enables in-process measurements since there is no sensor contact or near contact with the workpiece. Kercel et al. [87] used laser-generated plate (Lamb) waves to propagate from one plate into the weld nugget as a means of detecting defects.

Shao et al. [88] discussed various applications of acoustic, optical, visual, thermal, and ultrasonic sensing techniques for monitoring laser welding processes. Structure-borne AE sensor has been extensively investigated and appears to be a good choice for detecting material phase transformation and crack formation. However, the sensor needs contact installation with the workpiece. Airborne emission sensor can detect weld surface defects. But it is difficult to be used in noisy and hostile environments. In fact, very little in terms of applications of AE in microelectronics industry has been reported so far. Optical detectors have been widely used over a wide range of industries since they are relatively simple, cheap, and effective. Although some commercial systems are available, signal processing and classification would be crucial for the further development of this technique. CCD (charge-coupled device) and CMOS (complementary metal-oxide semiconductor) camera or array sensors have been reported to monitor the continuous welding processes. However, it is tricky to apply these techniques for monitoring pulsed laser spot welding processes as the process dynamics tends to have prolonged transients.

Luo et al. [69] studied acoustic sounds in keyhole and conduction laser welding process. Their research indicated that the acoustic sound frequency is strongly related to the welding quality. The characteristic signals representing good welding quality was from 10 to 20 kHz. Furthermore, keyhole shape also affected the acoustic signal intensities.

The main problem of AFM-based nanomanipulation is the lack of real-time visual feedback. Random drift aroused from an uncontrolled manipulation environment generates a position error between the manipulation coordinate and the true environment. Liu et al. [89] proposed a real-time fault detection and correction (RFDC) method to solve these problems by using the AFM tip as an end effector as well as a force sensor during manipulation. Based on the interaction force measured from the AFM tip, the validity of the visual feedback is monitored in real time by the developed Kalman filter. Once the faulty display is detected, it can be corrected online through a quick local scan without interrupting manipulation. In this way, the visual feedback keeps consistent with the true environment changes during manipulation, which makes it possible for several operations to finish without an image scan in between.

Gozen et al. [68] studied the 3D dynamic characterization of nanotool motions in the nanomilling process using a custom-designed laser Doppler vibrometer-based measurement system. The measurement system is also used to correlate the orientation of the nanotool motions and nanopositioning stage axes with the orientation of the sample surface. The control method for calculating the necessary inputs to the nanopositioning stage and the piezoelectric actuator to generate the desired nanomilling motions is then presented and experimentally evaluated to assess the capability to generate desired rotational nanotool motions.

Geese et al. [15] proposed an image feature extraction algorithm for a laser welding process. They extracted features from the recorded images (e.g., full penetration hole and spatter), connected those features to the quality of the laser welding process, and enabled a real-time control for defect defection.

Beeck et al. [90] investigated laser metrology as a diagnostic tool for analyzing and optimizing complex coupled processes inside and between automotive components and structures, such as the reduction of a vehicle's interior or outer acoustic noise and the combustion analysis for diesel and gasoline engines to further reduce fuel consumption and pollution. Pulsed electronic speckle pattern interferometry (ESPI) and holographic interferometry were used to study the knocking behavior of modern engines and positioning of knocking sensors. Holographic interferometry shows up the vibrational behavior of brake components and their interaction during braking and allows optimization for noise-free brake systems. Scanning laser vibrometry analyzes structure-borne noise of a whole car body for the optimization of its interior acoustical behavior.

Leopold et al. [91] reviewed various laser-based optical methods, such as projected fringes method, electronic speckle interferometry, and laser scanning methods for surface control. This laser measurement technique can enable fast detection during the manufacturing process.

Laser welding has been widely used in a variety of manufacturing processes, such as automobile production. However, the high dynamics of the process has made it impossible to construct a camera-based real-time quality and process control. Geese et al. [15] proposed an image feature extraction algorithm, running at a frame rate of 10 kHz, for a laser welding process. Many pictures of different welding processes were recorded during the experiments, so it is possible to match the recorded images to the theoretical concept of the welding process described earlier. A large number of quality features were observed in the images, and the most important quality feature is full penetration. They showed that the welding process had good quality when the full penetration hole became visible.

Considerable effort has been put into extracting geometric features from point clouds in laser scanner, which provides noncontact, high accuracy, and high-speed probing and has found many applications in manufacturing inspection. However, the research results for feature extraction from edge detection and computer vision do not meet the metrology requirement of high accuracy. Zhang et al. [92,93] studied the features extracted from a point cloud obtained from laser and improved the poor accuracy in laser scanning for geometric features.

Kumar et al. [94] presented a strategy to fuse information from two vision sensors and one infrared proximity sensor to obtain a 3D distance information to avoid single sensor uncertainty and thus improve sensing quality during the process.

Jager et al. [56] investigated principal component imagery (PCI) for the quality monitoring of dynamic laser welding processes by recording laser-induced plasma radiation with a high-speed camera. They employed appearance-based features to describe the relevant characteristics of the recorded images. The classification performance of geometric and appearance-based features is compared on a representative data set from an industrial laser-welding application. They demonstrated that a classification system based on appearance-based features can outperform geometric features.

Lu et al. [95] investigated an in situ surface roughness measurement method based on laser light scattering, which is obtained from the spatial distribution of the scattered light intensity. Root-mean-square (RMS) height of the surface roughness is extracted by means of image processing of the scattered light distribution in the direction parallel to the manufacturing mark.

Shinozaki et al. [96] developed a fast scanning system to extract 1D surface profile. The profile is measured by integration of a slope distribution of the surface obtained from angular deflection of a scanning laser beam. The scanning optical system consists principally of a spherical concave mirror and a rotating scanner mirror, and is insensitive to mechanical vibration because of its high-speed scanning, of the order of milliseconds.

Udupa et al. [97] investigated a confocal scanning optical microscope (CSOM) to measure and assess micro and macro surface irregularities of various machined planar and cylindrical surfaces. They extracted features, such as arithmetic mean deviation, maximum surface valley depth and surface height, and the mean of the maximum surface summit height and maximum surface peak-to-valley height, to characterize the 3D surface roughness and roundness measurements.

## Case Study

The use of high-resolution sensors and large-scale experiments has resulted in the availability of vast amount of a data to study laser-based nanomanufacturing processes. However, most of the research works reported were limited to studying a single snapshot of a system in time (e.g., temperature, distance), allowing only coarse description of the underlying dynamics of the underlying systems. Even if time-series data (e.g., acoustic signals) have been gathered from in situ sensors and instruments [69], conventional time and frequency domain features are mostly extracted for process control. The rich causal and dynamic information available in the time-series data from these sensors have seldom been explored to study and manage these complex processes and systems. Time-series informatics research is therefore considered to be the next frontier in advancing the prediction of the global states of complex dynamic systems [98]. In this case study, we demonstrate one of the first attempts of using time-series information for laser-based nanomanufacturing process simulation and quality control.

Advanced nanoscale processes are data intensive, and incorporation of informatics is anticipated to improve process and system control, increase productivity, and speed up innovation [99,100]. Unlike conventional manufacturing systems, nanomanufacturing systems exhibit complex dynamics due to the inherent nonlinear and nonstationary characteristics of the synthesis processes. Hence, time-series informatics provides a means to the modeling, simulation and control through accurate state prediction, model identification, structure reference, and parameter estimation.

We have been investigating nanoinformatics to study laser-based CNT synthesis process in our research lab at Oklahoma State University. The experimental procedures are aimed at discerning the evolution of the CNT growth profile distributions over time under different process conditions (defined as pressure, gas flow rates, temperature, catalyst type settings, and distributions). The as-grown CNT synthesis process consists of two main stages, namely, catalyst deposition and CNT growth. Fully aligned MWNTs will be grown in the microwave plasma-enhanced (PE)-CVD apparatus (see Figure 10.5) instrumented with multiple sensors on a substrate with a catalyst specially prepared using a PLD and plasma pretreatment processes [101–103]. Our prior experiments have provided the process parameter settings to consistently yield catalyst films with nanometric-size grains that serve as CNT growth sites in PE-CVD process [102–104].

The laser used in the PLD process is a Lambda Physik Compex (Model 201) krypton fluoride (KrF) excimer laser (with an average power of 4 W at 10 Hz, 30 KV, wavelength of 248 nm, pulse width of 25 ns) (Figure 10.5a). The PLD setup consists of a vacuum chamber that houses both the target and the substrate holder. The chamber will be evacuated to $10^{-3}$ torr using a turbo molecular pump backed by a mechanical pump. Depending on the thickness (2–10 μm) of the catalyst film required, the laser exposure time will be varied over 15–45 s. To generate patterned growth of CNTs, a mask with the required geometric patterns will be placed on top of the substrate. The catalyst film would be deposited only in the open, unmarked areas. Once the catalyst is deposited on to the wafer surface, the mask would be carefully removed and the sample transferred to the CVD reactor for plasma

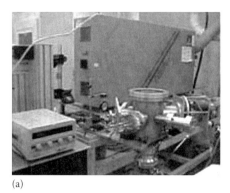
(a)

(b)

**FIGURE 10.5**
**(See color insert.)** (a) Excimer laser PLD setup, (b) plasma-assisted MW CVD setup. (From Raghavan, D., Synthesis of multi-walled carbon nanotubes by plasma enhanced microwave CVD using colloidal form of iron oxide as a catalyst, Oklahoma State University, Stillwater, OK, 2005; Ramakrishnan, M.P., Experimental study on microwave assisted CVD growth of carbon nanotubes on silicon wafer using cobalt as a catalyst, Oklahoma State University, Stillwater, OK, 2005; Nidadavolu, A.G.R., Synthesis of carbon nanotubes by microwave plasma enhanced CVD on silicon using iron as catalyst, Oklahoma State University, Stillwater, OK, 2005.)

pretreatment and CNT growth. The microstructure and distribution of the catalyst significantly affect the yield and quality of MWNT. The optimized pretreatment settings (e.g., substrate preparation, catalyst size/distribution, exposure time, and beam density) from our prior experiments have achieved catalyst properties (Figure 10.6) conducive for consistently realizing vertically aligned CNTs [102–104].

As we have stated in our study about CNT synthesis study [105], efforts have been made to optimize the CVD and other processes used for CNT synthesis, yet current production and yield rates remain too low for wider applications. Atomistic molecular dynamics (MD)/Monte Carlo (MC) modeling is essential to track the short-time (<1 ps) phenomena central to CNT synthesis and other nanomanufacturing processes, and the computation time required for MC simulations needs to be greatly reduced to allow timely adjustment of model parameters based on sensor and characterization data, as well as estimation and eventual intermittent control of nanostructure (CNT) geometry and other quality variables. Pertinently, 80%–95% of the computational overhead during an MC simulation is attributed to the relaxation procedure implemented at every growth step. Despite various attempts [106–108] to increase computational speed, the current MC models are not computationally tractable for simulation of CNT and other nanostructure growth well beyond the nucleation stage. One of the longest reported CNTs from atomistic/mesoscale simulations consists of some 10,000 carbon atoms with a length of ~150 nm [109] using

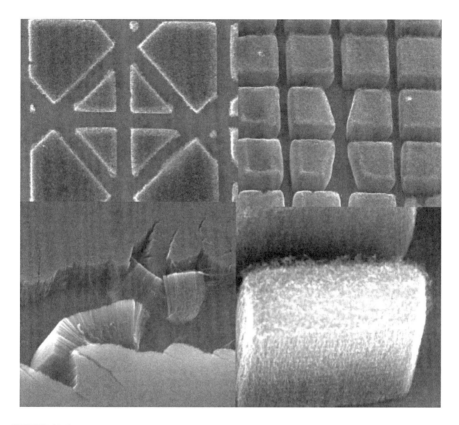

**FIGURE 10.6**
Highly aligned and patterned CNT structures. (From Raghavan, D., Synthesis of multi-walled carbon nanotubes by plasma enhanced microwave CVD using colloidal form of iron oxide as a catalyst, Oklahoma State University, Stillwater, OK, 2005; Ramakrishnan, M.P., Experimental study on microwave assisted CVD growth of carbon nanotubes on silicon wafer using cobalt as a catalyst, Oklahoma State University, Stillwater, OK, 2005; Nidadavolu, A.G.R., Synthesis of carbon nanotubes by microwave plasma enhanced CVD on silicon using iron as catalyst, Oklahoma State University, Stillwater, OK, 2005.)

the earth simulator. This is about an order of magnitude below the CNT sizes (500 nm to several μm in length) from experiments. A key to our approach is our recent finding that CNT growth rates from MC simulation exhibit a distinct nonlinear stochastic dynamics [105]. Nanoinformatics, that is, prediction of these growth increments, offers an opportunity to reduce computational time of MC nanostructure growth simulations and can consequently update the dynamic simulation models with sensor and characterization data for real-time monitoring.

We recently developed a time-series informatics-based fast and computationally efficient atomistic Monte Carlo simulation of CNT synthesis process

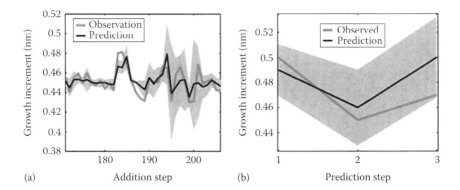

**FIGURE 10.7**
**(See color insert.)** Comparison of growth increment from simulation and LGP prediction model; the shaded region represents the 95% confidence interval: (a) one-step-ahead prediction; (b) three-step-ahead prediction. (From Cheng, C. et al., *Proc. NAMRI/SME*, 40, 371, 2012.)

in a CVD chamber. Among the predictive models, recurrence-based local Gaussian process (LGP) models [110] were found advantageous in terms of prediction accuracy and computational speed (Figure 10.7). The LGP can capture the drifts and variations in the CNT growth increment with high prediction accuracy. The results indicated that it can save over 70% of the simulation time for the conventional atomistic simulation, which led to one of the longest CNTs (~194 nm) from atomistic simulations.

Such speed-up simulations can be used for CNT in situ length control over a larger length and time scales that were possible earlier. The utility function plot at different synthesis process steps to realize CNTs of length 90 nm is shown in Figure 10.8. The red (light) curve shows the variation of the utility over time (addition steps) as one decides to continue synthesis, and the blue (dark) curve captures the variation of utility when the synthesis process is stopped at different times. When the two curves intersect, it is the recommended end point for realizing desired CNT length. Numerical studies indicate that CNTs generated through such time-series informatics-based procedures are within the 1 nm variation of the specifications. Thus, time-series informatics can be useful for advance simulation and monitoring and control of, plausibly laser-based, nanoprocesses.

## Summary and Future Directions

Evidently, laser applications are now mundane in everyday life. However, they are currently far from reaching the zenith of their capabilities. Ongoing research in a large swath of applications, such as laser-guided automotive

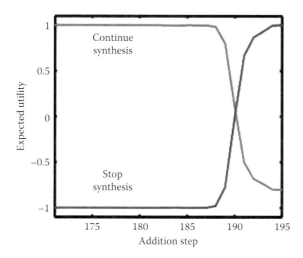

**FIGURE 10.8**
Utility function for the synthesis of CNT with desired lengths as 90 nm. (From Cheng, C. et al., *Proc. NAMRI/SME*, 40, 371, 2012.)

steering, laser-based terrain mapping, smart munitions, and nanomanufacturing, are promise revolutionary developments in the coming years. In the broadest terms, lasers allow the functions of energy transfer, measurement, positioning, manipulation, and wireless communication without the need for direct contact. There are ample opportunities for using information from lasers as well as various physical and chemical phenomena in laser-based processes for end-user applications and in-process control. The traditional penchant of the Industrial Engineering (IE) for data-driven decision making can be leveraged for analyzing information extracted from laser-based processes and measurement systems; a wealth of information remains to be harnessed in this area.

Although laser-based technologies are beginning to be widely used to affect various aspects of our life, many issues regarding their large-scale applications especially in the emerging areas of ultraprecision and nanomanufacturing still remain, from both processing and measurement and quality assurance standpoints. Assurance of real-time accurate monitoring of quality, reduction of the total manufacturing cost, and thus facilitating the large-scale production are impeded by various factors including limited understanding of the underlying physical phenomena, resolution limitations and confounding effects associated with the measurement instruments and systems, complexity and diversity of signals emanating from these processes, as well as the accuracy and scalability issues with conventional models used for process monitoring and control. For example, investigations are currently ongoing toward addressing issues pertaining to material grain uniformity and surface morphology in laser processing. Also, how would laser radiation affects the nanomaterial properties and structures is still a scientific challenge, and further understanding would open opportunities not just for laser processing and modification of functional nanomaterials but also for wider deployment of laser-based in situ measurement systems. More pertinently, modeling of laser beam interactions with different media at nanoscales is a very challenging task, particularly when multiple phases and multiscale problems are encountered. Atomistic MD/MC modeling and finite difference in time domain (FDTD) modeling [5] are particularly suitable for modeling the synthesis and transformations at micro-/nanoscale compared to conventional continuum and linear systems-based time-series analysis approaches. In specific, MD/MC models would be seen to have their increasing applications in understanding the nanomaterial behaviors in laser-based nanomanufacturing as they can capture the mechanisms at the picosecond range time scales over which the structural modifications (e.g., additions of atoms to the structure) take place in laser-based nanomanufacturing processes. It was also gleaned from the case study that the use of informatics can help in bridging the gap between these characteristic time scales of nanomanufacturing process mechanisms and the coarser time scales over which various measurements are available. Moreover, informatics techniques can be combined with MD, MC [111], and/or FDTD models to facilitate the understanding of the laser-based nanomanufacturing process and advance real-time quality control of these emerging manufacturing processes.

## References

1. Taniguchi, N., Current status in, and future trends of ultra-precision machining and ultra-fine materials processing. *CIRP Annals*, 1983. **32**: 573–582.
2. Bukkapatnam, S. et al., Nanomanufacturing systems: Opportunities for industrial engineers. *IIE Transactions*, 2012. **44**(7): 492–495.
3. Gao, W., *Precision Nanometrology: Sensors and Measuring Systems for Nanomanufacturing*, 2010. Springer: London, U.K.
4. Li, L. et al., Laser nano-manufacturing—State of the art and challenges. *CIRP Annals—Manufacturing Technology*, 2011. **60**(2): 735–755.
5. Wang, N., Y. Chen, and L. Zhang. Design of multi-agent-based distributed scheduling system for bus rapid transit. In *Intelligent Human-Machine Systems and Cybernetics (IHMSC), 2011 International Conference on*, Hangzhou, China, 2011.
6. Leggett, G.J., Scanning near-field photolithography-surface photochemistry with nanoscale spatial resolution. *Chemical Society Reviews*, 2006. **35**(11): 1150–1161.
7. Lo, S.C. and H.N. Wang, Near-field photolithography by a fibre probe. *Proceedings of First IEEE Conference on Nanotechnology*, Maui, Hi, 2001. 36–39.
8. DuBay, N.M., et. al, Nanolithography using high transmission nanoscale ridge aperture probe. *Applied Physics A: Materials Science and Processing*, 2008. **93**: 881–884.

9. Savas, T.A. et al., Properties of large-area nanomagnet arrays with 100 nm period made by interferometric lithography. *Journal of Applied Physics*, 1999. **85**(8): 6160–6162.

10. de Boor, J., D.S. Kim, and V. Schmidt, Sub-50 nm patterning by immersion interference lithography using a Littrow prism as a Lloyd's interferometer. *Optics Letters*, 2010. **35**(20): 3450–3452.

11. Alkaisi, M.M. et al., Sub-diffraction-limited patterning using evanescent near-field optical lithography. *Applied Physics Letters*, 1999. **75**(22): 3560–3562.

12. Martinez-Anton, J.C., Surface relief subwavelength gratings by means of total internal reflection evanescent wave interference lithography. *Journal of Optics A: Pure and Applied Optics*, 2006. **8**(4): S213.

13. Chua, J.K. and V.M. Murukeshan, UV laser-assisted multiple evanescent waves lithography for near-field nanopatterning. *Micro & Nano Letters, IET*, 2009. **4**(4): 210–214.

14. Venkatesh, V.C., Tool wear investigations on some cutting tool materials. *Wear of Materials*, 1979. Dearborn, MI, pp. 501–508.

15. Geese, M. et al. Feature extraction in laser welding processes. In *Cellular Neural Networks and Their Applications, 2008. CNNA 2008. 11th International Workshop on*, Compostela, Spain, 2008.

16. Cristoforetti, G. et al., Production of palladium nanoparticles by pulsed laser ablation in water and their characterization. *The Journal of Physical Chemistry C,* 2010. **115**(12): 5073–5083.

17. Burgess, D.S., Laser ablation generates nanoparticles. *Photonics Spectra*, 2001. **35**: 26–29.

18. Cheng, M.-D. et al., Formation studies and controlled production of carbon nanohorns using continuous in situ characterization techniques. *Nanotechnology*, 2007. **18**(18): 185604.

19. Henglein, A., Physicochemical properties of small metal particles in solution: "Microelectrode" reactions, chemisorption, composite metal particles, and the atom-to-metal transition. *The Journal of Physical Chemistry*, 1993. **97**(21): 5457–5471.

20. Tsuji, T., Preparation of nano-size particles of silver with femtosecond laser ablation in water. *Applied Surface Science*, 2003. **206**: 314–320.

21. Barcikowski, S. et al., Properties of nanoparticles generated during femtosecond laser machining in air and water. *Applied Physics A*, 2007. **87**(1): 47–55.

22. Favazza, C. et al., Nanomanufacturing via fast laser-induced self-organization in thin metal films. *SPIE Proceedings*, San Diego, California, USA, 2007. **6648**: 664809.

23. Hong, S. and S. Myung, Nanotube electronics: A flexible approach to mobility. *Nature Nanotechnology*, 2007. **2**(4): 207–208.

24. Guo, T. et al., Self-assembly of tubular fullerenes. *Journal of Physical Chemistry*, 1995. **99**: 10694.

25. Maser, W.K. et al., Production of high-density single-walled nanotube material by a simple laser-ablation method. *Chemical Physics Letters*, 1998. **292**(4–6): 587–593.

26. Klanwan, J. et al., Generation and size classification of single-walled carbon nanotube aerosol using atmospheric pressure pulsed laser ablation (AP-PLA). *Journal of Nanoparticle Research*, 2010. **12**(8): 2747–2755.

27. Morales, A.M. and C.M. Lieber, A laser ablation method for the synthesis of crystalline semiconductor nanowires. *Science*, 1998. **279**(5348): 208–211.
28. Wang, N. et al., Transmission electron microscopy evidence of the defect structure in Si nanowires synthesized by laser ablation. *Chemical Physics Letters*, 1998. **283**(5–6): 368–372.
29. Jia, T. et al., Femtosecond laser-induced ZnSe nanowires on the surface of a ZnSe wafer in water. *Solid State Communications*, 2007. **141**(11): 635–638.
30. Shen, Y. et al., A general approach for fabricating arc-shaped composite nanowire arrays by pulsed laser deposition. *Advanced Functional Materials*, 2010. **20**: 703–707.
31. Johannes de, B. et al., Sub-100 nm silicon nanowires by laser interference lithography and metal-assisted etching. *Nanotechnology*, 2010. **21**(9): 095302.
32. Muriel, L.A.R. et al., Fabrication and photoluminescence of hyperbranched silicon nanowire networks on silicon substrates by laser-induced forward transfer. *Nanotechnology*, 2008. **19**(24): 245303.
33. Ali, M. et al., Review of laser nanomachining. *Journal of Laser Applications*, 2008. **20**: 169–184.
34. Joglekar, A.P. et al., A study of the deterministic character of optical damage by femtosecond laser pulses and applications to nanomachining. *Applied Physics B*, 2003. **77**(1): 25–30.
35. Tan, B., A. Dalili, and K. Venkatakrishnan, High repetition rate femtosecond laser nano-machining of thin films. *Applied Physics A*, 2009. **95**(2): 537–545.
36. Chimmalgi, A. et al., Femtosecond laser aperturless near-field nanomachining of metals assisted by scanning probe microscopy. *Applied Physics Letters*, 2003. **82**(8): 1146–1148.
37. Iga, Y. et al., Characterization of micro-channels fabricated by in-water ablation of Fs-laser pulses. *Japanese Journal of Applied Physics*, 2004. **43**: 4207–4211.
38. Kamlage, G. et al., Deep drilling of metals by Fs-laser pulses. *Applied Physics A: Materials Science & Processing*, 2003. **77**: 307–310.
39. Hu, A. et al., Direct synthesis of sp-bonded carbon chains on graphite surface by Fs-laser irradiation. *Applied Physics Letters*, 2007. **91**: 131906.
40. Perry, T. et al., Pulsed laser micro polishing of microfabricated Ni and TI6AL4V samples. *Proceedings of International Manufacturing Science and Engineering Conference*, New York, 2008.
41. Kim, Y. et al., Microroughness reduction of tungsten films by laser polishing technology with a line beam. *Japanese Journal of Applied Physics*, 2004. **43**: 1315–1322.
42. Tonshoff, H. and H. Kappel, Surface modification of ceramics by laser machining. *CIRP Annals*, 1998. **47**: 471–474.
43. Bereznai, M. et al., Surface modifications induced by ns and sub-ps excimer laser pulses on titanium implant material. *Biomaterials*, 2003. **24**: 4197–4203.
44. Gumpenberger, T. et al., F2-laser polishing of polytetrafluoroethylene surfaces. *Europhysics Letters*, 2005. **70**: 831–835.
45. Lamikiz, A. et al., Laser polishing of parts built up by selective laser sintering. *International Journal of Machine Tools and Manufacture*, 2007. **47**(12–13): 2040–2050.
46. Mai, T. and G. Lim, Micromelting and its effects on surface topography and properties in laser polishing of stainless steel. *Journal of Laser Applications*, 2004. **16**: 221–228.

47. Shao, T. et al., An approach to modeling laser polishing of metals. *Surface and Coatings Technology*, 2005. **197**: 77–84.
48. Lee, K.-S. et al., Advances in 3D nano/microfabrication using two-photon initiated polymerization. *Progress in Polymer Science*, 2008. **33**(6): 631–681.
49. Ashkin, A. et al., Observation of a single-beam gradient force optical trap for dielectric particles. *Optics Letters*, 1986. **11**(5): 288–290.
50. Nedyalkov, N.N., H. Takada, and M. Obara, Nanostructuring of silicon surface by femtosecond laser pulse mediated with enhanced near-field of gold nanoparticles. *Applied Physics A*, 2006. **85**(2): 163–168.
51. Bechtold , P., S. Roth, and M. Schmidt, Precise subnanometer positional adjustments with laser-induced shock waves. *Photonics Spectra*, 2008. 658–663.
52. Liddle, J.A., *Measurement Challenges in Nanomanufacturing*. New England Nanomanufacturing Summit, 2010. Lowell, MA.
53. Klinger, P. et al. In process 3D-sensing for laser material processing. In *3-D Digital Imaging and Modeling, 2001. Proceedings. Third International Conference on 3-D Digital Imaging and Modeling,* Quebec City, Canada, 2001.
54. Sibillano, T. et al., A real-time spectroscopic sensor for monitoring laser welding processes. *Sensors*, 2009. **9**(5): 3376–3385.
55. Ancona, A. et al., Optical sensor for real-time monitoring of CO2 laser welding process. *Applied Optics*, 2001. **40**(33): 6019–6025.
56. Jager, M. and F.A. Hamprecht, Principal component imagery for the quality monitoring of dynamic laser welding processes. *Industrial Electronics, IEEE Transactions on*, 2009. **56**(4): 1307–1313.
57. Dagalakis, N.G., T. LeBrun, and J. Lippiatt. Micro-mirror array control of optical tweezer trapping beams. In *Nanotechnology, 2002. IEEE-NANO 2002. Proceedings of the 2002 Second IEEE Conference on Nanotechnology*, Washington, DC, USA, 2002.
58. Ferrara, M. et al., Online quality monitoring of welding processes by means of plasma optical spectroscopy. *Proceedings of SPIE*, 2000. **3888**: 750–758.
59. Kratzsch, C. et al., Coaxial process control during laser beam welding of tailored blanks. 2000, **3888**: 472–482.
60. Muller, R., Real time monitoring of laser weld plume temperature and species concentration. *Proceedings of the ICALEO*, 1996. **81**: B68–B75.
61. Anwander, M. et al., Noncontacting strain measurements at high temperatures by the digital laser speckle technique. *Experimental Mechanics*, 2000. **40**(1): 98–105.
62. Hu, D. and R. Kovacevic, Sensing, modeling and control for laser-based additive manufacturing. *International Journal of Machine Tools and Manufacture*, 2003. **43**(1): 51–60.
63. Chen , S.-L. et al., In-Process Laser Beam Position Sensing. *Proceedings of SPIE*, The Hague, Netherlands, 1991: pp. 123–134.
64. Meyer, G. and N.M. Amer, Novel optical approach to atomic force microscopy. *Applied Physics Letters*, 1988. **53**(12): 1045–1047.
65. Malshe, A.P. et al., Tip-based nanomanufacturing by electrical, chemical, mechanical and thermal processes. *CIRP Annals—Manufacturing Technology*, 2010. **59**(2): 628–651.
66. Ryabov, O. et al., An in-process direct monitoring method for milling tool failures using a laser sensor. *CIRP Annals—Manufacturing Technology*, 1996. **45**(1): 97–100.

67. Golnabi, H., Role of laser sensor systems in automation and flexible manufacturing. *Robotics and Computer-Integrated Manufacturing*, 2003. **19**(1–2): 201–210.
68. Gozen, B.A. and O.B. Ozdoganlar, Design and evaluation of a mechanical nano-manufacturing system for nanomilling. *Precision Engineering*, 2012. **36**(1): 19–30.
69. Luo, H. et al., Application of artificial neural network in laser welding defect diagnosis. *Journal of Materials Processing Technology*, 2005. **170**(1–2): 403–411.
70. Alok, S., M.S. Vijay, and W. Sy-Bor, The generation of nano-patterns on a pure silicon wafer in air and argon with sub-diffraction limit nanosecond laser pulses. *Journal of Physics D: Applied Physics*, 2010. **43**(14): 145301.
71. Ma, X. et al., Surface-enhanced Raman scattering sensor on an optical fiber probe fabricated with a femtosecond laser. *Sensors*, 2010. **10**(12): 11064–11071.
72. Li, L. and W.M. Steen, Non-contact acoustic emission monitoring during laser processing. *Proceedings of the ICALEO* 1992. **75**: 719–728.
73. Li, L. and N. Qi, On-line laser welding sensing for quality. *Proceedings of the ICALEO*, 1994: 411–421.
74. Li, L., Sensor development for in-process quality inspection and optimisation of high speed laser can welding process. *Proceedings of the LAMP*, Nagaoka, Japan, 1992. pp. 421–426.
75. D avies, M.A. et al., Application of precision diamond machining to the manufacture of microphotonics components. *Proceedings of SPIE*, San Diego, California, USA, 2003. **5183:** 94–108.
76. Okafor, A.C. and Y.M. Ertekin, Vertical machining center accuracy characterization using laser interferometer: Part 1. Linear positional errors. *Journal of Materials Processing Technology*, 2000. **105**(3): 394–406.
77. Jiang, X., Precision surface measurement. *Philosophical Transactions of the Royal Society A: Mathematical, Physical and Engineering Sciences*, 2012. **370**: 4089–4114.
78. Zobrist, T.L., J.H. Burge, and H.M. Martin, Accuracy of laser tracker measurements of the GMT 8.4 m off-axis mirror segments, 2010. **7739**: 77390S.
79. Wang, S. et al., Development of a laser-scattering-based probe for on-line measurement of surface roughness. *Applied Optics*, 2003. **42**(7): 1318–1324.
80. Wang, S.H. et al., Surface roughness measurement in the submicrometer range using laser scattering. *Optical Engineering*, 2000. **39**(6): 1597–1601.
81. Ancona, A. and T. Sibillano, Monitoring laser welding, in *Real-Time Monitoring of Welding Processes*, Zhang, Y.M., (Ed.) 2008. Woodhead Publishing Limited: Cambridge, U.K. pp. 260–287.
82. de Groot, P. et al., Optical interferometry for measurement of the geometric dimensions of industrial parts. *Applied Optics*, 2002. **41**(19): 3853–3860.
83. Wong, Y.S. et al., Tool condition monitoring using laser scatter pattern. *Journal of Materials Processing Technology*, 1997. **63**(1–3): 205–210.
84. Sárosi, Z. et al., Detection of surface defects on sheet metal parts using one-shot deflectometry in the infrared range. FLIR Technical Series Whie Paper, Application Note for Research & Science, ETH Zurich, IWF 2011.
85. Liu, J. and L. Li, In-time motion adjustment in laser cladding manufacturing process for improving dimensional accuracy and surface finish of the formed part. *Optics & Laser Technology*, 2004. **36**(6): 477–483.
86. Hu, D., H. Mei, and R. Kovacevic, Improving solid freeform fabrication by laser-based additive manufacturing. *Proceedings of the Institution of Mechanical Engineers, Part B: Journal of Engineering Manufacture*, 2002. **216**(9): 1253–1264.

87. Kercel, S.W. et al., In-process detection of weld defects using laser-based ultrasound. *Proceedings of SPIE*, Boston, Massachusetts, USA, 1999. 3852: 81–92.

88. Shao, J. and Y. Yan, Review of techniques for on-line monitoring and inspection of laser welding. *Journal of Physics: Conference Series*, 2005. **15**(1): 101.

89. Liu, L. et al., Sensor referenced real-time videolization of atomic force microscopy for nanomanipulations. *Mechatronics, IEEE/ASME Transactions on*, 2008. **13**(1): 76–85.

90. Beeck, M.-A. and W. Hentschel, Laser metrology—A diagnostic tool in automotive development processes. *Optics and Lasers in Engineering*, 2000. **34**(2): 101–120.

91. Leopold, J., H. Günther, and R. Leopold, New developments in fast 3D-surface quality control. *Measurement*, 2003. **33**(2): 179–187.

92. Zhang, S., J. Wootton, and A. Chisholm, Geometric feature extraction from point clouds obtained by laser scanning. *Proceedings of SPIE*, Dresden, Germany, 2005.6157: 61570R–61570R-9.

93. Nicolosi, L. et al. Multi-feature detection for quality assessment in laser beam welding: Experimental results. In *Cellular Nanoscale Networks and Their Applications (CNNA), 2012 13th International Workshop on Cellular Nanoscale Networks and Their Applications*, Turin, Italy, 2012.

94. Kumar, M., D.P. Garg, and R. Zachery. Multi-sensor fusion strategy to obtain 3-D occupancy profile. In *Industrial Electronics Society, 2005. IECON 2005. 31st Annual Conference of IEEE Industrial Electronics Society*, Raleigh, NC. 2005.

95. Lu, R.-S. and G.Y. Tian, On-line measurement of surface roughness by laser light scattering. *Measurement Science and Technology*, 2006. **17**(6): 1496.

96. Shinozaki, R., O. Sasaki, and T. Suzuki, Fast scanning method for one-dimensional surface profile measurement by detecting angular deflection of a laser beam. *Applied Optics*, 2004. **43**(21): 4157–4163.

97. Ganesha, U. et al., Characterization of surface topography by confocal microscopy: II. The micro and macro surface irregularities. *Measurement Science and Technology*, 2000. **11**(3): 315.

98. McKinney, B.A., Informatics approaches for identifying biologic relationships in time-series data. *Wiley Interdisciplinary Reviews: Nanomedicine and Nanobiotechnology*, 2009. **1**(1): 60–68.

99. Huang, Q., Physics-driven Bayesian hierarchical modeling of nanowire growth process at each scale. *IIE Transactions on Quality and Reliability*, 2011. **43**: 1–11.

100. Park, C. et al., A multistage, semi-automated procedure for analyzing the morphology of nanoparticles. *IIE Transactions*, 2012. **44**(7): 507–522.

101. Cui, H., O. Zhou, and B.R. Stoner, Deposition of aligned bamboo-like carbon nanotubes via microwave plasma enhanced chemical vapor deposition. *Journal of Applied Physics*, 2000. **88**: 6072–6074.

102. Raghavan, D., Synthesis of multi-walled carbon nanotubes by plasma enhanced microwave CVD using colloidal form of iron oxide as a catalyst, Master Thesis, 2005. Oklahoma State University: Stillwater, OK.

103. Ramakrishnan, M.P., Experimental study on microwave assisted CVD growth of carbon nanotubes on silicon wafer using cobalt as a catalyst, Master Thesis, 2005. Oklahoma State University, Stillwater, OK.

104. Nidadavolu, A.G.R., Synthesis of carbon nanotubes by microwave plasma enhanced CVD on silicon using iron as catalyst, Master Thesis, 2005. Oklahoma State University, Stillwater, OK.

105. Cheng, C. et al., Monte Carlo simulation of carbon nanotube nucleation and growth using nonlinear dynamic predictions. *Chemical Physics Letters*, 2012. **530**: 81–85.

106. Maksym, P.A., Fast Monte Carlo simulation of MBE growth. *Semiconductor Science and Technology*, 1988. **3**(6): 594.

107. Rocha, W.R. et al., An efficient quantum mechanical/molecular mechanics Monte Carlo simulation of liquid water. *Chemical Physics Letters*, 2001. **335**(1–2): 127–133.

108. Gentile, N.A., Implicit Monte Carlo diffusion—An acceleration method for Monte Carlo time-dependent radiative transfer simulations. *Journal of Computational Physics*, 2001. **172**(2): 543–571.

109. Tejima, S. et al., *Earth Simulator Research Projects: Epoch-making Simulation*, Annual report of the Earth Simulator Center, Yokohama, Japan 2008.

110. Bukkapatnam, S.T.S. and C. Cheng, Forecasting the evolution of nonlinear and nonstationary systems using recurrence-based local Gaussian process models. *Physical Review E*, 2010. **82**(5): 056206.

111. Cheng, C. et al., Towards control of CNT synthesis process using prediction-based fast Monte Carlo simulations. *Proceedings of NAMRI/SME*, 2012. **40**: 371–378.

# 11

## *System Optimization for Laser and Photonic Applications*

Longfei Wang and Leyuan Shi

**CONTENTS**

The integration of different processes in manufacturing, healthcare, and communication imposes new challenges. The service-centric platform requires that physical processes of different nature be systematically integrated so that the resources, for example, manufacturing resources, can be effectively drawn and assembled in a timely manner when needed. In addition, dynamic time sharing among different people, assets, and processes in both remote and proximal environments should be handled automatically by monitoring programs, enabling optimal allocation of critical resources and effort on the tasks with different levels of priorities.

There are many optimization problems that exist in laser and photonic application areas. Advanced optimization theories and solution techniques are needed to solve these problems. In this chapter, three optimization problems in laser and photonic applications are reviewed and then an effective and flexible optimization framework—the nested partitions (NP) method— is introduced.

## Optimization Problems in Laser and Photonic Applications

### Chemical Laser Modeling

The typical chemical oxygen iodine laser (COIL) utilizes an energy transfer from the singlet delta excited state of oxygen $O_2(^1\Delta)$ directly and indirectly to $I_2$ to dissociate the iodine molecule (Carroll, 1996). There are many issues that should be investigated about the operation of COIL systems, such as the effects of increased total pressure on mixing, kinetics, gain, and power. These effects should be analyzed through reliable computer models. Crowell and Plummer (1993) have demonstrated a premixed model for the laser cavity calculations; however, the results were not in good agreement with data. Carroll (1995) used a modified mixing Blaze II computer model, which is originally developed to be a generic chemical laser model. In the Blaze II model, there are five unknown parameters: yield, water vapor flow rate, $k_{18}$ multiplier, the jet expansion factor, and $DCM_{exp}$. It is very important to find a set of appropriate parameters that best match modeling predictions with experimental data. Carroll (1996) used a genetic algorithm (GA) technique to solve this problem. The results showed that the GA method performed very well and in a cost-effective and time-efficient manner. This study also indicated that it may be possible to optimize the performance of any chemical laser systems through the use of GA.

### Remote Laser Welding Path Optimization

Recent laser developments such as the fiber laser have increased the focal length of a focused laser beam. When applied in welding, this development has increased the welding distances between optics and the workpiece. This new technology is called remote laser welding (RLW). The main advantage of RLW is reduced cycle time because the positioning speed of the focus spot on the workpiece increases by small deflections of the laser beam. But these developments have created the challenge of programming an optimal robot movement with respect to cycle time. A useful programming and optimization system is needed for finding a good solution for a given welding problem. The path planning and optimization problem is very complicated, so Reinhart et al. (2008) decomposed the problem into several subproblems, of which the two central problems are welding sequence and the variation of inclination angles. In the welding sequence problem, it is assumed that the shortest Cartesian path between all seams results in the shortest cycle time for a welding problem. With this assumption, the problem becomes analogous to the travelling salesman problem (TSP), which is a typical optimization problem. Reinhart et al. (2008) used a simple heuristic method to solve this problem.

### Maximum A Posteriori Estimate

Bayesian object recognition can be applied in many areas, such as the analysis of complex forest object configurations. When using the Bayesian object recognition method, a probabilistic model of the active sensing process and a prior probability model on object configurations are incorporated. And in Bayesian analysis, all inferences are based upon the posterior distribution of the object configuration: $p(x|y) \propto l(y|x)p(x)$, where $p(x)$ is the prior distribution, $p(x|y)$ is the conditional distribution, and $l(y|x)$ is the likelihood function. The typical objective of Bayesian object recognition is to estimate the true configuration of object $x$, given the observed data $y$. In particular, the maximum a posteriori (MAP) estimate, representing the mode of the posterior distribution (a Markov object process), is of primary interest in the context of object recognition. Finding the object configuration that maximizes the posterior probability is a combinatorial optimization problem that requires optimization techniques to be applied. In Andersen et al. (2002), a simulated annealing algorithm was used to find the object configuration.

### NP Method

For many deterministic optimization problems, it is notoriously difficult to find out an optimal or even near-optimal solution. The main reason is that the size of problems is too large and the structure of the solution space is too complex to deal with. In general, stochastic optimization problems are more difficult to solve because of the existence of randomness.

Solving large-scale optimization problems is a subject of intensive research and many research achievements thus far. For deterministic optimization problems, the methods that have been used to solve these problems can be classified into two categories: exact algorithms and heuristics. One of the most important exact algorithms is the branch and bound method. This method is usually used in solving integer or mixed integer programming (Wolsey, 1998; Nemhauser and Wolsey, 1999). Decomposition methods are also powerful to solve optimization problems, and relaxation methods approach these problems in other intelligent ways. For example, when using the Lagrangian methods, one or more complicated constraints will be moved into the objective function (Lemarechal, 2001). Dynamic programming is a class of exact methods to deal with sequential decision-making problems. These exact methods are very powerful to solve deterministic optimization problems as they can assure that the optimal solution will be obtained. But for large-scale optimization problems, the computation time and effort that is needed to solve problems by these methods is too large to implement. In this case, heuristic algorithms should be used for some large-scale problems.

Heuristic methods cannot guarantee that the optimal solution will be obtained, but they can find good solutions in a very short time. Heuristics have been proven to be very effective and efficient in solving many practical problems. One simple heuristic algorithm is greedy search (Cormen et al., 1990), which often gives a locally optimal solution. Tabu search (Cvijović and Klinowski, 1995; Glover and Laguna, 1997) improves the efficiency of the search by prohibiting the repeated visits to the same solution within a certain short period of time. There is a class of meta-heuristics that are called evolutionary algorithms. These algorithms start with some feasible solutions and improve the quality of the feasible solutions iteratively. These algorithms include GA (Goldberg, 1989; Mitchell, 1998), ant colony optimization (Dorigo, 1992; Dorigo et al., 1999), and particle swarm optimization (Kennedy and Eberhart, 1995; Poli et al., 2007). Some meta-heuristics utilize probability distributions to improve the efficiency of the search process such as simulated annealing (Kirkpatrick et al., 1983; van Laarhoven and Aarts, 1987) and the cross-entropy method (Rubinstein and Kroese, 2004; De Boer et al., 2005). Meta-heuristics are flexible and can be applied to solve many different types of problems. What is more, other useful techniques and domain knowledge can be integrated into these methods.

Stochastic optimization problems are more difficult to solve because the objective function value can only be estimated. The study about the methods and techniques to solve these problems is in an immature stage. But there are some algorithms that can be applied to these problems such as the stochastic ruler (Yan and Mukai, 1992), the COMPASS method (Hong and Nelson, 2006), and ordinal optimization (Ho et al., 1992, 2000).

The NP method (Shi and Ólafsson, 2000) is a newly developed and novel optimization framework for solving large-scale optimization problems. It can be applied to many complex problems in both manufacturing and service industries. In the following sections, the NP method is discussed in detail, including the basic elements of the NP method, the global convergence property, and the integration of the NP method with other algorithms.

## NP for Deterministic Optimization

Consider the following optimization problem:

$$\min_{\theta \in \Theta} f(\theta)$$

The set of feasible solutions, called the feasible solution space, is denoted as $\Theta$ and $\theta$ denotes the feasible solution. The problem can be discrete or continuous. And $f:\Theta \to \mathbb{R}$ denotes the objective function, which can be linear or nonlinear. What we want to do is to find out the solution that minimizes the objective function value. For most problems, it is not easy to solve by using exact algorithms due to the large size or complicated structure of the problems. However, we can solve these problems with the NP method.

During each iteration of the NP method, we assume that there is a region that is considered to be the *most promising* region, that is, the probability that the optimal solution is contained within the region is considered to be the largest. The most promising region is divided (partitioned) into a fixed number of $M$ subregions, and the entire complementary region (surrounding region) is aggregated into one region. So the entire solution space is partitioned into $M+1$ regions, and the regions are disjoint and their union is equal to $\Theta$. Using some random sampling scheme, sample solutions will be collected from each of the regions. And based on the objective values of these randomly generated solutions, we can calculate the promising index for each region. The promising index can be used to determine which region is the most promising region in the next iteration. If one of the subregions of the current most promising region is found to be the best, then this subregion will be the next most promising in the next iteration. If the complementary region is found to be the best, then the process will backtrack to a larger region that contains the previous most promising region. The larger region will be the most promising region in the next iteration. The process will continue until the terminating condition is satisfied.

To describe the NP method in detail, notations and terminologies should be defined. If region $\eta$ is a subregion of region $\sigma$, then we call $\sigma$ a superregion of $\eta$. And let $\sigma(k)$ denote the most promising region of the $k$th iteration, and let $d(k)$ denote the depth of $\sigma(k)$. The depth of $\Theta$ is 0, the depth of the subregion of $\Theta$ is 1, and so forth. If the feasible solutions in $\Theta$ is finite, then there will be regions that contain only one solution. The singleton regions are called regions of maximum depth. If the feasible solutions in $\Theta$ are infinite, we define the maximum depth to correspond to the smallest desired sets. The maximum depth is denoted as $d^*$. Based on Shi and Ólafsson (2008), the *generic NP algorithm* or *pure NP* is shown as follows:

Generic NP Algorithm

1. *Partitioning.* Partition the most promising region $\sigma(k)$ into $M$ subregions $\sigma_1(k), \ldots, \sigma_M(k)$ and aggregate the complimentary region $\Theta \backslash \sigma(k)$ into one region $\sigma_{M+1}(k)$.

2. *Random sampling.* Randomly generate $N_j$ sample solutions from each of the regions $\sigma_j(k), j = 1, 2, \ldots, M+1$:

$$\theta_1^j, \theta_2^j, \ldots, \theta_{N_j}^j, j = 1, 2, \ldots, M+1$$

   Calculate the corresponding performance values:

$$f\left(\theta_1^j\right), f\left(\theta_2^j\right), \ldots, f\left(\theta_{N_j}^j\right), j = 1, 2, \ldots, M+1$$

3. *Calculate promising index.* For each region $\sigma_j, j = 1, 2, \ldots, M+1$, calculate the *promising index* as the best performance value within the region:

$$I\left(\sigma_j\right) = \min_{i=1,2,\ldots,N_j} f\left(\theta_i^j\right), j = 1, 2, \ldots, M+1$$

4. *Move.* Calculate the index of the region with the best performance value:

$$\hat{j}_k^* = \arg\min_{j=1,\ldots,M+1} I\left(\sigma_j\right), j = 1, 2, \ldots, M+1$$

If more than one region is equally promising, the tie can be broken arbitrarily. If this index corresponds to a region that is a subregion of $\sigma(k)$, that is, $\hat{j}_k^* \leq M$, then let this be the most promising region in the next iteration:

$$\sigma(k+1) = \sigma_{\hat{j}_k^*}(k)$$

Otherwise, if the index corresponds to the complementary region, that is, $\hat{j}_k^* = M+1$, backtrack to the superregion of the current most promising region (previous most promising region):

$$\sigma(k+1) = \sigma(k-1)$$

or backtrack to the entire solution space:

$$\sigma(k+1) = \Theta$$

For the special case of $d(k) = 0$, the steps are identical except there is no complementary region, and the algorithm generates feasible sample solutions from the subregions and in the next iteration moves to the subregion with the best promising index. For the special case of $d(k) = d^*$, there are no subregions. And the algorithm generates feasible sample solutions from the complementary region and either backtracks or stays in the current most promising region.

## Partitioning

In the generic NP algorithm, partitioning is not discussed in detail because it is very flexible and problem-specific. Partitioning is very important and can affect the efficiency of the NP method significantly. The main principle of partitioning is that good solutions should be clustered together. Clustering can make it easy to identify the next promising region, that is, make correct moves.

## Sampling

Sampling is also an important factor in determining the efficiency of the NP method. In the generic NP algorithm, sampling is done randomly, but the flexibility of the NP method allows us to use other methods to generate samples and determine how many random samples should be obtained.

Many techniques can be used to improve the efficiency of the NP method by increasing the probability of making a correct move. To make the correct move frequently, good solutions should be sampled in each region. We can use three ways to increase the probability of making a correct move when sampling: (1) biasing the sampling distribution so that good solutions are more likely to be selected, (2) using heuristic methods to search for good solutions, and (3) using a sufficiently large sample.

When solving problems, we can obtain some information about good solutions based on domain knowledge, linear programming (LP) relaxation, or other methods. Using these methods, the sampling distribution can be biased and good solutions are more likely to be selected. As further discussion of this method is problem-specific, we will not cover it in detail here.

Heuristic methods can also be used to generate good solutions. We can obtain one or several random samples using either uniform or weighted sampling and then use heuristics to improve the solution set. Hybrid NP algorithms will be addressed later in more detail.

When sampling feasible solutions, there is a very important parameter that should be determined, total sampling effort, that is, how many sample solutions are needed to make the correct move with a sufficiently large probability. The number of feasible sample solutions required to assure this minimum probability depends on the variance of the performance of the generated solutions. For example, if the solutions that are generated have the same or similar performance, then it is unnecessary to generate more than one solution. On the other hand, if the solutions have greatly variable performance, then we should generate more solutions to obtain a sufficiently good estimate of the overall performance of the region.

Based on observations, a two-stage sampling approach can be designed and used to sample solutions. The first stage would consist of generating a small number of feasible solutions using uniform sampling, weighted sampling, or other methods. Then the variance of the performance (objective values) for these solutions is calculated and then a statistical selection technique could be used to determine how many total samples are needed to ensure the probability of making a correct move.

## Backtracking and Initialization

In the NP method, if the best feasible solution is found in the complimentary region, or the best promising index corresponds to the complementary region, backtracking is necessary. If the best feasible solution is found in the complimentary region, this indicates that an incorrect move was made when $\sigma(k)$ was selected as the most promising region. Generating solutions from the complementary region and backtracking ensure that the NP method converges to an optimal solution in finite time.

Even though backtracking is very important, excessive backtracking indicates that the NP method is inefficient. By monitoring the number of backtracks, it is possible to design an adaptive NP algorithm. When the number of backtracks is excessive, this indicates that more effort is needed to improve the decision of which region is the next most promising region, that is, more and higher-quality solutions should be sampled and used to calculate the promising index for each region.

In the generic NP algorithm, we let the entire feasible solution space be the most promising region for the initial state. But if time is very limited and the quality of the solution is very important, a heuristic or domain knowledge method should be used to generate a partial solution and set it as the most promising region in the initial state.

## Global Convergence Analysis

Global convergence is a very important property of the NP framework. In this section, two important theorems are proposed. The first is that the NP method converges to an optimal solution in finite time and the second is that the time until convergence can be bounded in terms of the size of the problem and the probability of making the correct move.

In this section, we discuss the basic ideas of global convergence and give some propositions and theorems. For details about the proofs, and more theoretical properties of the NP method, we refer to Shi and Ólafsson (2000).

When using the NP method, the sequence of the most promising regions is an absorbing Markov chain and the set of optimal solutions corresponds exactly to the absorbing states. Because the state space of the Markov chain is finite, it is absorbed in finite time. To describe the ideas with mathematical notations, the following propositions and theorems are presented:

## Proposition 11.1

Assume that the partitioning of the feasible region is fixed and $\Sigma$ is the set of all valid regions. The stochastic process $\{\sigma(k)\}_{k=1}^{\infty}$, defined by the most promising region in each iteration of the pure NP algorithm, is a homogeneous Markov chain with $\Sigma$ as state space.

## Proposition 11.2

Assume that the partitioning of the feasible region is fixed and $\Sigma$ is the set of all valid regions. A state $\eta \in \Sigma$ is an absorbing state for the Markov chain $\{\sigma(k)\}_{k=1}^{\infty}$ if and only if $d(\eta) = d^*$ and $\eta = \{x^*\}$, where $x^*$ is a global minimizer of the original problem.

## Theorem 11.1

The NP algorithm converges almost surely to a global minimum of the optimization problem. In mathematical notation, the following equation holds:

$$\lim_{k \to \infty} \sigma(k) = \{x^*\} \ a.s.,$$

where

$$x^* \in \arg\min_{x \in \Theta} f(\theta)$$

After proving that the Markov chain of the most promising regions will converge to the global optimum, we also need to study how many iterations and consequently how many function evaluations are required before the global optimum is found. Assume that $\Sigma$ is finite and the global optimum is unique. Defined in the following are some notations that are needed when studying the expected number of iterations until absorption.

Let $Y$ denote the number of iterations until the Markov chain is absorbed and let $Y_\eta$ denote the number of iterations spent in state $\eta \in \Sigma$. Define $T_\eta$

to be the hitting time of state $\eta \in \Sigma$, that is, the first time that the Markov chain visits the state. Let $E$ denote an arbitrary event and let $\eta \in \Sigma$ be a valid region, $P_\eta[E]$ the probability of event $E$ given that the chain starts in state $\eta \in \Sigma$. Define $\Sigma_1 = \{\eta \in \Sigma \backslash \{\sigma^*\} \mid \sigma^* \subseteq \eta\}$ and $\Sigma_2 = \{\eta \in \Sigma \mid \sigma^* \not\subseteq \eta\}$. Then $\Sigma = \{\sigma^*\} \cup \Sigma_1 \cup \Sigma_2$ and these three sets are disjoint. Using these notations, we can state the following theorems:

**Theorem 11.2**

The expected number of iterations until the NP Markov chain is absorbed is given by

$$E[Y] = 1 + \sum_{\eta \in \Sigma_1} \frac{1}{P_\eta[T_{\sigma^*} < T_\eta]} + \sum_{\eta \in \Sigma_2} \frac{P_X[T_\eta < \min\{T_X, T_{\sigma^*}\}]}{P_\eta[T_X < T_\eta] \cdot P_X[T_{\sigma^*} < \min\{T_X, T_\eta\}]}.$$

Assume that $P^*$ is a lower bound on the probability of selecting the correct region. Then, we can get an upper bound for the expected time until the NP algorithm converges in terms of $P^*$ and the size of the feasible region as measured by $d^*$.

**Theorem 11.3**

Assume that $P^* > 0.5$. The expected number of iterations until the NP Markov chain is absorbed is bounded by

$$E[Y] \le \frac{d^*}{2P^* - 1}.$$

Because $E[Y]$ grows only linearly in $d^*$, the expected number of iterations required by the NP algorithm grows very slowly with the size of the problem. Because of this, the NP method is very effective to solve large-scale optimization problems.

When $P^* \to \frac{1}{2}$, $E[Y]$ grows exponentially, that is, the number of expected iterations increases rapidly as the success probability decreases. The most important way to improve the efficiency of the NP method is to increase $P^*$, that is, increase the probability of making correct moves. We can use intelligent partitioning, efficient sampling, and other methods discussed in this chapter to achieve this goal.

## NP for Stochastic Optimization

This section will discuss using the NP framework for stochastic optimization. For stochastic optimization problems, the objective function is noisy, that is, there are some random effects when evaluating the objective function value. This makes the problems more complicated because there is no simple analytical model that can be specified for the system being optimized. The challenge of evaluating the objective function exactly exacerbates the difficulty of the optimization problem. Consider the following stochastic problem:

$$\min_{\theta \in \Theta} E[f_w(\theta)],$$

where $w$ represents the randomness in function evaluation. The goal of the optimization problem is to minimize the expected objective function value.

When solving a stochastic optimization problem using the NP method, some special issues should be considered, but some implementation aspects remain unchanged. All changes in the implementation and analysis of the NP method refer to how the algorithm moves on the fixed state space, such as how much more additional computational effort is needed to make the correct moves. In stochastic optimization problems, the NP method no longer generates an absorbing Markov chain, so convergence of the NP method will be revisited.

The global convergence analysis of the NP method for stochastic optimization problems is very complicated. Definitions, assumptions, propositions, and theorems are discussed in this section. For more details, refer to Shi and Ólafsson (2008).

When using the NP method for stochastic optimization problems, the set performance function $I: \Sigma \rightarrow R$ is defined, and the original performance function $f: X \rightarrow R$ and the set performance function agree on singletons. So we can define a new problem that is to find an element $\sigma^* \in S$, where

$$S = \arg \min_{\sigma \in \Sigma_0} I(\sigma).$$

This new problem is equivalent to solving the original problem.

For any $\sigma, \eta \in \Sigma$, we use $P(\sigma, \eta)$ to denote the transition probability of the Markov chain moving from state $\sigma$ to state $\eta$ and $P^n(\sigma, \eta)$ to denote the $n$-step transition probability. Because the next promising region $\sigma(k+1)$ in the $k$th iteration only depends on the current most promising region $\sigma(k)$ and the sampling solutions obtained in the $k$th iteration, the sequence of the most promising regions $\{\sigma(k)\}_{k=1}^{\infty}$ is a Markov chain with state space $\Sigma$. And let $\Sigma_0 \subset \Sigma$ denote all of the maximum depth (singleton) regions, $\mathcal{H}(\sigma)$ denote the set of subregions of a valid region $\sigma \in \Sigma \setminus \Sigma_0$, and $b(\sigma)$ denote the parent region of each valid region $\sigma \in \Sigma\{X\}$.

**Assumption 11.1**

For all $\sigma \in \Sigma \backslash \{X\}$, $P(\sigma, b(\sigma)) > 0$.

**Proposition 11.3**

If Assumption 11.1 holds, then the NP Markov chain has a unique stationary distribution $\{\pi(\sigma)\}_{\sigma \in \Sigma}$.

**Theorem 11.4**

Assume that Assumption 11.1 holds. The estimate of the best region $\hat{\sigma}^*(k)$ converges to a maximum of the stationary distribution of the NP Markov chain, that is,

$$\lim_{k \to \infty} \hat{\sigma}^*(k) \in \arg \max_{\sigma \in \Sigma_0} \pi(\sigma), w.p.1.$$

**Definition 11.1**

Let $\sigma$ and $\eta$ be any valid regions. Then there exists some sequence of regions $\sigma = \xi_0, \xi_1, \ldots, \xi_n = \eta$ along which the Markov chain can move to get from state $\sigma$ to state $\eta$. We call the shortest such sequence the shortest path from $\sigma$ to $\eta$. We also define $\kappa(\sigma, \eta)$ to be the length of the shortest path.

**Assumption 11.2**

The set

$$S_0 = \{\xi \in \Sigma_0 : P^{\kappa(\eta, \xi)}(\eta, \xi) \geq P^{\kappa(\xi, \eta)}(\xi, \eta), \forall \eta \in \Sigma_0\}$$

satisfies $S_0 \subseteq S$, that is, it is a subset of the set of global optimizers.

**Theorem 11.5**

Assume that the NP I algorithm is applied and Assumptions 11.1 and 11.2 hold. Then

$$\arg\max_{\sigma\in\Sigma_0}\pi(\sigma)\subseteq S$$

and consequently the NP I algorithm converges with probability one to a global optimum.

**Theorem 11.6**

Assume that the NP II algorithm is applied and Assumptions 11.1 and 11.2 hold. Then

$$\arg\max_{\sigma\in\Sigma_0}\pi(\sigma)\subseteq S$$

and consequently the NP II algorithm converges with probability one to a global optimum.

**Definition 11.2**

Let $\Sigma_g = \{\sigma\in\Sigma : \sigma^*\subseteq\sigma\}$ denote all the regions containing the unique global optimum $\sigma^*$ and $\Sigma_b = \Sigma\backslash\Sigma_g$ denote the remaining regions. Define a function $Y : \Sigma_g\times\Sigma_b\to R$ by

$$Y(\sigma_g,\sigma_b) = J\left(x^{[1]}_{\sigma_b}\right) - J(x^{[2]}_{\sigma_g}),$$

where $x^{[1]}_{\sigma}$ denotes the random sample point from $\sigma\in\Sigma$ that has the estimated best performance and is generated in Step 2 of the NP algorithm. Furthermore, let

$$\hat{Y}(\sigma_g,\sigma_b) = L\left(x^{[1]}_{\sigma_b}\right) - L(x^{[1]}_{\sigma_g})$$

denote the corresponding simulation estimate.

**Theorem 11.7**

Assume that $\sigma^*$ is unique and

$$P\left[\hat{Y}(\sigma_g,\sigma_b)>0\,|\,Y(\sigma_g,\sigma_b)=y\right] > \frac{y}{2E[Y(\sigma_g,\sigma_b)]}$$

for all $\min_{x\in\sigma_b} J(x) - \max_{x\in\sigma_g} J(x) \leq y \leq \max_{x\in\sigma_b} J(x) - \min_{x\in\sigma_g} J(x), \sigma_g \in \Sigma_g$ *and* $\sigma_b \in \Sigma_b$. Then the NP I algorithm converges to a global optimum.

---

## Enhancements and Advanced Developments

### Intelligent Partitioning

Partitioning is very important for the efficiency of the NP method as the selected partition imposes a structure on the feasible region. After partitioning, if good solutions are clustered together, then these subregions will be selected easily, yet if good solutions are not clustered together, that is, the good solutions are surrounded by poor-quality solutions, then the probability that the algorithm makes a move toward those subregions is small, that is, the probability that finding the optimal solution is small. So we should use an intelligent partitioning method to make good solutions clustered together.

Intelligent partitioning methods are usually application dependent, but there are also some general intelligent partitioning methods that are suitable for a large class of problems. Here, we introduced a method that is based on information theory.

In this intelligent partitioning method, solutions are divided into different classes based on their objective value, that is, if the objective values of two solutions are the same or there is little difference between the two values, then these two solutions are considered to be the same. If a valid subregion contains many solutions that are different from each other, then this subregion is considered diverse. When using the NP method, diverse subregions are undesirable because it will be difficult to determine which subregion should be selected as the promising region in the next iteration as there are many diverse subregions.

The following procedure is from Shi and Ólafsson (2008), and using the procedure an intelligent partitioning can be constructed:

1. Use random sampling to generate a set of $M_0$ sample solutions.
2. Evaluate the performance $f(x)$ of each one of these sample solutions and record the average standard error $\bar{s}^2$.
3. Construct $g(\bar{s}^2)$ intervals or categories for the sample solutions.
4. Let $S_l$ be the frequency of the $l$th category in the sample set and $q_l = \dfrac{S_l}{M_0}$ be the relative frequency.
5. Let $i = 1$.

6. Fix $x_i = x_{ij}$, $j = 1, 2, \ldots, m(x_i)$.

7. Calculate the proportion $p_l$ of solutions that falls within each category and use this to calculate the corresponding entropy value:

$$E(i) = \sum_{l=1}^{g(\bar{s}^2)} q_l(i) \cdot I_l(i),$$

where

$$I_l(i) = -\sum_{l=1}^{g(\bar{s}^2)} p_{ij} \cdot \log_2(p_{ij}),$$

where $p_{ij}$ is the proportion of samples with $x_i = x_{ij}$.

8. If $i = n$, stop; otherwise, let $i = i + 1$ and go back to Step 6.

Because a high entropy value equates to high diversity, it is desirable to partition by fixing the lowest entropy dimension first. By using the ordering of entropy values of these dimensions to guide the partitioning, good solutions will be clustered together and the efficiency of the NP method will be improved.

## Sampling

When using the NP method, some samples should be collected from each subregion in each iteration. The samples determine the value of the promising index and determine which subregion is the next promising region. It is very important to collect high-quality solutions to create samples. In the traditional NP method, samples are collected randomly according to a uniform probability. To improve the efficiency of the NP method, a more advanced sampling method should be designed and proposed.

When solving some problems using the NP method, we can use mathematical programming to sample feasible solutions. There are two main approaches to implement this idea.

The first is to use the solutions of a relaxation of the original problem to bias the sampling. The idea behind this approach is that if a solution is similar to the optimal solutions for the relaxation problem, then the solution should be sampled with higher probability. For example, consider the following problem:

$$z = \min_x cx$$

$$Ax \leq b$$

$$x \in \{0,1\}^n$$

Solve the LP relaxation problem, and we can obtain the optimal solution $(x_1^{LP}, x_2^{LP}, \ldots, x_n^{LP})$. We can use the solution to bias the sampling. For example, for any $x_i$, the sampling distribution should be

$$P[x_i = 1] = x_i^{LP},$$

$$P[x_i = 0] = 1 - x_i^{LP}.$$

If a particular variable $x_i$ is close to a solution in the LP relaxation solution, then it should be selected with high probability in the sample solution.

The second approach can be used when the problem can be decomposed into two parts, one part that is easy to solve from a mathematical programming perspective and another part that is harder to solve. Because it is impossible to solve the entire problem using mathematical programming in a reasonable time, we can use mathematical programming and heuristics together to generate feasible solutions. The idea behind this approach is that we can use sampling to generate partial solutions that fix the hard part of the problem and then use a mathematical programming method to solve the partial problem to complete the solution. Because the mathematical programming output is optimal given the partial sample solution, it can be expected that there are higher-quality feasible solutions than the solutions that are obtained by sampling entirely. Thus, the efficiency and effectiveness of NP method is improved.

There are also some variants of these two basic approaches that are problem-specific. In Pi et al. (2008) and Chen et al. (2009), these approaches are applied to a local pickup and delivery problem and a facility location problem in the intermodal industry, and the results show that these approaches are very effective in solving these kinds of large-scale discrete optimization problems.

## Promising Index

In the NP methods, the promising index is used to determine the quality of each subregion. In each iteration, the next most promising region is selected based on the promising index. The promising index has a significant impact on the performance of NP method. In the standard NP framework, the promising index is defined as a minimum/maximum sample of each region. In Shi et al. (1999), the promising index is defined as the sample mean of each region. To increase the probability of making a correct move, that is, to improve the performance of the NP method, other promising index definitions will be discussed here.

When defining a more suitable promising index, the promising index should contain more information. Let $\Sigma$ denote the space of all valid regions, $\sigma$ a valid region and $\sigma \in \Sigma$, and $D_\sigma$ the set of feasible solutions in $\sigma$. Assume that we obtain a lower bound $z(\sigma)$ on the objective function for $\sigma \in \Sigma$, and by solving $\min_{x \in D_\sigma} cx$, the upper bound can be obtained. So the promising index can be calculated as

$$I(\sigma) = \alpha_1 z(\sigma) + \alpha_2 \min_{x \in D_\sigma} cx,$$

where $\alpha_1, \alpha_2 \in R$ are the weights given to the lower bound and upper bound, respectively. We can use LP relaxation, Lagrangian dual, or other methods to solve the problem and obtain the lower bound.

Based on extreme value theory, we can propose another promising index. When solving large-scale optimization problems, it is difficult to find a good feasible solution in the solution space. Using the extreme value theory, we can get the limiting results for large samples. Based on the extreme value theory, the limiting distribution of the minimum (or maximum) of a set of variables approaches one of the three extreme value distributions given as a large set of independent and identically distributed random variables (Coles, 2001). The three extreme value distributions are Gumbel, Fréchet, and Weibull distribution. For a minimization problem, the objective value of the global optimal solution is an implicit lower bound of the objective value, and extreme values follow the Weibull distribution.

At each iteration of the NP method, for every region $\sigma_i$, we can randomly sample a set of feasible solutions of which the objective values are independent and identically distributed. We can divide the feasible solutions into $n_i$ groups, each of which has the same number $l$ of samples. Let $\theta_{jk}$ denote the $k$th solution in the $j$th group, and let $y_{jk}$ denote the objective value of $\theta_{jk}$, that is, $y_{jk} = f(\theta_{jk})$, and $z_j$ the best objective value in the $j$th group, that is, $z_j = \min\{y_{j1}, \ldots, y_{jl}\}, \forall j = 1, 2, \ldots, n_i$. $y_{jk}$ is an individual sample and $z_j$ is a supersample. We assume that the individual samples $y$'s are independent and identically distributed. Based on the extreme value theory, if the sample size $l$ is large enough, the limit distribution of supersamples $z$'s follows a Weibull distribution since minimization problems are bounded from below (Derigs, 1985). The cumulative distribution function of Weibull distribution is

$$F(z; \alpha, \beta, \gamma) = \begin{cases} 0 & \text{for } z < \alpha \\ 1 - \exp\left[-\left(\dfrac{z - \alpha}{\beta}\right)^\gamma\right] & \text{for } z \geq \alpha' \end{cases}$$

where
  $\alpha$ is the location parameter
  $\beta$ ($>0$) is the scale parameter
  $\gamma$ ($>0$) is the shape parameter

We can use the maximum likelihood estimation (MLE) to fit the Weibull parameters. By finding the values that maximize the following log-likelihood function, we can obtain the MLEs of the three Weibull parameters $\tilde{\alpha}, \tilde{\beta}, \tilde{\gamma}$:

$$\ln L\left(z_1, \ldots, z_n; \alpha, \beta, \gamma\right) = n\left(\ln\gamma - \gamma\ln\beta\right) + (\gamma - 1)\sum_{i=1}^{n}\ln\left(z_i - \alpha\right) - \beta^{-\gamma}\sum_{i=1}^{n}\left(z_i - \alpha\right)^{\gamma},$$

where the parameters $\alpha$, $\beta$, $\gamma$ should satisfy the constraints $\alpha < \min z_i$, $\beta > 0, \gamma > 0$.

Because the Weibull distribution is bounded by the location parameter $\alpha$, we can use the estimator $\tilde{\alpha}$ to indicate the optimal value of each region. For each region, we can use the estimator $\tilde{\alpha}$ to be the promising index of the region. This kind of promising index contains more information than the traditional promising index, so the probability of making a correct move increases and the performance of NP method can be improved.

## Hybrid NP Algorithms

The inherent flexibility of the NP method makes it possible to incorporate other heuristic algorithms for generating good feasible solutions. As we discussed earlier in the chapter, it is very important to make a correct move when using the NP method, that is, move either to a subregion containing the global optimal solution or backtrack. In order to increase the probability of making a correct move, high-quality feasible solution should be generated or sampled from each region being considered. Because heuristic algorithms have the advantage of generating an improving sequence of feasible solutions, they can be combined with NP method resulting in hybrid NP algorithms. By incorporating these heuristic algorithms to generate feasible solutions, the effectiveness and efficiency of the NP method can be improved.

### *Greedy Heuristics in the NP Framework*

Greedy heuristics are the simplest heuristics in terms of implementation. There are two kinds of greedy heuristics: construction heuristics and improvement heuristics. Construction heuristics build up a single feasible solution by determining the values of the decision variables one by one. Improvement heuristics improve the feasible solution iteratively by making some relatively small change to the solution. So a sequence of improving solutions can be obtained by an improvement heuristics.

When using a greedy construction heuristic, the value of one of the variables is fixed using a single variable objective function, which correlates in some way with the original objective function defined for all variables. In most situations, the solution obtained by the use of the greedy construction heuristic is not the optimal solution though it can be obtained quickly and easily.

Greedy construction heuristics can be incorporated into the NP framework in two ways: to bias the sampling distribution or to define an intelligent partitioning. We will discuss how to bias the sampling distribution first.

Let $x_i, i = 1, 2, \ldots n$ denote the decision variables and the single variable objective function $g_i(x_i)$ corresponds to the variable $x_i$. Assume for simplicity, but without loss of generality, that the objective function is nonnegative, that is, $g_i(x_i) \geq 0, \forall x_i$. Then we can define a probability distribution

$$P\left[x_i = x_i^0\right] = \frac{g_i(x_i^{\max}) - g_i(x_i^0)}{g_i(x_i^{\max})},$$

where $x_i^{\max} = \arg\max_{x_i} g_i(x_i)$.

The intuition behind this method is very simple. Good values of each variable, that is, values that are closer to the optimal values, will be sampled with a higher probability. So the sampling distribution is biased and the quality of sample solutions can be improved, and the probability of making correct move increases.

Construction heuristics can also be incorporated into the NP framework in another way, that is, to define an intelligent partitioning. Construction heuristics always use an ordering of the importance of the variables to construct good solutions. This idea also can be used to impose a structure on the search space when partitioning in the NP method. The ordering of the importance of the variables should be determined first. When partitioning the solution space, the most important variables should be used at the top of the partitioning tree.

With greedy improvement heuristics, a good solution can be obtained by transforming a feasible solution. It can be incorporated into the NP framework to generate good feasible solutions quickly and the efficiency of the NP method can be improved.

### Random Search in the NP Framework

Random search methods usually use one or more local improvement moves at their core but use randomization to escape local optima and explore a larger part of the feasible region and so a sequence of good feasible solutions can be generated.

There are two basic approaches to utilizing random search methods for generating good feasible solutions in the NP methods, depending on if the random search is point-to-point or population-based. The point-to-point random search method starts from a single solution $x^i$ and explores a neighborhood $N(x^i) \subset X$ of this solution, then selects a candidate $x^c \in N(x^i)$ and either accepts it or rejects it. Population-based methods start with a set of solutions $D \subset X$ or the entire population and then improve the set. So in the NP framework, we can generate an initial population from each region being considered and improve these populations by search methods and get the final population from each region to choose the next most promising region.

*NP with Genetic Algorithm*

The GA is a population-based random search method based on the concept of natural selection. Starting from an initial population, a better population can be obtained by some operations such as reproduction, crossover, and mutation. The GA can be incorporated into the NP algorithm as follows: start with the sample solution sets from each region as an initial population and then apply GA to these sets to generate better solution sets, and then the promising index for each region can be estimated based on the final solution sets.

*NP with Tabu Search*

The tabu search method is a point-to-point method. The main idea behind the tabu search is defining certain solutions or moves to be tabu, that is, these solutions and moves cannot be visited as long as they are on the tabu list. After each iteration, the tabu list is updated, that is, the latest solution or move is added into the tabu list and the oldest solution or move is removed from the list. In every iteration, the best non-tabu solution from the neighborhood of the current solution is selected, even if it is worse than the current solution. This mechanism allows the search to escape from the local optima and the tabu list ensures that the search does not revert back. Similar to the hybrid NP/GA mentioned earlier, the tabu search method can be incorporated into the NP framework to generate good feasible sample solutions from each region and the promising index can be estimated from these solutions.

*NP with Ant Colony Optimization*

An interesting phenomenon has been observed by scientists that ants can find the shortest distance between their nests and food sources. Ant colony optimization algorithms imitate this natural behavior to find good solutions. To find and mark the good routes from their nests and food sources, ants can exchange information about good routes through a chemical substance called pheromone, which can accumulate for short routes and evaporate for long routes. So the ant colony optimization algorithms identify what solutions should be visited based on some pheromone values $\tau$, which are updated according to solution quality, and an evaporation rate $\rho < 0$. The ant colony optimization algorithm can be incorporated in the NP framework in similar ways to GA and tabu search, that is, using ant colony optimization algorithm to generate good feasible solutions for each region.

**Domain Knowledge in the NP Framework**

For many complicated, large-scale optimization problems in the real world, there is a lot of expertise or domain knowledge that can help to improve the performance of the real systems. If the domain knowledge can be properly used, the effectiveness and efficiency of the NP method will be improved.

Many methods to incorporate domain knowledge into the NP framework are problem-specific, so this section will briefly introduce two main strategies here. The first strategy is to use domain knowledge to determine the order of importance for decision variables and use this information to define an intelligent partition. The second strategy is that domain knowledge can be applied to generate partial solutions that will make the optimization problem easier to solve.

## Application of NP Method in the Chemical Laser Modeling Problem

In this section, we will discuss the application of the NP method in the chemical laser modeling problem, which demonstrates how the NP method can be used to solve optimization problems in laser and photonic applications.

### Problem Description

For the typical COIL system, the performance can be affected by many factors, such as the ratio of $I_2$ to $O_2$ flow rate and the total pressure of the flow. These effects can be analyzed with reliable computer models. The Blaze II model is one such model, and it has been used extensively to predict the performance of COIL systems. But the experimental results showed that it is very difficult to produce good agreement with experimental data for different mass flow rates. Based on previous research, to produce good agreement with data, five unknown parameters in the Blaze II model must be varied simultaneously until the best combination of these parameters is identified. The five parameters are yield, the amount of vapor, $k_{18}$ multiplier (a multiplier of the reaction rate constant $k_{18}$), jet expansion factor, and $DCM_{exp}$ (the exponent of diffusion coefficient multiplier ($DCM$), that is, $DCM$ is increased by the exponent: $DCM = (DCM)^{DCMexp}$).

In Carroll (1996), the range and increment of the five parameters are selected based on experimental data. The details are shown in Table 11.1.

**TABLE 11.1**

Parameter Space for the Blaze II Modeling

| Parameter | Yield | $H_2O$, Moles/s | $k_{18}$ Multiplier | Jet Expansion Factor, $J_{exp}$ | $DCM_{exp}$ |
|---|---|---|---|---|---|
| Range | 0.34–0.65 | 0.01–0.16 | 1–16 | 1.0–2.5 | 1.0–2.5 |
| Increment | 0.01 | 0.01 | 1 | 0.1 | 0.1 |

The objective of the chemical laser modeling problem is to find the best combination of the five parameters that best predict the experimental data. Let $\theta^1$ denote the first parameter, that is, yield; $\theta^2$ the second parameter, water vapor; $\theta^3$ the third parameter, $k_{18}$ multiplier; $\theta^4$ the fourth parameter, jet expansion factor; and $\theta^5$ the fifth parameter, $DCM_{exp}$. And let $\Theta$ denote the solution space. Define the objective function as

$$f(\theta^1,\theta^2,\theta^3,\theta^4,\theta^5) = \sum_m | (power)_{Blaze\,II} - (power)_{data} | ,$$

where

$(power)_{Blaze\,II}$ is the result obtained by the Blaze II model, which can be affected by the values of the five parameters

$(power)_{data}$ is the experimental result

$m$ is the number of Blaze II calculations being compared to $m$ different data points, and it must be set as an appropriate value

If $m$ is too big, much computational effort will be taken to make the Blaze II calculations, and if $m$ is too small, it cannot cover a broad range of flow rates and resulting powers. Based on the previous research, $m = 5$ is a good choice, because we want to find the combination of the five parameters that provide good overall agreement with experimental data, so the problem can be formulated as

$$\min_{\left(\theta^1,\theta^2,\theta^3,\theta^4,\theta^5\right)\in\Theta} f(\theta^1,\theta^2,\theta^3,\theta^4,\theta^5).$$

The solution space is very large, which makes the optimal solution hard to find. What makes the situation more complicated is that the solution space is highly multimodal, that is, there are a great deal of local optimal solutions in the solution space, so it is very likely to be trapped in local optimum for many algorithms. These difficulties can be overcome effectively by using the NP method.

### Partitioning

To develop an NP algorithm for the chemical laser modeling problem, it is first necessary to consider how to partition the solution space. Figure 11.1 shows a simple example of the partitioning. First, divide the solution space into 32 subregions by fixing $\theta^1$. And then further partition each subregion by fixing $\theta^2$. This procedure can be repeated until a singleton region is reached, when all five parameters are fixed.

To improve the performance of the NP algorithm, intelligent partitioning can also be designed and implemented based on the techniques that are introduced in the "Intelligent Partitioning" section.

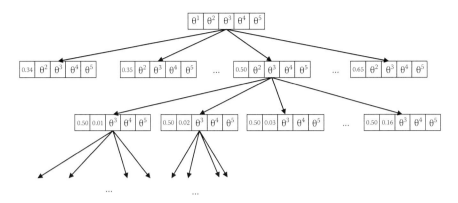

**FIGURE 11.1**
Partition of the solution space.

## Generating Feasible Solutions

Generally, there are two ways to generate feasible solutions. One way is to randomly select the samples from the feasible region with equal probability. Another way is to use biased sampling. Using domain knowledge or experimental experience, some solutions can be considered to be close to the optimal solution, so they should be selected with higher probability. By selecting samples in this way, high-quality solutions are more likely to be selected, so the performance of NP method will be improved.

## Defining the Promising Index

We can define the promising index as the minimum objective function value of samples in the subregion. Different promising index estimators can also be used to improve the efficiency of NP method.

When calculating the promising index, other methods, such as the meta-heuristics, can be used to increase the probability of making a correct move. As discussed in the "Hybrid NP Algorithms" section, meta-heuristic algorithms have the advantage of generating an improving sequence of feasible solutions. So after sampling feasible solutions, meta-heuristics can be used to improve the quality of these samples. Then the promising index can be calculated based on these better solutions. Carroll (1996) used a GA to solve the chemical laser modeling problem. The main steps of the GA used by Carroll (1996) are shown in the following:

Step 1: Generate the initial population randomly, that is, select $n$ individual samples randomly from the feasible solution space. Denote the initial population as $P(0)$ and set $i = 0$.

Step 2: Calculate the fitness for each individual in population $P(i)$ based on the following function:

$$Fitness = \frac{1}{\sum_{m} |(power)_{Balze\,II} - (power)_{data}|}$$

Step 3: Based on the fitness of each individual, select the most fitted individuals and let them be parents.

Step 4: Using the parents selected, perform the crossover operations and obtain the children.

Step 5: Select some children with a low probability and perform the mutation operation.

Step 6: Repeat Steps 3–5 until the new population size reaches $n$.

Step 7: If the terminating conditions are satisfied (such as $i = 20$), stop; otherwise, $i = i + 1$. Go back to Step 2 and continue the process.

In the algorithm, the population size $n$ can be set as different values, such as 100 or 200. The solutions can be encoded as chromosomes by different encoding schemes, such as floating point coding (the value of the chromosome is stored as a floating point number) and binary coding (the chromosome consists of a string of 0 and 1). The selection operation, crossover operation, and mutation operation can be implemented in many different ways. Some popular methods include expected value selection, tournament selection, single-point crossover, uniform crossover, jump mutation, and creep mutation. In Carroll (1996), computation experimental results showed that the tournament selection, uniform crossover, and jump and creep mutation schemes are the best combination of methods for the application of GA in the chemical laser modeling problem.

It is possible to design a hybrid NP algorithm based on the NP and GA methods. The hybrid NP algorithm is expected to perform better than other traditional methods. The main idea is that in each iteration we can use the sample solution set from each region as an initial population, then apply GA to these samples to generate better solution sets, and then estimate the promising index for each region based on the final solution sets. Let $\theta = (\theta^1, \theta^2, \theta^3, \theta^4, \theta^5)$ denote the solution. The details of the hybrid algorithm are shown in the following:

Step 1: Let $k = 1$, and let the entire solution space $\Theta$ be the most promising region $\sigma(k)$.

Step 2: In the iteration $k$, partition the most promising region $\sigma(k)$ into $M$ subregions $\sigma_1(k), \ldots, \sigma_M(k)$ (the partition scheme introduced in the "Partitioning" section can be used) and aggregate the complimentary region $\Theta \setminus \sigma(k)$ into one region $\sigma_{M+1}(k)$.

Step 3: Randomly generate $N_j$ (based on the size of the subregion $\sigma_j(k)$, $N_j$ can be chosen as one value from 1 to 100) sample solutions from each of the subregions $\sigma_j(k)$, $j = 1, 2, \ldots, M + 1$:

$$\theta_1^j, \theta_2^j, \ldots, \theta_{N_j}^j, j = 1, 2, \ldots, M+1$$

Let $j = 1$.

Step 4: Use the $N_j$ sample solutions from the region $\sigma_j(k)$ as the initial population; use the binary coding scheme to encode the solutions, for example, the solution $\theta = (\theta^1, \theta^2, \theta^3, \theta^4, \theta^5) = (0.35, 0.16, 2, 1.2, 1.5)$ can be represented as 00010′1000′0010′0011′0110 (0.35 is the second number in the range of [0.34,0.65] with increment 0.01, so it should be represented as 2 by binary digit, i.e., 00010; 0.16, 2, 1.2, and 1.5 are represented in the same way); and denote the population as $P_j^0(k)$, let $h = 0$.

Step 5: Calculate the fitness for each individual in population $P_j^h(k)$ by the following fitness function:

$$Fitness = \frac{1}{\sum_m |(power)_{Balze \, II} - (power)_{data}|}$$

Step 6: Based on the fitness of each individual, select the most fitted individuals and let them be parents. The expected value selection scheme may be used, that is, for each individual $l \in P_j^h(k)$, the expected probability of selection is based on the fitness of the individual divided by the total fitness of the individuals in population $P_j^h(k)$, that is, the probability that the individual is selected is $p_l = \dfrac{f_l}{\sum_{l \in P_j^h(k)} f_l}$.

Step 7: Use the single-point crossover scheme to perform the crossover operations on the parents and obtain the children. For example, the chromosomes of parents are 00010′1000′0010′0011′0110 and 01110′1010′0100′0101′0110, and the position between the 11th number and the 12th number is selected as the crossover point, so the chromosome of the child is 00010′1000′0000′0101′0110, in which the first 11 binary numbers are succeeded from the first individual of the parents and the last 10 binary numbers are from the second individual.

Step 8: Select some children with a low probability and perform the mutation operation, that is, change one binary number in the chromosome from 0 to 1 or from 1 to 0 with a probability, the binary number that is changed is selected randomly.

Step 9: Repeat Steps 6–8 until the new population size reaches $N_j$.

Step 10: If the terminating conditions are satisfied (such as $h = 10$), stop and decode the chromosomes and get the final sample solutions in the region $\sigma_j(k)$; otherwise, $h = h + 1$. Go back to Step 5 and continue the process.

Step 11: If $j = M+1$, go to Step 12; otherwise, $j = j+1$. Go back to Step 4 and use GA to improve the sample solutions in the next region.

Step 12: Calculate the corresponding performance values based on the individual samples in the final population of the region $\sigma_j(k), j = 1, 2, \ldots, M+1$ using the following function:

$$f(\theta) = \sum_m |(power)_{Blaze\ II} - (power)_{data}|$$

Step 13: For each region $\sigma_j, j = 1, 2, \ldots, M+1$, calculate the promising index as the best performance value within the region:

$$I(\sigma_j) = \min_{i=1,2,\ldots,N_j} f(\theta_i^j), j = 1, 2, \ldots, M+1$$

Step 14: To find the next most promising region, calculate the index of the region with the best performance value:

$$j_k^* = \arg\min_{j=1,\ldots,M+1} I(\sigma_j), j = 1, 2, \ldots, M+1$$

If more than one region is equally promising, the tie can be broken arbitrarily. If this index corresponds to a region that is a subregion of $\sigma(k)$, that is, $j_k^* \leq M$, then let this be the most promising region in the next iteration:

$$\sigma(k+1) = \sigma_{j_k^*}(k)$$

Otherwise, if the index corresponds to the complementary region, that is, $j_k^* = M+1$, backtrack to the superregion of the current most promising region (previous most promising region):

$$\sigma(k+1) = \sigma(k-1)$$

or backtrack to the entire solution space:

$$\sigma(k+1) = \Theta$$

Step 15: If the terminating conditions are satisfied (such as the most promising region contains only one solution), stop; otherwise, $k = k+1$. Go back to Step 2 and continue the process.

Except for GA, other meta-heuristics, such as simulated annealing, tabu search, and ant colony algorithm, can also be combined with the NP method to improve the efficiency.

Based on the techniques earlier, an NP algorithm for the chemical laser modeling problem can be developed. Using the algorithm, the difficulties of optimizing the parameters can be solved. And this method can also be used to design experiments to improve the performance of chemical laser systems.

## Conclusions

In this chapter, some operation and system optimization problems in laser and photonic applications are discussed. There are numerous similar large-scale and complex problems in the laser and photonic application areas that require powerful optimization frameworks and methods. A newly developed optimization framework—the NP method—is introduced. It has been shown that the NP method and its variants are very effective and efficient in solving complex real problems. It is expected that many optimization problems in the laser and photonic systems design and operation could also be solved via NP optimization framework.

## References

Andersen, H. E., Reutebuch, S. E., and Schreuder, G. F. 2002. Bayesian object recognition for the analysis of complex forest scenes in airborne laser scanner data. *ISPRS Commission III Symposium*, Graz, Austria, p. A-73.

Carroll, D. L. 1995. Modeling High-Pressure Chemical Oxygen-Iodine Lasers, *AIAA Journal* 33(8): 1454–1462.

Carroll, D. L. 1996. Chemical laser modeling with genetic algorithms, *AIAA Journal* 34(2): 338–346.

Chen, W., Pi, L., and Shi, L. 2009. Nested partitions and its applications to the intermodal hub locations problem. *Optimization and Logistics Challenges in the Enterprise*. Springer, New York, pp. 229–251.

Coles, S. 2001. *An Introduction to Statistical Modeling of Extreme Values*. Springer, New York.

Cormen, T. H., Leiserson, C. E., Rivest, R. L., and Stein, C. 1990. *Introduction to Algorithms*. MIT Press and McGraw-Hill, Cambridge, MA and New York.

Crowell, P. G. and Plummer, D. N. 1993. Simplifies chemical oxygen iodine laser (COIL) system model, *Proceedings of Intense Laser Beams and Applications*, 1871: 148–180.

Cvijović, D. and Klinowski, J. 1995. Tabu search: Approach to the multiple minima problem, *Science* 267(5198): 664–666.

De Boer, P. T., Kroese, D. P., Mannor, S., and Rubinstein, R. Y. 2005. A tutorial on the cross-entropy method, *Annals of Operations Research* 134: 19–67.

Derigs, U. 1985. Using confidence limits for the global optimum in combinatorial optimization, *Operations Research* 33(5): 1024–1049.

Dorigo, M. 1992. Optimization, learning and natural algorithms, PhD thesis, Politecnico di Milano, Milan, Italy.

Dorigo, M., Caro, G. D., and Gambardella, L. M. 1999. Ant algorithms for discrete optimization, *Artificial Life* 5(2): 137–172.

Glover, F. W. and Laguna, M. 1997. *Tabu Search*. Kluwer Academic Publishers, Boston, MA.

Goldberg, D. E. 1989. *Genetic Algorithm in Search, Optimization and Machine Learning*. Addison-Wesley, Reading, MA.

Ho, Y. C., Cassandras, C. G., Chen, C. H., and Dai, L. Y. 2000. Ordinal optimization and simulation, *Journal of Operations Research Society* 51: 490–500.

Ho, Y. C., Sceenivas, R. S., and Vakili, P. 1992. Ordinal optimization of DEDS, *Discrete Event Dynamic Systems: Theory and Applications* 2(1): 61–88.

Hong, L. J. and Nelson, B. L. 2006. Discrete optimization via simulation using compass, *Operations Research* 54: 115–129.

Kennedy, J. and Eberhar, R. 1995. Particle swarm optimization. In *Proceedings of 1995 IEEE International Conference on Neural Networks*, New York, Vol. 4, pp. 1942–1948.

Kirkpatrick, S., Gelatt, C. D., and Vecchi, M. P. 1983. Optimization by simulated annealing, *Science* 220(4598): 671–680.

Lemarechal, C. 2001. Lagrangian relaxation. In M. Jünger and D. Nadder (eds.), *Computation Combinatorial Optimization, Lecture Notes in Computer Science*. Vol. 2241. Springer-Verlag, Berlin, Germany.

Mitchell, M. 1998. An Introduction to Genetic Algorithms. MIT Press, Cambridge, MA.

Nemhauser, G. L. and Wolsey, L. A. 1999. *Integer and Combinatorial Optimization*. Wiley Series in Discrete Mathematics and Optimization, New York.

Pi, L., Pan, Y., and Shi, L. 2008. Hybrid nested partitions and mathematical programming approach and its applications, *IEEE Transactions on Automation Science and Engineering* 5(4): 573–586.

Poli, R., Kennedy, J., and Blackwell, T. 2007. Particle swarm optimization, *Swarm Intelligence* 1(1): 33–57.

Reinhart, G., Munzert, U., and Vogl, W. 2008. A programming system for robot-based remote-laser-welding with conventional optics, *CIRP Annals—Manufacturing Technology* 57(1): 37–40.

Rubinstein, R. Y. and Kroese, D. P. 2004. *The Cross-Entropy Method: A Unified Approach to Combinatorial Optimization, Monte-Carlo Simulation, and Machine Learning*. Springer, New York.

Shi, L. and Ólafsson, S. 2000. Nested Partitions method for global optimization, *Operations Research* 48(3): 390–407.

Shi, L. and Ólafsson, S. 2008. *Nested Partitions Method, Theory and Applications*. Springer, New York.

Shi, L., Ólafsson, S., and Sun, N. 1999. New parallel randomized algorithms for the traveling salesman problem, *Computers and Operations Research* 26(4): 371–394.

van Laarhoven, P. J. M. and Aarts, E. H. L. 1987. *Simulating Annealing: Theory and Applications*. Kluwer Academic Publisher, Norwell, MA.

Wolsey, L. A. 1998. *Integer Programming*. Wiley Series in Discrete Mathematics and Optimization, New York.

Yan, D. and Mukar, H. 1992. Stochastic discrete optimization, *SIAM Journal of Control and Optimization* 30: 594–612.

# 12

## Network Models and Operations of Laser and Photonics Systems

Xin W. Chen

### CONTENTS

Laser and photonics systems have been developed for new medical processes and new manufacturing processes, with promising new solutions. They enable new techniques of better and faster communications. They are used for material processing to create unique characteristics for more efficient and economical energy systems, such as solar energy. The US Department of Energy is striving to deliver laser fusion power stations with Laser Inertial Fusion Energy (LIFE). The integration of all the technologies required for a power station is planned for the mid-2020s. Laser technology is an inseparable part of computer-created virtual realities and advanced personalized learning that aim at furthering engineering's contributions to the joy of living. A major motivation to integrate laser technology and photonic systems is the inherent need of scientific discoveries to investigate the vastness of the cosmos or the inner intricacy of life and atoms. The laser and photonics technologies must be seamlessly integrated at the system and network level in order to fully achieve their promise. This chapter addresses the emerging approaches and challenges in network models and operations of laser and photonics systems.

## Collaboration Network for Laser and Photonic Systems

As laser technology and various photonic systems continue to mature, their coherence and integration with humans and other systems and tools pose significant challenges. For instance, current laser technology is limited by the useful transmission range of nonlinear optic materials; longer wavelength materials have been developed but are incompatible with mature laser systems. To fully explore their potential, scientific and engineering challenges must be addressed (Li, 2010). Some of the challenges stem from the development of laser technology, while others are related to its implementation and integration in various manufacturing and service systems.

### Challenges

#### *System-Oriented Structure*

The first challenge is the local versus networked system-oriented structure. Hundreds of thousands of laser- and photonic-based systems have already been developed and more advanced ones are emerging. In certain cases, it is costly to maintain these stand-alone systems and an untold number of system-to-system interfaces, both within local facilities and between facilities. The local system-oriented structure is challenged by the need to rationally scale up application solutions, network them on a broader basis where it is useful, and eliminate unnecessary duplications. A service-centric platform is needed, which enables collaboration between product and service providers and between providers and customers.

#### *Inherent Sequential Process*

The second challenge is the inherent sequential process. Even with concurrent design, product development must undergo several sequential, repetitive steps, for example, design, prototyping, and testing. The sequential process is vulnerable to changes in a product life cycle and in technology innovation. For instance, a relatively small design change discovered during the testing stage can be costly and may not be accommodated. A network-centric process will increase parallelism and help flatten the manufacturing, development, and integration process. Innovative and flexible aspects of the laser and photonic technologies would promote parallelism, and it is important to carefully explore this opportunity from a network-centric perspective.

### Integration of Laser-Based and Traditional Processes

The third challenge is the integration of laser- and photonic-based processes with traditional manufacturing, communication, and health-care processes. Both types of processes can be applied to a variety of products and services. In general, laser- and photonic-based manufacturing is more energy efficient and provides better quality than traditional manufacturing. In practice, however, many objectives such as affordability, process time, safety, system sustainability and vulnerability, and recycling must also be considered. While emerging laser and photonic systems are still maturing, the characteristics of certain laser-based processes are yet difficult to predict. Therefore, to overcome such difficulties, the integration needs to be planned in an open architecture (Figure 12.1). For instance, the integration of different manufacturing processes is emerging as being enabled by an open manufacturing environment supported by a library of physical processes, simulators, and virtual manufacturing tools.

### System Modeling and Analysis

Successful integration of laser and photonic technologies into manufacturing and service systems through the open architecture (Figure 12.1) gives rise to new system models that may not yet have been investigated. New photonics-based processes can introduce different behaviors and interactions within and across processes and systems. For instance, the closer and faster integration of information and decision flows with processing steps and control are becoming feasible. Therefore, the fourth challenge is to define the system models emerging from the photonics and laser technologies, thereby developing tools and techniques of optimally designing and controlling the systems. Though photonics technologies have a great potential in enhancing the performance of systems in terms of cost, timeliness, and quality, if these processes are not integrated from a broad systems perspective, the promise they can provide would be diminished. On the other hand, successful integration at the system and network level will prove the usefulness of the technology, enabling its fruitful diffusion into society at an accelerated rate.

### Emerging Approaches

Figure 12.1 depicts a service- and network-centric collaboration platform for laser and photonic systems. A hub-enhanced collaborative intelligence

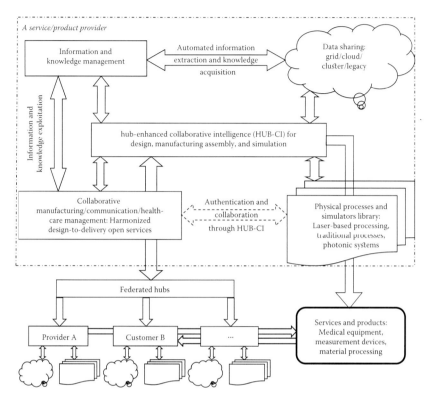

**FIGURE 12.1**
Integration architecture (framework) of a collaboration network for laser- and photonic-based services and products.

(HUB-CI) integrates systems for a provider and aligns their functions to services requested by customers. Service and product providers and customers collaborate with each other through a network of cyber-hubs. The objective is to establish a network of sustainable, resilient, and collaborative laser and photonic technology providers and customers for better solution of societal problems. Table 12.1 summarizes the five key components of the collaboration network for laser- and photonic-based services and products.

## Network Models

Interactions, integration, and collaboration of laser and photonics systems must be supported by appropriate network models. Network structures

**TABLE 12.1**

Key Components of the Collaboration Network for Laser- and Photonic-Based Services and Products (Figure 12.1)

| | Data Sharing | Information and Knowledge Management | HUB-CI | Collaborative Manufacturing/ Communication/Health-Care Management | Physical Processes and Simulators Library |
|---|---|---|---|---|---|
| Objectives | Provide basic services and computing resources for data and simulation sharing. | Build a fusion center that integrates CAD, CAM, and process planning software systems to enhance design for manufacturability and enable rapid manufacturing. | Integrate services to support all the steps pertinent to laser manufacturing. | Provide a collaborative workspace where people and computational agents can share data and analyses synchronously or asynchronously. | Allow engineers to predict and select desirable processing parameters without going through expensive and time-consuming experiments with the actual hardware. |
| Example functions | Cloud computing; federated hubs. | Mechanisms and application programming interfaces to manage manufacturing expertise. | Creating ad hoc networks of best-matching manufacturing entities. | Service architecture and workflow enabling seamless data and simulation sharing. | A virtual manufacturing system integrating computer simulators for laser-based manufacturing processes. |
| Major challenges | Data integrity; authentication. | Intellectual property control; conflict of interest. | Duplicate services; incompatible protocols. | Errors in design and manufacturing; conflicts between materials and processes. | Model verification and validation. |
| Network structures | Distributed clusters connected through hubs. | Distributed clusters connected through hubs. | Heterogeneous networks of designers, engineers, and software systems. | Homogeneous network with virtual point-to-point connection between many pairs of collaborative entities. | Three-layer network of laser processes: direct deposition and machining, joining, and surface enhancement. |

illustrated in Table 12.1 indicate that there are at least three types of emerging networks:

1. Knowledge networks, including data and information networks, for example, the American Society for Laser Medicine and Surgery, the Institute of Electrical and Electronics Engineers (IEEE) Photonics Society, and the Laser Institute of America. In these networks, each member is a data or knowledge source. Some sources are connected to each other. They may form a random network, a scale-free network, or other types of networks. When two members are connected, there are also different types of relationships. For instance, members may form a subgroup, also known as a circle in a knowledge network, and each member in the group is aware of what knowledge other members in the same group have or do not have. Knowledge networks play an important role in disseminating knowledge, acquiring expertise, and stimulating new discoveries that are essential to the learning and research of laser and photonics technology.

2. Manufacturing and logistics networks that comprise service providers and customers. Providers are characterized by the type of services they deliver, price, quantity, lead time, due-date performance, and other features for which customers are concerned. Traditionally, manufacturing and logistics networks have been hierarchical networks in which a major provider delivers services or products to layers of customers or multiple layers of providers interact with a major customer. The emerging trend is to flatten the network to keep up with innovations in laser and photonics technology. A node in the network may be both a customer and a provider and interacts directly with many other nodes; the manufacturing and logistics network does not have to be a strict bipartite network (Newman et al., 2001).

3. Cyber-enabled collaboration networks, for example, the HUB-CI (Seok and Nof, 2011) and HUBzero® (McLennan and Kennell, 2010). The Internet and World Wide Web (WWW) enable data and information sharing, which is the foundation for various collaboration tools that advance laser and photonics technology. The web pages on the WWW and the underlying physical structure of the Internet are well-known scale-free networks (Albert et al., 1999). Structures of cyber-enabled collaboration networks need to be thoroughly studied to help design and engineer more effective and efficient collaboration networks.

Network models capture the structure of laser and photonics systems. These models may be analyzed to predict system behaviors and enable robust collaborative control. The study of network models has had a long history.

The classic model, random network, was first discussed in the early 1950s (Solomonoff and Rapoport, 1951) and was rediscovered and analyzed in a series of papers published in the late 1950s and early 1960s (Erdös and Rényi, 1959, 1960, 1961). The degree of a node, $d$, is the number of links connected to it. In a random network with $n$ nodes and probability $p$ to connect any pair of nodes, the maximum number of links in the network is $\frac{1}{2}n(n-1)$. The probability $p_d$ that a node has degree $d$ is $\binom{n-1}{d}p^d(1-p)^{n-1-d}$, which is also the fraction of nodes in the network that have degree $d$. The mean degree $\bar{d} = (n-1)p$. One of the important properties of a random network is phase transition or bond percolation (Solomonoff and Rapoport, 1951; Newman et al., 2006; Angeles Serrano and De Los Rios, 2007). There is a phase transition from a fragmented random network for the mean degree $\bar{d} \le 1$ to a random network dominated by a giant component for the mean degree $\bar{d} > 1$.

Two other network models have been studied extensively and capture the topology of many laser and photonics systems; these are (1) scale-free network (Price, 1965; Albert et al., 1999; Barabasi and Albert, 1999; Broder et al., 2000) (in a scale-free network, the probability $p_d$ that a node has $d$ degree follows a power law distribution, i.e., $p_d = cd^{-\gamma}$, where $\gamma$ is between 2.1 and 4 for real-world systems (Barabasi and Albert, 1999) and $c$ is a normalization factor) and (2) Bose–Einstein condensation network (Bianconi and Barabasi, 2001a), which was discovered in an effort to model the competitive nature of networks. A fitness model (Bianconi and Barabasi, 2001b) was proposed to assign a fitness parameter $\eta_i$ to each node $i$. A node with higher $\eta_i$ has higher probability to obtain links. $\eta_i$ is randomly chosen from a distribution $\rho(\eta)$.

Another network model that has been gaining attention recently is hierarchical network, which describes many systems in biology and supply chains. Biological networks can be decomposed into modular components that recur across and within organisms. The top level in a hierarchical biological network is comprised of interacting regulatory motifs consisting of groups of 2–4 components such as proteins or genes (Lee et al., 2002; Shen-Orr et al., 2002; Zak et al., 2003). Motifs are small subnetworks and pattern-searching techniques can be used to determine the frequency of occurrence of these simple motifs (Shen-Orr et al., 2002). At the lowest level in this hierarchy is the module describing transcriptional regulation (Barkai and Leibler, 2000). In supply chains, a customer may have multiple suppliers, some of which may have their suppliers. The top level in a hierarchical supply chain is comprised of customers only whereas the lowest level is comprised of suppliers only. All other levels in the middle are both customers and suppliers. The Bose–Einstein condensation network is a special case of the hierarchical network, in which there are only two levels, the fitness node and all other nodes. Analytical models need to be developed to describe the structure and reveal properties of hierarchical networks. Table 12.2 summarizes the four network models with examples.

**TABLE 12.2**

Network Models and Examples

| | Random | Scale-Free | Bose–Einstein Condensation | Hierarchical |
|---|---|---|---|---|
| Degree distributions | $\binom{n-1}{d} p^d (1-p)^{n-1-d}$ | $cd^{-\gamma}$ | The fittest node acquires a finite fraction of links | To be developed |
| Structures | Homogeneous | Hierarchical with a few hubs | Winer-takes-all | Hierarchical with a hub |
| Structure examples | | | | |
| Real-world examples | US highway | Power grid | HUBzero® | Supply chain |

*Source:* Chen, X.W. and Nof, S.Y., *Automatica*, 48(5), 770, 2012a.

# Interactive Prognostics and Diagnostics Network for Errors and Conflicts

A universal challenge (Table 12.1) for integration of laser- and photonic-based services and products is to effectively and efficiently detect, prevent, and resolve conflicts and errors (CEs) that emerge within and between networked systems and human operators. A conflict is an inconsistency between collaborative units' (Co-Us') goals, dependent tasks, associated activities, or plans of sharing resources (Chen and Nof, 2010). A conflict occurs whenever an inconsistency between two or more units in a system occurs. An error is any input, output, or intermediate result that does not meet system specification, expectation, or comparison objective (Klein, 1997). Other terms such as fault, failure, exception, and flaw are used to describe CEs. Table 12.3 shows typical CEs in laser- and photonic-based manufacturing.

CE prevention and detection (CEPD) has been studied in many applications where centralized algorithms are often used. Centralized algorithms have a central unit that performs CEPD sequentially and controls system information. Decentralized algorithms have distributed agents that prevent and detect CEs in parallel through collaboration. Centralized algorithms have been more prevalent because CEs are often detected using system output.

**TABLE 12.3**

CE Examples in Laser- and Photonic-Based Services and Products

| Conflict | Error |
|---|---|
| Output of a virtual manufacturing process cannot be duplicated in real manufacturing environment. | Low surface quality due to incorrect frequency used in laser penning |
| Parameters retrieved from a cloud storage cannot be interpreted to control processes. | Communication interruption due to failures in optic fibers |

*Source:* Chen, X.W. and Nof, S.Y., *Automatica*, 48(5), 770, 2012a.

Detection algorithms identify CEs after they occur whereas prevention algorithms identify potential CEs before they occur. Two main challenges of CEPD are (1) system modeling for integrated prevention and detection of CEs and (2) the use of relationships between constraints for prevention and detection. Table 12.4 summarizes various CEPD methods.

A conflict or an error is prevented or detected if and only if a predefined constraint is not satisfied (Chen and Nof, 2012c). A constraint describes what must occur with a Co-U or a collaborative network (Co-net) of Co-Us by a specific time. $u_i(t)$ represents Co-U $i$ in a system at time $t$. $i$ is the index of Co-Us. Let $con_r(t)$ denote constraint r in the system at time $t$. $r$ is the index of constraints. $u_i(t)$ needs to satisfy one or more constraints. $n_r(t)$ is Co-net $r$ in the system at time $t$. $r$ is the index of both Co-nets and constraints. Constraints and Co-nets have a one-to-one relationship. Each $con_r(t)$ has one and only one corresponding $n_r(t)$. Each $n_r(t)$ satisfies one and only one $con_r(t)$.

Constraints are necessary for CEPD. For instance, faulty components must be identified when laser equipment fails. Either centralized or decentralized algorithms may be applied to identify faulty components. To determine if a component is faulty, however, desired states or conditions of each component in the laser equipment must be defined. These states or conditions are represented by constraints, which might be related to each other. For instance, a failure in component X might indicate that component Y has a failure. A constraint network is comprised of nodes, each of which represents a constraint, and links, each of which represents the relationship between constraints.

A constraint describes the first-order dependence between Co-Us. Constraints are related through Co-Us. Relationships between constraints are relationships between CEs and reflect high-order dependences between Co-Us. Two constraints may be dependent or independent. There are at least two types of dependences: inclusive and mutually exclusive. $con_{r_1}(t_1)$ is inclusive of $con_{r_2}(t_2)$, that is, $con_{r_1}(t_1) \subset con_{r_2}(t_2)$ if the probability that $con_{r_2}(t_2)$ is not satisfied is 1 given that $con_{r_1}(t_1)$ is not satisfied. $con_{r_1}(t_1) \oplus con_{r_2}(t_2)$ indicates $con_{r_1}(t_1)$ and $con_{r_2}(t_2)$ are mutually exclusive, that is, the probability that $con_{r_2}(t_2)$ is not satisfied is 0 given that $con_{r_1}(t_1)$ is not satisfied. A constraint $con_{r_1}(t_1)$ is independent of any other constraint if the probability that $con_{r_1}(t_1)$

**TABLE 12.4**

Summary of CEPD Methods

| Comparison | CEPD | | | | |
|---|---|---|---|---|---|
| Methods | Analytical | Data-Driven | Knowledge-Based | Diagnostics Algorithms | Decentralized Algorithms |
| Error detection | Yes | Yes | Yes | Yes | Yes |
| Conflict detection | No | No | No | No | Yes |
| Diagnostics | No | No | Yes | Yes | Yes |
| Prediction/prevention | Yes | Yes | Yes | No | Yes |
| Centralized/decentralized | Centralized | Centralized | Centralized | Centralized | Decentralized |
| Main advantages | Accurate and reliable | Can process large amount of data | Do not require detailed system information | Accurate and reliable | Accurate, reliable, and efficient |
| Main disadvantages | Require mathematical models that are often not available | Rely on the quantity, quality, and timeliness of data | Results are subjective and may not be reliable | Time-consuming for large systems | Require network modeling and analysis |
| Bibliographies | Raich and Cinar (1996), Gertler (1998), Chiang et al. (2001) | | | Pattipati and Alexandridis, (1990), Pattipati and Dontamsetty (1992), Deb et al. (1995), Raghavan et al. (1999a,b), Shakeri et al. (2000), Tu and Pattipati, (2003), Tu et al. (2003) | Chen and Nof (2007, 2009, 2010, 2012a–c) |

*Source:* Adapted from Table 30.5 in Chen and Nof, 2009.

is not satisfied given that another constraint is not satisfied is the same as the probability that $con_{r_n}(t_1)$ is not satisfied.

Let $E(u_{r,i}(t))$ and $C(n_r(t))$ represent an error and a conflict, respectively. Figure 12.2 depicts the centralized CEPD algorithm, which includes three parts: detection, diagnostics, and prognostics. Dashed lines "- - -" in Figure 12.2 indicate the border of each part. The algorithm starts with checking the first-in-first-out (FIFO) queue. If there are tasks in the queue, the algorithm executes

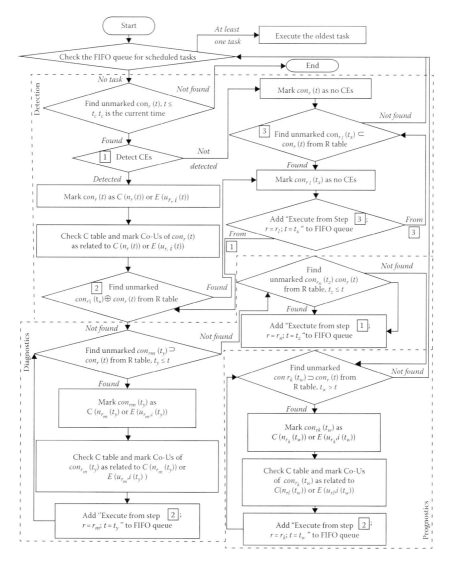

**FIGURE 12.2**
Centralized CEPD algorithm. (From Chen, X.W. and Nof, S.Y., *Eng. Optim.*, 44(7), 821, 2012c.)

the oldest task, the task that enters the queue the earliest. If there is no task in the queue, the algorithm moves to the detection part. Two tables, constraint (C) table and constraint relationship (R) table, are used in both the centralized and decentralized algorithms. The C table describes constraints and specifies first-order dependences between Co-Us. The R table describes relationships between constraints and specifies high-order dependences between Co-Us.

## Detection

In the detection part (Figure 12.2), the algorithm first randomly identifies a constraint that needs to be satisfied before or at the current time $t_c$. If no constraint is found, the algorithm ends, indicating all constraints have been checked and no tasks are in the FIFO queue. If a constraint $con_r(t)$ is found, detection techniques are applied (step $\boxed{1}$) to determine if $con_r(t)$ is satisfied. If $con_r(t)$ is not satisfied, that is, a CE is detected, the algorithm marks $con_r(t)$ and its related Co-Us. The next step, step $\boxed{2}$ in the algorithm is designed to find all constraints that are mutually exclusive with $con_r(t)$. According to the exclusive relationship between constraints, all constraints that are mutually exclusive with $con_r(t)$ are satisfied and do not have CEs because $con_r(t)$ is not satisfied. Step $\boxed{2}$ repeats until all multiple exclusive constraints are found. After that, the algorithm moves to the diagnostics part.

If a CE is not detected at step $\boxed{1}$, the algorithm marks $con_r(t)$ and moves to step $\boxed{3}$, which is designed to find all constraints that are inclusive of $con_r(t)$. Because $con_r(t)$ is satisfied and does not have a CE, all constraints that are inclusive of $con_r(t)$ must not have CEs. Step $\boxed{3}$ repeats until all constraints that are inclusive of $con_r(t)$ are found. After that, the algorithm goes back to check the FIFO queue.

The detection part of the algorithm detects CEs, marks constraints, and identifies constraints that do not have CEs through constraint relationships. To mark constraints that do not have CEs reduces detection time. The step that has two branches going to either step $\boxed{2}$ or $\boxed{3}$ is designed to reduce detection time. This step guarantees that before the next constraint is detected for CEs, all constraints that have been marked as not having CEs go through step $\boxed{2}$ or $\boxed{3}$ to find their related constraints that do not have CEs. Depending on whether a CE is detected in the detection part, the algorithm moves to two different directions. If a CE is detected, the algorithm moves to the diagnostics part to find causes of the CE. If a CE is not detected, the algorithms goes back to check the FIFO queue.

## Diagnostics

The diagnostics part identifies causes of CEs detected in the detection part. The algorithm in this part starts with finding constraints of which $con_r(t)$ is inclusive $(con_{r_m}(t_y) \supset con_r(t), t_y \leq t)$. Because $C(n_r(t))$ or $E(u_{r,i}(t))$ exists, constraints of which $con_r(t)$ is inclusive are not satisfied. Since $t_y \le t, C(n_{r_m}(t_y))$ or $E\left(u_{r_m,i}(t_y)\right)$ is the cause of $C\left(n_r(t)\right)$ or $E\left(u_{r,i}(t)\right)$. After all these constraints are found and marked, the algorithm moves to find constraints that are inclusive of $con_r(t)$ $(con_{r_n}(t_z) \subset con_r(t), t_z \le t)$. If $con_{r_n}(t_z)$ is found, it will go through the algorithm from step $\boxed{1}$ to be detected for CEs. $con_{r_n}(t_z)$ may or may not be satisfied. The objective is to find potential causes of $C\left(n_r(t)\right)$ or $E\left(u_{r,i}(t)\right)$. After all potential causes are found, the algorithm moves to the prognostics part.

## Prognostics

The prognostics part identifies constraints of which $con_r(t)$ is inclusive $(con_{r_n}(t_w) \supset con_r(t), t_w > t)$. Similar to the reasoning in the diagnostics part, constraints of which $con_r(t)$ is inclusive are not satisfied. Since $t_w > t, C\left(n_r(t)\right)$ or $E\left(u_{r,i}(t)\right)$ is the cause of $C\left(n_{r_k}(t_w)\right)$ or $E\left(u_{r_k,i}(t_w)\right)$ that has not occurred. After all future CEs have been found and marked, the algorithm goes back to check the FIFO queue.

The first four steps in the diagnostics part (lower left part in Figure 12.2) are essentially the same as the four steps in the prognostics part (lower right part in Figure 12.2). The only difference is $t_y \le t$ for diagnostics and $t_w > t$ for prognostics. The purpose of presenting them separately is to illustrate the difference between diagnostics and prognostics. Detection, diagnostics, and prognostics must be integrated for CEPD.

The centralized CEPD algorithm utilizes constraint relationships for effective CEPD. It is a sequential algorithm and requires a large database for systems with many constraints. An alternative is to use intelligent agents to perform CEPD in parallel. In the decentralized CEPD algorithm (Figure 12.3), each constraint employs a prevention and detection agent (PDA) that detects, diagnoses, and prognoses CEs. PDAs send and receive *CE* and *No CE* messages for prognostics and diagnostics. Each PDA controls information for one constraint. PDAs initiate CEPD in three events: (1) $PDA_r(t)$ receives a *CE* message; (2) current time $t_c \ge t$, the time $con_r(t)$, must be satisfied; and (3) $PDA_r(t)$ receives a *No CE* message. PDAs send and receive messages for collaborative CEPD. This differentiates the decentralized algorithm from the centralized algorithm. Other steps in Figure 12.3 are part of the steps in Figure 12.2. The decentralized algorithm is less complex than the centralized algorithm because prognostics and diagnostics are performed by activating collaborative PDAs.

An interactive prognostics and diagnostics network (IPDN) is comprised of nodes, each of which represents a constraint, and links that describe constraint relationships. Two types of links, inclusive and exclusive links, are used to describe the two types of relationships, inclusive and exclusive, respectively. The inclusive link has directions, for example, $con_{r_1}(t_1) \subset con_{r_2}(t_2)$ is different from $con_{r_1}(t_1) \supset con_{r_2}(t_2)$, whereas the exclusive link is undirected, for

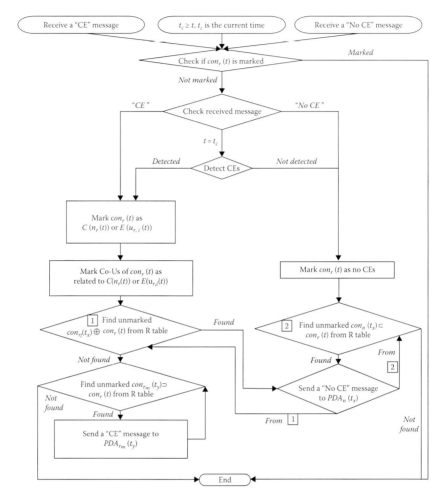

**FIGURE 12.3**
Decentralized CEPD algorithm. (From Chen, X.W. and Nof, S.Y., *Eng. Optim.*, 44(7), 821, 2012c.)

example, $con_{r_1}(t_1) \oplus con_{r_2}(t_2)$ is the same as $con_{r_2}(t_2) \oplus con_{r_1}(t_1)$. An IPDN may have both directed and undirected links. Figure 12.4 shows how relationships between constraints are represented by links.

Various network models described in the *Network Model* section, for example, random, scale-free, Bose–Einstein condensation, and hierarchical networks, may be used to model the structure of IPDNs. To analyze the CEPD algorithms, four performance metrics are defined:

1. *CA* (coverage ability): the quotient of the number of CEs that are detected, diagnosed, and prognosed, divided by the total number of CEs. $0 \leq CA \leq 1$.

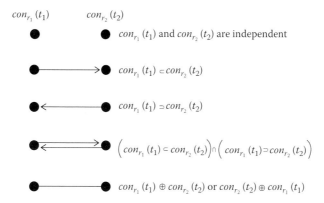

**FIGURE 12.4**
Relationships in an IPDN. (From Chen, X.W. and Nof, S.Y., *Eng. Optim.*, 44(7), 821, 2012c.)

2. *PA* (prognostics ability): the quotient of the number of CEs that are prognosed, divided by the total number of CEs. $0 \leq PA \leq 1$.
3. *TT* (total CEPD time): the time between the CEPD algorithm stops and begins. *TT* is the time required to check all nodes for CEs.
4. *TD* (total CE damage): the total damage caused by CEs, which is the sum of the damage caused by all CEs.

Tables 12.5 and 12.6 summarize the results of 1200 experiments conducted using AutoMod (AutoMod 1998–2003) on IPDNs, each of which has 150 nodes. $\overline{CA}$, $\overline{PA}$, $\overline{TT}$, and $\overline{TD}$ are the mean of *CA*, *PA*, *TT*, and *TD*, respectively. DR-IPDN, UR-IPDN, DF-IPDN, UF-IPDN, DB-IPDN, and UB-IPDN represent directed random IPDN, undirected random IPDN, directed scale-free IPDN, undirected scale-free IPDN, directed Bose–Einstein condensation IPDN, and undirected Bose–Einstein condensation IPDN, respectively.

**TABLE 12.5**

Performance of the Centralized CEPD Algorithm

| Network | $\overline{TT}$ | $\overline{CA}$ | $\overline{PA}$ | $\overline{TD}$ |
|---|---|---|---|---|
| DR-IPDN | 50 | 0.14 | 0.03 | 35,929 |
| UR-IPDN | | 0.26 | 0.08 | 29,893 |
| DS-IPDN | | 0.10 | 0.02 | 36,438 |
| US-IPDN | | 0.16 | 0.04 | 34,324 |
| DB-IPDN | | 0.23 | 0.07 | 29,828 |
| UB-IPDN | | 0.29 | 0.10 | 26,862 |

*Source:* Adapted from Figure 13 in Chen, X.W. and Nof, S.Y., *Automatica*, 48(5), 770, 2012a.

**TABLE 12.6**

Performance of the Decentralized
CEPD Algorithm

| Network | $TT$ | $\overline{CA}$ | $\overline{PA}$ | $\overline{TD}$ |
|---------|------|------|------|--------|
| DR-IPDN | 49.95 | 0.44 | 0.12 | 14,524 |
| UR-IPDN | 49.82 | 0.44 | 0.14 | 14,271 |
| DS-IPDN | 49.63 | 0.44 | 0.13 | 14,440 |
| US-IPDN | 41.54 | 0.45 | 0.17 | 14,329 |
| DB-IPDN | 49.96 | 0.43 | 0.10 | 14,700 |
| UB-IPDN | 49.89 | 0.43 | 0.13 | 14,503 |

*Source:* Adapted from Chen, X.W. and
Nof, S.Y., *Automatica*, 48(5), 770,
2012a, Figure 14.

For all six IPDNs, the decentralized algorithm outperforms the centralized algorithms with smaller $\overline{TT}$ and $\overline{TD}$ and larger $\overline{CA}$ and $\overline{PA}$. The performance of the centralized algorithm (Table 12.5) is DB-IPDN > DR-IPDN > DS-IPDN and UB-IPDN > UR-IPDN > US-IPDN in terms of $\overline{CA}$ and $\overline{PA}$ and DB-IPDN < DR-IPDN < DS-IPDN and UB-IPDN < UR-IPDN < US-IPDN in terms of $\overline{TD}$. The performance of the decentralized algorithm (Table 12.6) is not sensitive to the network structure. The decentralized algorithm performs slightly better over the scale-free IPDNs than the respective random IPDNs and Bose–Einstein condensation IPDNs because of the existence of large components in the scale-free IPDNs, which improve the performance of the decentralized but not the centralized algorithm.

Table 12.5 indicates that the centralized algorithm performs better over the undirected than the corresponding directed IPDN with larger $\overline{CA}$ and $\overline{PA}$ and smaller $\overline{TD}$. The centralized algorithm has the same $\overline{TT}$ over all IPDNs. The decentralized algorithm has larger $\overline{PA}$ and smaller $\overline{TD}$ over the undirected than the corresponding directed IPDN. The $\overline{TT}$ and $\overline{CA}$ of the decentralized algorithm are the same for the corresponding undirected and directed IPDNs. These results together with the analytical studies (Chen and Nof, 2012a,b,c) suggest the following:

1. Use the decentralized CEPD algorithm to minimize the $TT$ and damage and maximize the CA and preventability.

2. The performance of the centralized and decentralized algorithms improves and converges as the size of the giant component, the largest component in an IPDN, increases in the undirected networks.

3. The performance of the CEPD algorithms improves when IPDNs and algorithms are aligned. The centralized algorithm performs the best over the Bose–Einstein condensation IPDNs and performs better over the random IPDNs than scale-free IPDNs. The decentralized algorithm performs the best over the scale-free IPDNs. The

order from the most centralized to the most decentralized is Bose–Einstein condensation IPDNs, random IPDNs, and scale-free IPDNs. To optimize the performance of the centralized algorithm, it shall be applied to the most centralized IPDNs, that is, Bose–Einstein condensation IPDNs. To optimize the performance of the decentralized algorithm, it shall be applied to the most decentralized IPDNs, that is, scale-free IPDNs.

## Network Control with Limited Resources

In addition to CEPD, another great challenge inherent in many complex systems, such as laser and photonics systems, is to control a large system with limited resources. The ability to select key nodes in a collaboration network, for example, an IPDN, for protection, enhancement, improvement, or CEPD, is critical for the integration of laser and photonics solutions. For a network of nodes, that is, developers, operators, devices, equipment, and facilities to organize and carry out various functions, nodes must communicate to coordinate their actions. The goal is to facilitate lines of communication and collaboration by answering three questions:

1. What are network measures of communication and collaboration capabilities? An appropriate network measure may be used to optimally control networks and prevent disruptions. For instance, if three key nodes are to be identified in a network of 100 nodes, there are total $\binom{100}{3} = 161{,}700$ different combinations. A network measure may be calculated for each of the 161,700 networks, each of which has 97 nodes. The network with the worst network measure, that is, the network with the minimum communication or collaboration capability, reveals the three key nodes that require protection, enhancement, or improvement.

2. What are node measures for selecting key nodes? Network measures may be used to determine key nodes, but they are computationally inefficient because the number of combinations increases exponentially as the number of nodes increases. For example, if three key nodes are to be identified in a network of 1000 nodes, there are $\binom{1000}{3} = 166{,}167{,}000$ different combinations. The network measure must be calculated for all 166,167,000 networks, each of which has 997 nodes. Compared to the network of 100 nodes, the number of combinations increases over 1000 times when the network is only 10 times larger. A node measure is calculated for each node and indicates the node's communication

or collaboration capability. The number of times a node measure is calculated is at most the number of nodes in a network.

3. What algorithms can effectively and efficiently select key nodes? The node measure alone is not sufficient for selecting key nodes. For instance, removing nodes with the highest communication or collaboration capability does not guarantee that the remaining network has the lowest communication or collaboration capability. Correlations between network and node measures must be explored to develop effective and efficient algorithms.

Two measures that describe communication and collaboration capabilities at both network and node levels are diffusion speed and collaboration scale. Structural search algorithms may be designed to select key nodes (Dawande et al., 2012; Jahanpour and Chen, 2013). A structural search algorithm is a procedure to search networks for useful subsets of nodes and links and is equipped with unambiguous network measures, computationally efficient node measures, and effective and efficient search processes.

### Diffusion Speed

Diffusion speed ($S^n$; Equation 12.1) indicates how fast information diffuses in a network:

$$S^n = \frac{\sum_{i=1}^{n}\sum_{j=1, i<j}^{n} \frac{1}{d_{ij}}}{n(n-1)/2} \tag{12.1}$$

$n$ is the total number of nodes in a network, that is, network order
$d_{ij}$ is the geodesic distance between nodes $i$ and $j$ in the network

$0 \le S^n \le 1$. $S^n = 0$ if there is no link in the network. $S^n = 1$ if the network is a clique. The larger $S^n$ is, the higher diffusion speed the network has. Suppose two key nodes are to be identified from a network (Figure 12.5). There are totally $\binom{5}{2} = 10$ different combinations. $S^3_{\backslash(1\wedge2)\vee(4\wedge5)\vee(1\wedge5)} = \dfrac{\frac{1}{1}+\frac{1}{2}+\frac{1}{1}}{3} = 0.833$ when nodes 1 and 2, 4 and 5, or 1 and 5 are removed; $S^3_{\backslash(1\wedge3)\vee(1\wedge4)\vee(3\wedge5)\vee(2\wedge5)\vee(2\wedge3)\vee(3\wedge4)} = \dfrac{\frac{1}{1}}{3} = 0.333$ when nodes 1 and 3, 1 and 4, 3 and 5, 2 and 5, 2 and 3, or 3 and 4 are removed;

**FIGURE 12.5**
A network of five nodes.

and $S^3_{2 \wedge 4} = \dfrac{0}{3} = 0$ when nodes 2 and 4 are removed. Removing nodes 2 and 4 minimizes $S^3$, and they are therefore the key nodes.

Since the calculation of a network measure often involves the calculation of the same property for each node, it is intuitive to use Equation 12.1 to calculate the node diffusion speed: $S_1 = S_5 = \dfrac{\frac{1}{1} + \frac{1}{2} + \frac{1}{3} + \frac{1}{4}}{4} = 0.521$;

$S_2 = S_4 = \dfrac{\frac{1}{1} + \frac{1}{1} + \frac{1}{2} + \frac{1}{3}}{4} = 0.708$; $S_3 = \dfrac{\frac{1}{1} + \frac{1}{1} + \frac{1}{2} + \frac{1}{2}}{4} = 0.750$. If only one key node is to be identified, it can be shown that removing the node with the highest diffusion speed (here is node 3) produces the lowest diffusion speed for the network in most cases. To prove this, let $k_1$ and $k_2$ represent two

nodes in a network of $n$ nodes. $S_{k_1} = \dfrac{\sum\limits_{j=1, j \neq k_1}^{n} \dfrac{1}{d_{k_1 j}}}{n-1}$ and $S_{k_2} = \dfrac{\sum\limits_{j=1, j \neq k_2}^{n} \dfrac{1}{d_{k_2 j}}}{n-1}$.

Without losing generality, suppose that $S_{k_1} > S_{k_2}$. Let $S^{n-1}_{\backslash k_1}$ and $S^{n-1}_{\backslash k_2}$ represent the network diffusion speed after removing nodes $k_1$ and $k_2$, respectively.

$S^{n-1}_{\backslash k_1} = \dfrac{\sum\limits_{i=1, i \neq k_1}^{n} \sum\limits_{j=1, j \neq k_1, i<j}^{n} \dfrac{1}{d_{ij}}}{\dfrac{(n-1)(n-2)}{2}}$ and $S^{n-1}_{\backslash k_2} = \dfrac{\sum\limits_{i=1, i \neq k_2}^{n} \sum\limits_{j=1, j \neq k_2, i<j}^{n} \dfrac{1}{d_{ij}}}{\dfrac{(n-1)(n-2)}{2}}$. It is nec-

essary to prove that $S^{n-1}_{\backslash k_1} < S^{n-1}_{\backslash k_2}$.

If neither $k_1$ nor $k_2$ is on the only shortest path between any two nodes $i$ and $j$ in the network, removing $k_1$ or $k_2$ only affects the diffusion speed between $k_1$ and all other nodes in the network or between $k_2$ and all other nodes in the network, respectively.

$$\sum_{i=1, i \neq k_1}^{n} \sum_{j=1, j \neq k_1, i<j}^{n} \frac{1}{d_{ij}} = \sum_{i=1}^{n} \sum_{j=1, i<j}^{n} \frac{1}{d_{ij}} - \sum_{j=1, j \neq k_1}^{n} \frac{1}{d_{k_1 j}}$$

and $$\sum_{i=1, i \neq k_2}^{n} \sum_{j=1, j \neq k_2, i<j}^{n} \frac{1}{d_{ij}} = \sum_{i=1}^{n} \sum_{j=1, i<j}^{n} \frac{1}{d_{ij}} - \sum_{j=1, j \neq k_2}^{n} \frac{1}{d_{k_2 j}}.$$

$$S^{n-1}_{\backslash k_1} = \frac{\sum\limits_{i=1, i \neq k_1}^{n} \sum\limits_{j=1, j \neq k_1, i<j}^{n} \dfrac{1}{d_{ij}}}{\dfrac{(n-1)(n-2)}{2}} = \frac{\sum\limits_{i=1}^{n} \sum\limits_{j=1, i<j}^{n} \dfrac{1}{d_{ij}} - \sum\limits_{j=1, j \neq k_1}^{n} \dfrac{1}{d_{k_1 j}}}{\dfrac{(n-1)(n-2)}{2}}$$

$$= \frac{\dfrac{n(n-1)}{2} S^n - (n-1) S_{k_1}}{\dfrac{(n-1)(n-2)}{2}} = \frac{n S^n - 2 S_{k_1}}{n-2}$$

Similarly, $S_{\backslash k_2}^{n-1} = \dfrac{nS^n - 2S_{k_2}}{n-2}$. Since $S_{k_1} > S_{k_2}$, $S_{\backslash k_1}^{n-1} < S_{\backslash k_2}^{n-1}$. It is necessary that $n > 2$.
When $n = 2$, it is always true that $S_{k_1} = S_{k_2}$.

In a more general case, removing $k_1$ or $k_2$ decreases the diffusion speed between other nodes. There are three situations in which removing $k_1$ or $k_2$ affects the diffusion speed between nodes $i$ and $j$: (1) every shortest path between $i$ and $j$ covers $k_1$ but not $k_2$. The geodesic distance between $i$ and $j$ is $d_{ij}^{k_1}$, which is $k_1$-dependent. After removing $k_1$, the geodesic distance between $i$ and $j$ becomes $d_{ij}^{\backslash k_1}$. $d_{ij}^{k_1} < d_{ij}^{\backslash k_1}$ and $\dfrac{1}{d_{ij}^{k_1}} > \dfrac{1}{d_{ij}^{\backslash k_1}}$. Removing $k_1$ increases the geodesic distance between $i$ and $j$ and therefore decreases the diffusion speed between $i$ and $j$. Removing $k_2$ does not affect the diffusion speed between $i$ and $j$; and (2) every shortest path between $i$ and $j$ covers $k_2$ but not $k_1$. $d_{ij}^{k_2} < d_{ij}^{\backslash k_2}$ and $\dfrac{1}{d_{ij}^{k_2}} > \dfrac{1}{d_{ij}^{\backslash k_2}}$. $d_{ij}^{k_2}$ is $k_2$-dependent. Removing $k_2$ decreases the diffusion speed between $i$ and $j$. Removing $k_1$ does not affect the diffusion speed between $i$ and $j$; and (3) every shortest path between $i$ and $j$ covers both $k_1$ and $k_2$. After removing either $k_1$ or $k_2$, the geodesic distance between $i$ and $j$ increases, that is, $d_{ij}^{k_1 k_2} < d_{ij}^{\backslash k_1}$ and $d_{ij}^{k_1 k_2} < d_{ij}^{\backslash k_2}$. $d_{ij}^{k_1 k_2}$ is $k_1$-dependent and $k_2$-dependent. Removing either $k_1$ or $k_2$ decreases the diffusion speed between $i$ and $j$.

There might be multiple pairs of nodes $i$ and $j$ whose diffusion speed is affected by removing nodes $k_1$ or $k_2$.

$$S_{\backslash k_1}^{n-1} = \dfrac{\displaystyle\sum_{i=1}^{n}\sum_{j=1, i<j}^{n} \dfrac{1}{d_{ij}} + \sum_{i,j}\left(\dfrac{1}{d_{ij}^{\backslash k_1}} - \dfrac{1}{d_{ij}^{k_1}}\right) - \sum_{j=1, j\neq k_1}^{n} \dfrac{1}{d_{k_1 j}}}{\dfrac{(n-1)(n-2)}{2}} \quad \text{where} \quad \dfrac{1}{d_{ij}^{\backslash k_1}} - \dfrac{1}{d_{ij}^{k_1}}$$

is negative and represents the decrease in diffusion speed between $i$ and $j$ when every shortest path between $i$ and $j$ covers $k_1$ (situations (1) and (2)) and $k_1$ is removed. $\displaystyle\sum_{i,j}\left(\dfrac{1}{d_{ij}^{\backslash k_1}} - \dfrac{1}{d_{ij}^{k_1}}\right)$ calculates the summation of speed decrease between multiple pairs of nodes $i$ and $j$. Further simplification shows that $S_{\backslash k_1}^{n-1} = \dfrac{nS^n + \dfrac{2}{n-1}\displaystyle\sum_{i,j}\left(\dfrac{1}{d_{ij}^{\backslash k_1}} - \dfrac{1}{d_{ij}^{k_1}}\right) - 2S_{k_1}}{n-2}$. Similarly,

$S_{\backslash k_2}^{n-1} = \dfrac{nS^n + \dfrac{2}{n-1}\displaystyle\sum_{i,j}\left(\dfrac{1}{d_{ij}^{\backslash k_2}} - \dfrac{1}{d_{ij}^{k_2}}\right) - 2S_{k_2}}{n-2}$. To prove that $S_{\backslash k_1}^{n-1} < S_{\backslash k_2}^{n-1}$, it is equivalent to show that $S_{\backslash k_1}^{n-1} - S_{\backslash k_2}^{n-1} < 0$, that is, Equation 12.2 must be true:

$$\sum_{i,j}\left(\frac{1}{d_{ij}^{k_2}}-\frac{1}{d_{ij}^{\backslash k_2}}\right)-\sum_{i,j}\left(\frac{1}{d_{ij}^{k_1}}-\frac{1}{d_{ij}^{\backslash k_1}}\right)<(n-1)\left(S_{k_1}-S_{k_2}\right) \qquad (12.2)$$

If neither $k_1$ nor $k_2$ is on the only shortest paths between any two nodes $i$ and $j$, the left-hand side of Equation 12.2 becomes zero. If $S_{k_1}>S_{k_2}, 0<(n-1)\left(S_{k_1}-S_{k_2}\right)$. This indicates that removing a node with higher diffusion speed always results in lower diffusion speed for the network. Also note that if $S_{k_1}=S_{k_2}$, both sides of Equation 12.2 become zero. This indicates that $S_{\backslash k_1}^{n-1}=S_{\backslash k_2}^{n-1}$ when $S_{k_1}=S_{k_2}$; removing either $k_1$ or $k_2$ results in the same diffusion speed for the network. The difficulty is to prove Equation 12.2 when removing $k_1$ or $k_2$ decreases the diffusion speed between other nodes. Equation 12.3 rewrites Equation 12.2:

$$\left(\sum_{i,j}\frac{1}{d_{ij}^{k_2}}-\sum_{i,j}\frac{1}{d_{ij}^{k_1}}\right)-\left(\sum_{i,j}\frac{1}{d_{ij}^{\backslash k_2}}-\sum_{i,j}\frac{1}{d_{ij}^{\backslash k_1}}\right)<(n-1)\left(S_{k_1}-S_{k_2}\right) \qquad (12.3)$$

$\displaystyle\sum_{i,j}\frac{1}{d_{ij}^{k_2}}-\sum_{i,j}\frac{1}{d_{ij}^{k_1}}$ is the difference in summations of $k_2$-dependent geodesic distances and $k_1$-dependent geodesic distances. $\displaystyle\sum_{i,j}\frac{1}{d_{ij}^{\backslash k_2}}-\sum_{i,j}\frac{1}{d_{ij}^{\backslash k_1}}$ is the difference in summations of corresponding distances after removing $k_2$ and $k_1$. $i$'s and $j$'s in $\displaystyle\sum_{i,j}\frac{1}{d_{ij}^{k_2}}$ are the same as $i$'s and $j$'s in $\displaystyle\sum_{i,j}\frac{1}{d_{ij}^{\backslash k_2}}$. $i$'s and $j$'s in $\displaystyle\sum_{i,j}\frac{1}{d_{ij}^{k_1}}$ are the same as $i$'s and $j$'s in $\displaystyle\sum_{i,j}\frac{1}{d_{ij}^{\backslash k_1}}$. Additional analytical studies or simulation experiments may be conducted to provide further insight into the relationship between node diffusion speed and network diffusion speed.

If two nodes are to be removed, either nodes 2 and 3 or nodes 4 and 3 (Figure 12.5) shall be removed according to the node diffusion speed. According to the network diffusion speed, however, removing nodes 2 and 4 minimizes the network diffusion speed. The discrepancy between node and network measures is due to the structural dependency between nodes 2, 3, and 4 in Figure 12.5. A sequential search process using only the node measure is not sufficient for identifying multiple key nodes. Structural search processes need to be developed to identify key nodes and help improve network communication capability.

## Collaboration Scale

One of the objectives for laser and photonics systems integration is to identify key nodes that affect the scale of a collaboration network. Figure 12.6

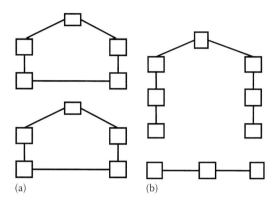

**FIGURE 12.6**
Two networks with different diffusion speeds and collaboration scales, (a) a network of two 5-node components, (b) a network of one 7-node component and one 3-node component.

shows two networks, both of which have two (disconnected) components and 10 nodes. Using Equation 12.1, $S^{10} = \dfrac{2\sum\limits_{i>j} \dfrac{1}{d_{ij}}}{10 \times 9} = 0.333$ for Figure 12.6a and $S^{10} = \dfrac{2\sum\limits_{i>j} \dfrac{1}{d_{ij}}}{10 \times 9} = 0.303$ for Figure 12.6b; network Figure 12.6a has a higher diffusion speed than Figure 12.6b. Network Figure 12.6b may, however, enable collaboration among at most seven nodes whereas network Figure 12.6a may support collaboration among at most five nodes; network Figure 12.6b is a larger collaboration network than network Figure 12.6a.

Collaboration scale of a network indicates the expected scale of collaboration. Equation 12.4 calculates the expected collaboration scale *ES*:

$$ES = \sum_{i=1}^{m} \frac{c_i}{n} \times c_i = \sum_{i=1}^{m} \frac{c_i^2}{n} \tag{12.4}$$

$c_i$ is the size of component $i$
$c_i / n$ is the probability that a source node belongs to component $i$
$m$ is the total number of components

*ES* for Figure 12.6a and b are 5 and 5.8, respectively. On average, Figure 12.6b is able to support larger-scale collaboration than Figure 12.6a. In reality, however, collaboration is time sensitive. Suppose a collaboration network must begin its operations in $l$ time units and it takes $t$ time units to diffuse information between any two directly linked nodes. $t$ is fixed, that is, deterministic diffusion time. If $\dfrac{l}{t} < max(c_i)$, the maximum component size,

Equation 12.5 calculates the expected time-sensitive collaboration scale with deterministic diffusion times:

$$ES(l,t) = \frac{\sum_{i=1}^{n}\sum_{j=1,j\neq i}^{n} q_{i,j}}{n} = \frac{2\sum_{i=1}^{n}\sum_{j=i+1}^{n} q_{i,j}}{n} \tag{12.5}$$

$$q_{i,j} = \begin{cases} 1, & if\ d_{i,j} \leq \dfrac{l}{t} \\ 0, & \text{otherwise} \end{cases} \qquad i \neq j$$

If $\dfrac{l}{t} \geq max(c_i)$, $ES = ES(l,t)$. If the diffusion time between two directly linked nodes is a random variable $T$ described by a probability density function $p(T)$. Replacing $t$ with $T$ in Equation 12.5, Equation 12.5 may be used to calculate the expected time-sensitive collaboration scale with stochastic diffusion times.

Many other domain-specific algorithms have been developed to optimize network operations and performance, for example, model checking for software design (Clarke et al., 2000), testing algorithms for hardware design (Tu et al., 2003), and decentralized algorithms for CEPD (Chen and Nof, 2007, 2010). Emerging research is expected to integrate network modeling and analysis and design of heuristic algorithms to effectively and efficiently control complex systems.

## Conclusions

The integration of laser and photonics solutions is a cross-disciplinary topic and requires systems perspective. A major challenge is to design and engineer a platform that effectively and efficiently supports seamless collaboration among humans and machines. Recently, collaboration control and network science have taken the lead in modeling complex systems and interactions, performance analysis, and efficient and optimal control of systems. Emerging solutions to integration and collaboration of laser and photonics systems must bridge the gap between systems engineering, which may be inefficient in dealing with numerous systems states, and the traditional divide-and-conquer approach, in which important system properties become lost. A balanced approach is needed to achieve optimal system performance with efficient algorithms through distributed and collaborative control.

# References

Albert, R., Jeong, H., and Barabasi, A.L. (1999). Internet: Diameter of the World-Wide Web. *Nature*, 401(6749), 130–131.

Angeles Serrano, M. and De Los Rios, P. (2007). Interfaces and the edge percolation map of random directed networks. *Physical Review E—Statistical, Nonlinear, and Soft Matter Physics*, 76(5), 56–121.

AutoMod Version 11.1. (1998–2003). Brooks Automation, Inc, Chelmsford, MA.

Barabasi, A.L. and Albert, R. (1999). Emergence of scaling in random networks. *Science*, 286(5439), 509–512.

Barkai, N. and Leibler, S. (2000). Circadian clocks limited by noise. *Nature*, 403, 267–268.

Bianconi, G. and Barabasi, A.L. (2001a). Bose-Einstein condensation in complex networks. *Physical Review Letters*, 86(24), 5632–5635.

Bianconi, G. and Barabasi, A.L. (2001b). Competition and multiscaling in evolving networks. *Europhysics Letters*, 54(4), 436–442.

Broder, A., Kumar, R., Maghoul, F., Raghavan, P., Rajagopalan, S., Stata, R., Tomkins, A., and Wiener, J. (2000). Graph structure in the Web. *Computer Networks*, 33(1), 309–320.

Chen, X.W. and Nof, S.Y. (2007). Error detection and prediction algorithms: Application in robotics. *Journal of Intelligent and Robotic Systems*, 48(2), 225–252.

Chen, X.W. and Nof, S.Y. (2009). Automating errors and conflicts prognostics and prevention. In: S.Y. Nof, ed. *Handbook of Automation*. Heidelberg, Germany: Springer, pp. 503–525.

Chen, X.W. and Nof, S.Y. (2010). A decentralized conflict and error detection and prediction model. *International Journal of Production Research*, 48(16), 4829–4843.

Chen, X.W. and Nof, S.Y. (2012a). Conflict and error prevention and detection in complex networks. *Automatica*, 48(5), 770–778.

Chen, X.W. and Nof, S.Y. (2012b). Interactive, constraint-network prognostics and diagnostics to control errors and conflicts. Pending U.S. Patent Application, Purdue Research Foundation (PRF) 65241.00.US.

Chen, X.W. and Nof, S.Y. (2012c). Constraint-based conflict and error management. *Engineering Optimization*, 44(7), 821–841.

Chiang, L.H., Braatz, R.D., and Russell, E. (2001). *Fault Detection and Diagnosis in Industrial Systems*. London, U.K.: Springer.

Clarke, E.M., Grumberg, O., and Peled, D.A. (2000). *Model Checking*. Cambridge, MA: The MIT Press.

Dawande, M., Mookerjee, V., Sriskandarajah, C., and Zhu, Y. (2012). Structural search and optimization in social networks. *INFORMS Journal on Computing*, 24, 611–623.

Deb, S., Pattipati, K.R., Raghavan, V., Shakeri, M., and Shrestha, R. (1995). Multi-signal flow graphs: A novel approach for system testability analysis and fault diagnosis. *IEEE Aerospace and Electronics Systems Magazine*, 10(5), 14–25.

Erdös, P. and Rényi, A. (1959). On random graphs. *Publicationes Mathematicae Debrecen*, 6, 290–291.

Erdös, P. and Rényi, A. (1960). On the evolution of random graphs. *Publications of the Mathematical Institute of the Hungarian Academy of Sciences* 5, 17–61.

Erdös, P. and Rényi, A. (1961). On the strength of connectedness of a random graph. *Acta Mathematica Academiae Scientiarum Hungaricae*, 12, 261–267.

Gertler, J. (1998). *Fault Detection and Diagnosis in Engineering Systems*. New York: Marcel Dekker.

Jahanpour, E. and Chen, X.W. (2013). Analysis of complex network performance and heuristic node removal strategies. *Communications in Nonlinear Science and Numerical Simulation*, 18, 3458–3468.

Klein, B.D. (1997). How do actuaries use data containing errors: Models of error detection and error correction. *Information Resources Management Journal*, 10(4), 27–36.

Lee, T.I. et al. (2002). Transcriptional regulatory networks in Saccharomyces cerevisiae. *Science*, 298, 799–804.

Li, L. (2010). The challenges ahead for laser macro, micro and nano manufacturing. In: J. Lawrence et al., eds. *Advances in Laser Materials Processing: Technology, Research and Applications*. Oxford, U.K.: Woodhead Publishing.

Mclennan, M. and Kennell, R. (2010). HUBzero: A platform for dissemination and collaboration in computational science and engineering. *Computing in Science and Engineering*, 12(2), 48–52.

Newman, M.E.J., Barabasi, A.L., and Watts, D.J. (2006). *The Structure and Dynamics of Networks*. Princeton, NJ: Princeton University Press.

Newman, M.E.J., Strogatz, S.H., and Watts, D.J. (2001). Random graphs with arbitrary degree distributions and their applications. *Physical Review E—Statistical, Nonlinear, and Soft Matter Physics*, 64(2), 0261181–02611817.

Pattipati, K.R. and Alexandridis, M.G. (1990). Application of heuristic search and information theory to sequential fault diagnosis. *IEEE Transactions on Systems, Man and Cybernetics*, 20(4), 872–887.

Pattipati, K.R. and Dontamsetty, M. (1992). On a generalized test sequencing problem. *IEEE Transactions on Systems, Man and Cybernetics*, 22(2), 392–396.

Price, D.J. (1965). Networks of scientific papers. *Science*, 149, 510–515.

Raghavan, V., Shakeri, M., and Pattipati, K. (1999a). Optimal and near-optimal test sequencing algorithms with realistic test models. *IEEE Transactions on Systems, Man, and Cybernetics Part A: Systems and Humans*, 29(1), 11–26.

Raghavan, V., Shakeri, M., and Pattipati, K. (1999b). Test sequencing algorithms with unreliable tests. *IEEE Transactions on Systems, Man, and Cybernetics Part A: Systems and Humans*, 29(4), 347–357.

Raich, A. and Cinar, A. (1996). Statistical process monitoring and disturbance diagnosis in multivariable continuous processes. *AIChE Journal*, 42(4), 995–1009.

Seok, H.S. and Nof, S.Y. (2011). The HUB-CI initiative for cultural, education and training, and healthcare networks. *Proceedings of 21st ICPR*, Stuttgart, Germany.

Shakeri, M., Raghavan, V., Pattipati, K.R., and Patterson-Hine, A. (2000). Sequential testing algorithms for multiple fault diagnosis. *IEEE Transactions on Systems, Man and Cybernetics Part A: Systems and Humans*, 30(1), 1–14.

Shen-Orr, S.S., Milo, R., Mangan, S., and Alon, U. (2002). Network motifs in the transcriptional regulation network of Escherichia coli. *Nature Genetics*, 31, 64–68.

Solomonoff, R. and Rapoport, A. (1951). Connectivity of random nets. *Bulletin of Mathematical Biophysics*, 13, 107–117.

Tu, F. and Pattipati, K.R. (2003). Rollout strategies for sequential fault diagnosis. *IEEE Transactions on Systems, Man, and Cybernetics Part A: Systems and Humans*, 33(1), 86–99.

Tu, F., Pattipati, K.R., Deb, S., and Malepati, V.N. (2003). Computationally efficient algorithms for multiple fault diagnosis in large graph-based systems. *IEEE Transactions on Systems, Man, and Cybernetics Part A: Systems and Humans*, 33(1), 73–85.

Zak, D.E., Gonye, G.E., Schwaber, J.S., and Doyle, F.J.III. (2003). Importance of input perturbations and stochastic gene expression in the reverse engineering of genetic regulatory networks: Insights from an identifiability analysis of an in silico network. *Genome Research*, 13, 2396–2405.

# 13

## Dynamic Resource Allocation in Human-Centered Service Robot Applications

Seokcheon Lee

### CONTENTS

### Introduction

Service robotics has become one of the research areas attracting major attention for over a decade and the social aspect of service robots has been increasingly being emphasized. Service robots not only navigate in an environment but also have to be able to interact with people to provide services (upon request or by identification of need). The human-centered robotics is currently among the vital research areas in mobile robotics with a variety of potential applications in medical care, surveillance, and domestic settings (Burgard et al., 2005; Liu et al., 2005; Bellotto and Hu, 2009). People tracking is a key technology to the human-centered service robots that enables to be aware of the presence of adjacent people, identify their current positions, and keep track of them (as needed). People tracking is a very challenging task since people move around and their behaviors are often uncertain and unpredictable.

The core technology that enables people tracking is the laser sensors (e.g., scanning laser range finders) that are usually used in combination with cameras (Bellotto and Hu, 2009; Luber et al., 2011). Laser ranging is based on measurements performed on laser light reflected from target objects, and compared to vision, laser data are sparse (thus, much less processing is required) but more accurate while being less sensitive to noise and requiring almost no calibration (Fod et al., 2002). Therefore, laser sensors are ideal candidates to the people tracking problem. In laser-based people tracking, a person is represented, based on the laser sensor data, as a single state of torso (Kluge et al., 2001; Fod et al., 2002; Kleinhagenbrock et al., 2002; Schulz et al., 2003; Cui et al., 2005; Topp and Christensen, 2005; Mucientes and Burgard, 2006) or by the states of two legs (Fritsch et al., 2003; Scheutz et al., 2004; Taylor and Kleeman, 2004; Cui et al., 2006; Arras et al., 2008; Bellotto and Hu, 2009), which can be fused then with the face detected by a camera.

In order to fully leverage the benefits of the human-centered robotics and its enabling technologies, however, it is necessary to pose the system in a more generalized context where multiple service robots in an environment work collaboratively to provide services (see Figure 13.1). In such a context, it is essential to thoroughly address several issues of dynamic resource allocation by which the activities of service robots are dynamically coordinated among them to best serve the people in need. There are various types of resource allocation decisions that have to be coordinated, including searching, tracking, and service assignment (detailed in the "Generic Dynamic Resource Allocation Problem" section). These decisions are, in particular, difficult due to the unpredictable and uncertain nature of human behaviors.

Various service robot systems can be defined and corresponding dynamic resource allocation solutions can be developed. However, if individual solutions are dedicated to specific problems with certain characteristics, it will be hard to transfer such solutions to the problems with different characteristics. Therefore, when a new system emerges or the characteristics of an existing system change, the development has to be carried out somewhat from the ground up, consuming a significant amount of time and resources and consequently preventing this important area of research from being advanced enough to fulfill the need of our society.

This chapter therefore aims to propose to develop a unified decision framework for the dynamic resource allocation that can be flexibly and effectively applied to a wide range of possible applications and contexts. The development of such a decision framework will greatly alleviate the barriers against the introduction of complex service robot systems that have a great potential to enhance the safety and welfare of our society. For this purpose, in the "Generic Dynamic Resource Allocation Problem" section,

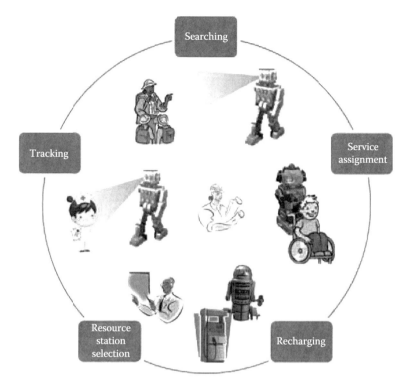

**FIGURE 13.1**
Dynamic resource allocation in human-centered service robot applications.

a preliminary, generic dynamic resource allocation problem is defined in the context of human-centered, service robot systems, from which various individual problems can be instantiated. The "Procedure in Developing Dynamic Resource Allocation Framework" section suggests a procedure for developing the dynamic resource allocation framework. The decision framework is based on the synthesis of various fundamental decision principles, and two important principles and their characteristics are introduced in the "Parallelism Principle, Centrality Principle, and Algorithmic Parameters" sections. Finally, the "Discussion" section concludes this research.

## Generic Dynamic Resource Allocation Problem

In this section, a preliminary description of the generic dynamic resource allocation problem is presented in the context of human-centered service robot applications. A team of service robots is deployed in an environment to provide people with certain services. People moving around in the service area request certain types of services. The service robots search and track people according to a certain search strategy and using laser-based tracking technology. Once service requests are identified, the robots allocate the requests among them according to a certain service assignment strategy. A robot, once allocated, moves to the designated person and provides the requested service. There can be various types of services provided by the robots, and a person may require on-site service only, transportation service only, or both.

Given a service request, a robot may or may not be able to perform the request depending on the robot's capabilities available at the time. There are two types of capabilities: consumable and inconsumable. Consumable capabilities are those that consume resources as the robot performs services (e.g., food), and inconsumable capabilities are those that are persistent (e.g., sensing). Robots spend energy for traveling and performing services and a robot has to have enough energy to be eligible to perform a service. The consumable resources and energy resources have to be recharged in designated resource stations to continue the service operations, and a robot, for recharge, has to choose a station along with the types and quantities to recharge in the limit of its capacity. Each robot is equipped with a laser-based people tracking system that is able to identify, to a certain degree of accuracy, people within its sensing range. Also, a robot has a certain range of communication and it can reach other robots directly within this range or indirectly by a sequence of communication links. The robots may be heterogeneous in various aspects including capabilities, traveling speed, capacity of energy resources, energy efficiency, sensing range, and communication range.

The objective of human-centered service robot system is to minimize total weighted response (waiting) time that is mathematically defined as

$$\min \sum_i w_{S(i)} \left( \tau_i - t_i \right)$$

where

$S(i)$ is the service type of service $i$

$w_{S(i)}$ is the relative importance (weight) of service type of service $i$

$\tau_i - t_i$ is the response time for the service request generated at time $t$ and fulfilled at time $\tau$

Various resource allocation decisions are associated with this objective: (1) searching (locating unidentified people), (2) tracking (keeping track of identified people), (3) service assignment (allocating robots to service requests), (4) resource station selection (choosing a resource station for recharge), and (5) recharging (types and quantities to recharge). The goal of service robots is to optimally and dynamically manage these decisions such that the total weighted response time is minimized even when the service environment is highly unpredictable and uncertain (especially due to the nature of human behaviors).

## Procedure in Developing Dynamic Resource Allocation Framework

This section proposes a procedure for developing a unified decision framework for the dynamic resource allocation problems in human-centered service robot systems (see Figure 13.2). A preliminary, descriptive definition of the generic resource allocation problem was presented in the "Generic Dynamic Resource Allocation Problem" section, and the first task in developing the decision framework is to concretely and mathematically define the generic problem. Various meaningful factors have to be taken into account in this formalism, including characteristics of services and robots, rechargeability, characteristics of resource stations, and properties of dynamic events and information uncertainty. This task will help understand the characteristics of and relationships between individual problems

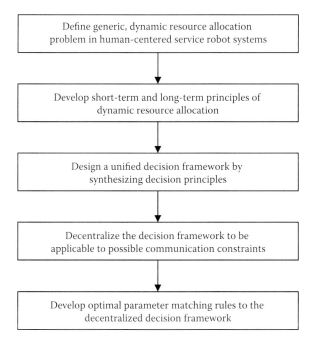

**FIGURE 13.2**
Procedure for developing dynamic resource allocation framework.

(e.g., complexity and reducibility), thereby systematically identifying the generic problem from which various meaningful service robot systems can be instantiated.

In order to develop a decision framework for the generic problem, fundamental principles that govern individual decisions and their interactions are first developed, and then they are mathematically and procedurally synthesized into a unified decision framework. The core and initiative of this approach is the generation of decision principles. Two types of principles can be incorporated: long-term principles and short-term principles. Long-term principles are those that enable to secure long-term performance, while short-term principles are those greedily pursuing immediate performance. The long-term and short-term principles can be explored not only by researching on the fundamental principles behind various (self-organizing) physical and biological systems (e.g., foraging and task allocation behaviors of insect colonies) but also by analyzing the patterns over optimal (or near-optimal) decisions on simplified (deterministic) scenarios. The built-in decision principles found in literature can also be utilized for this purpose as appropriate, such as resource welfare (Kim and Lee, 2010; Kim et al., 2011, 2012), preparedness (Lee, 2011), centrality (Lee, 2012, 2013), parallelism (Lee and Lee, 2011), and the collaboration control theory (CCT) principles: collaboration requirement planning (Berman and Nof, 2011), cyber parallelism (Ceroni and Nof, 2002, 2005), conflict and error detection and prevention (Anussornnitisarn et al., 2002), fault tolerance by teaming (Jeong and Nof, 2008, 2009), join/leave/remain association–dissociation (Yoon and Nof, 2011), and lines of command and collaboration (Tkach et al., 2011, 2012).

The decision framework, involving both types of principles, will enable to secure both long-term performance with respect to (uncertain) future events and, at the same time, short-term performance from (certain) current services. A decision framework can be developed by mathematically and procedurally translating and synthesizing the individual principles, and it will essentially contain several calibration parameters in terms of weights on different principles and parameters within each principle. The decision framework then has to be decentralized such that it is applicable to various possible communication constraints. As discussed before, a service robot has a certain range of communication, and therefore, a decision has to be made through the communication among the robots that are reachable to each other (by a sequence of communication links), based on their knowledge limited by sensing and communication capability.

The decentralized decision framework will have calibration parameters that will enable to achieve portable performance in various operational scenarios. However, the best parameter values have to be identified by calibration, which is quite laborious and computationally demanding. The parameters strongly depend on each problem instance at hand; thus, the calibration process has to be repeated for each problem instance. Therefore, optimal parameter matching rules have to be analyzed and constructed through theoretical and experimental analysis that relate the characteristics of operational scenarios to the best parameter values, so that users can be easily and directly guided by the rules in choosing appropriate parameter values for the scenarios of interest.

Two decision principles, parallelism and centrality, are introduced in next sections to evidence the impact of resource allocation on performance, provide exemplary cases of principle development, and envision a small picture of the procedure proposed for developing the dynamic resource allocation framework. The two principles are recently developed in the context of ambulance logistics (Lee and Lee, 2011; Lee, 2012, 2013); however, they are general and transferable to the dynamic resource allocation problem at hand, specifically to the service assignment between service requests and robots. The principles are introduced after being adapted to the service robot systems.

## Parallelism Principle

### Parallelism

A service assignment decision assigns service requests to appropriate robots such that the response time is minimized. A service assignment decision can be either service-initiated or robot-initiated. In service-initiated decisions, a new service request finds idle robots, thus initiating the decision of selecting a robot among the idle ones. On the other hand, if the requests cannot be immediately assigned, they start being queued and a robot that has just got freed has to choose a service request among those waiting, thus initiating the decision. The relevance of the two types of service assignment decisions depends on the busyness of the system. Service-initiated decisions are more relevant in scenarios where the system load is relatively low, while in high load conditions, the robot-initiated decisions play the primary role.

Greedy policies can be used for the service assignment decisions since they are popularly adopted in various applications due to their efficiency and at the same time capability of achieving a certain level of effectiveness (Egbelu and Tanchoco, 1984; Bertsimas and van Ryzin, 1991; Mantel and Landeweerd, 1995; Østergaard et al., 2001; Gerkey and Martarić, 2002; de Koster et al., 2004; Dias, 2004; Grundel, 2005; Lin and Zheng, 2005; Sujit and Ghose, 2005; Jones et al., 2006; Vig and Adams, 2006, 2007; Sujit and Beard, 2007). Greedy policies employ short-term principles only without any consideration of long-term consequences, and the greedy policy when applied to the resource allocation problem at hand is to choose the closest robot in case of service-initiated decisions and to choose the closest request in case of robot-initiated decisions.

The greedy policy, however, takes into account only the idle robots despite the possibility that a busy robot can respond more quickly, even after the completion of the currently assigned service, to the request that is otherwise assigned to an idle robot. The parallelism principle lets involve both idle and busy robots at the same time, leading to an assignment problem between multiple (idle and busy) robots and multiple unassigned requests. The parallelism policy that incorporates the assignment problem is presented in four steps as follows.

### *Parallelism Policy*

1. When either a service-initiated or robot-initiated decision has to be made, identify all unassigned service requests $U$ and all idle/busy robots $V = V_{idle} \cup V_{busy}$.

2. Compute fitness $f_{vu}$ between a robot $v \in V$ and a request $u \in U$ based on the expected response time $t_{vu}$ for the robot $v$ to reach the request location $u$, including, if any, the time to be spent on the already assigned request:

$$f_{vu} = \frac{1}{(1 + t_{vu})}$$

3. Establish a one-to-one assignment by prioritizing the matches with larger fitness.

4. Assign and dispatch the idle robots to their matched requests (if any) and leave unassigned the requests matched to busy robots.

This parallelism policy is exactly the same as the greedy policy when only idle robots are taken into account, that is, $V = V_{idle}$.

### Performance Evaluation

In this section, the parallelism policy is evaluated in various scenarios implemented in a discrete event simulator, by the performance enhancement over the greedy policy. The service area is represented in a $5 \times 5$ square grid as shown in Figure 13.3. Each vertex generates service requests at a certain rate and service robots move from vertex to vertex through edges each with 1 min of traveling time. Once assigned to a request, a robot serves the person with a service time that is exponentially distributed with mean service time = 0.5 min. The robot then stays there until it is assigned to another request.

Two factors are taken into account in generating different test conditions: (1) service request arrival pattern and (2) size of robot team. A total of 12,500 requests are generated at a rate of 1 request/min following an

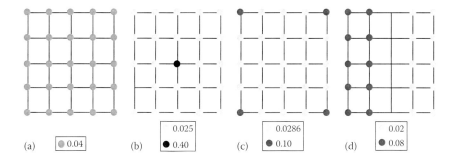

**FIGURE 13.3**
Service request patterns. (a) Uniform. (b) Centered. (c) Cornered. (d) Bipartite.

exponential distribution, and they are distributed to the vertices according to one of four request arrival patterns as shown in Figure 13.3: (a) uniform, (b) centered, (c) cornered, and (d) bipartite. A value in the figure of each pattern represents the probability for an arriving request to be allocated to a corresponding vertex. For example, in the cornered pattern, each vertex located in a corner gets an arriving request with probability 0.1. The size of the robot team is among {1, 2, 3, 4, 5, 6, 7} and the robots in each simulation run are initially located at random positions. As a result, 28 test conditions (4 arrival patterns × 7 sizes of robot team) are established. For each test condition, the two service assignment policies are applied: greedy policy and parallelism policy. Fifty simulation runs are replicated for each test scenario.

Figure 13.4 shows the average reduction in response time by the parallelism policy, that is, *reduction in response time (%) = (response time in greedy policy – response time in parallelism policy)/(response time in greedy policy)*. As can be observed in Figure 13.4, the parallelism considerably reduces the response time of the greedy policy in all different test conditions by up to 34% in uniform pattern, 28% in centered pattern, 43% in cornered pattern, and 35% in bipartite pattern, well supporting the potential of parallelism in reducing response time. Note that the parallelism policy is equivalent to the greedy policy when the size of robot team is one; therefore, there is always 0% reduction in such cases.

## Centrality Principle

The parallelism policy introduced in the previous section is computationally efficient, thus appropriate to the time-critical applications under consideration. However, it still tries to optimize the immediate performance without

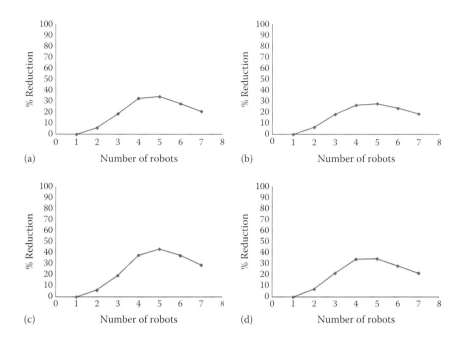

**FIGURE 13.4**
Average reduction in response time by parallelism over greedy policy. (a) Uniform. (b) Centered. (c) Cornered. (d) Bipartite.

taking into account the long-term consequences. A service assignment principle is designed in this section based on the notion of *centrality* from the complex network research, with which the parallelism policy becomes equipped with the global perspective of the geographical request distribution over the service area and thereby the capability of incorporating the long-term consequences.

## Centrality

The study of complex networks has emerged over the past decade as a theme of a wide range of disciplines, providing macroscopic characteristics of large-scale real networks such as the World Wide Web, the Internet, metabolic networks, and movie actor collaboration networks (see Newman, 2003, for a detailed review of this area). Node centrality in a network indicates the importance of a node in the operational efficiency of the network, that is, more central nodes contribute more to the efficiency of the network. There exist various measures of the node centrality available in literature, such as degree (Albert and Barabasi, 2002; Dorogovtsev and Mendes, 2002) and transitivity (Ravasz et al., 2002) in binary networks (where an edge has a binary state—present or absent), weighted degree (Barrat et al., 2004; Newman, 2004) and weighted transitivity (Barrat et al., 2004) in weighted networks (where an edge has a weight representing capacity or strength), and distance centrality (Sabidussi, 1996; López-Fernández et al., 2006) and betweenness centrality (Freeman, 1977; Barthelemy, 2004) applicable to both binary and weighted networks.

The node centrality is used as a decision principle for the service assignment problem under consideration. When a service robot gets freed, a network of service requests can be constructed where a node represents a waiting request (i.e., the request that has not been assigned to any robot) and an edge between every pair of requests has a weight value corresponding to the distance between the two request locations connected by the edge. The centrality of a request computed upon this service request network can be interpreted as the efficiency of the request in reaching out other requests or the density of requests around the request with respect to the geographical distribution of requests over the service area. When service requests are prioritized by the centrality and a robot is dispatched to the most central request, the robot will be given the opportunity, after the completion of the immediate service, to serve the other requests around it at the maximum rate of completion.

However, if requests are prioritized only by the centrality, the robots might keep jumping to the central requests without enough exploitation of requests in the vicinity, potentially leading to the excessive traveling just for repositioning purpose. On the other hand, the greedy policy, which assigns the freed robot to the request that is closest to the robot in an effort to minimize each current response time, has the strength in efficient exploitation of local regions. The robots under this greedy policy, however, are subject to getting stuck in sparse regions due to the lack of the global view of service area. Now, if service requests are prioritized by both closeness and centrality at the same time, the benefits of both principles can be incorporated while avoiding their inefficiencies. That is, the robots would smoothly proceed toward dense regions (global exploration) while continuously pursuing the exploitation of requests (local exploitation), thereby keeping the completion rate at maximum.

The complementary effects between the centrality and greediness lead to designing the centrality policy. The centrality measure $c_u$ represents the centrality of request $u$, in the following form:

$$c_u = \sum_{i \in U, i \neq u} \frac{1}{\left(1 + \tau_{ui}\right)}$$

where the centrality is represented by the weighted degree on the network of unassigned requests $U$ with each edge having a value of distance (in time) $\tau_{ui}$ between two requests $u$ and $i$. Among various centrality measures available in literature, the weighted degree is adopted here due to its capability of capturing the notion of centrality in weighted networks and computational efficiency appropriate to the real-time decisions. The weighted degree of a node is computed by the sum of the weights of connected edges when higher weight values are preferred (e.g., capacity and strength). However, the weight in the service request network represents distance; thus, lower weight values are preferred and the weighted degree in this case is computed by the sum of the reciprocals of weights. This measure is applied to the parallelism policy introduced in the "Parallelism Principle" section in computing fitness $f_{vu}$, giving rise to the centrality policy as follows.

### Centrality Policy

1. When either a service-initiated or robot-initiated decision has to be made, identify all unassigned service requests $U$ and all idle/busy robots $V = V_{idle} \cup V_{busy}$.

2. Compute centrality $c_u$ of each request $u \in U$ upon the network of service requests $U$ with the edge between every pair of requests having the value of distance $\tau_{ui}$ between them:

$$c_u = \sum_{i \in U, i \neq u} \frac{1}{(1 + \tau_{ui})}$$

3. Compute fitness $f_{vu}$ between a robot $v \in V$ and a request $u \in U$ based on the centrality $c_u$ and expected response time $t_{vu}$ for the robot $v$ to reach the request location $u$, including, if any, the time to be spent on the already assigned request:

$$f_{vu} = \frac{c_u}{(1 + t_{vu})}$$

4. Establish a one-to-one assignment by prioritizing the matches with larger fitness.

5. Assign and dispatch the idle robots to their matched requests (if any) and leave unassigned the requests matched to busy robots.

### Performance Evaluation

The centrality policy is evaluated in the same scenarios used in the "Parallelism Principle" section. Figure 13.5 shows the average reduction in response time by the centrality policy and parallelism policy over the greedy policy. As can be observed, the centrality policy considerably reduces the response time by up to 63% in uniform pattern, 35% in centered pattern, 52% in cornered pattern, and 36% in bipartite pattern, over the greedy policy. The improvements by the centrality are made especially when the parallelism alone is not very useful, that is, when the size of the robot team is small (since requests are queued everywhere and idle robots are likely to be assigned to the requests in the vicinity regardless of whether busy robots are considered

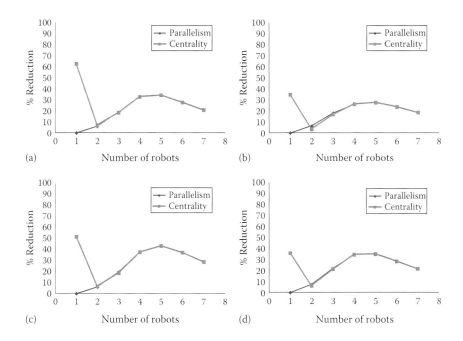

**FIGURE 13.5**
Average reduction in response time by centrality over greedy policy. (a) Uniform. (b) Centered. (c) Cornered. (d) Bipartite.

or not). In such high load conditions, on the contrary, more decisions are robot-initiated and the centrality can play an important role. Since the centrality complements the parallelism, the centrality policy can be effective in various load conditions.

## Algorithmic Parameters

The centrality policy can be more generalized by including two calibration parameters in order to provide portable performance in various possible operational scenarios: one parameter $\alpha$ for parallelism and another $\beta$ for centrality. The parallelism parameter $\alpha$ is used to determine the activation threshold of the parallelism in terms of the number of unassigned service requests. If there are too many requests waiting, the optimality of the solution provided by the parallelism becomes ineffective considering the uncertain future dynamics. Therefore, the parallelism may be better to be deactivated when there are unassigned service requests over a certain threshold, and in such a case, only idle robots are considered. Centrality parameter $\beta$ is the weight on centrality and it is used to trade off between centrality (global exploration capability) and closeness (local exploitation capability). When $\beta = 0$, the decision is based on the closeness only; however, when the weight is positive, the centrality is incorporated into the decision by the extent corresponding to the weight.

The centrality policy presented in the "Centrality Policy" section is transformed into the generalized service assignment policy with these two algorithmic parameters as follows.

### Generalized Service Assignment Policy

1. When either a service-initiated or robot-initiated decision has to be made, identify all unassigned service requests $U$, and if $|U| \le \alpha$, all idle/busy robots $V = V_{idle} \cup V_{busy}$; otherwise, only idle robots $V = V_{idle}$.
2. Compute centrality $c_u$ of each request $u \in U$ upon the network of service requests $U$ with the edge between every pair of requests having the value of distance $\tau_{ui}$ between them:

$$c_u = \sum_{i \in U, i \neq u} \frac{1}{(1 + \tau_{ui})}$$

3. Compute fitness $f_{vu}$ between a robot $v \in V$ and a request $u \in U$ based on the centrality $C_u$ weighted by $\beta$ and expected response time $t_{vu}$ for the robot $v$ to reach the request location $u$, including, if any, the time to be spent on the already assigned request:

$$f_{vu} = \frac{c_u^\beta}{(1 + t_{vu})}$$

4. Establish a one-to-one assignment by prioritizing the matches with larger fitness.
5. Assign and dispatch the idle robots to their matched requests (if any) and leave unassigned the requests matched to busy robots.

Note that the generalized service assignment policy is exactly the same as the centrality policy if $\alpha = \infty$ and $\beta = 1$, the parallelism policy when $\alpha = \infty$ and $\beta = 0$, and the greedy policy if $\alpha = 0$ and $\beta = 0$. Two calibration parameters are now associated with the policy and the right choice of the parameter values will be affected by various factors, and the optimal matching that relates the characteristics of individual operational scenarios to the optimal parameter values has to be thoroughly investigated. For example, Figure 13.6 shows that the effects of parallelism parameter $\alpha$ and centrality parameter $\beta$ on the performance in some test conditions. The nonlinear behaviors with the parameters in Figure 13.6 give rise to the need for carefully choosing the right values of parameters according to the operating environment, in order to maximize

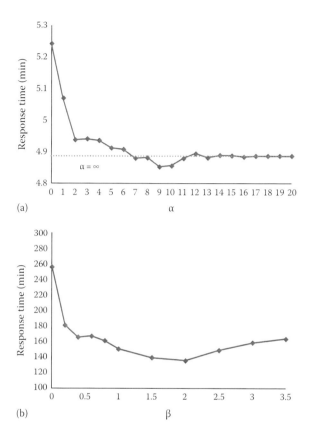

(a)

(b)

**FIGURE 13.6**
Effect of the calibration parameters. (a) Parallelism parameter. (b) Centrality parameter.

the benefit of the parallelism and centrality considerations. The matching process needs to utilize statistical analysis and data mining techniques that have been proven useful while referring to the research on parameter calibration (Grefenstette, 1986; Eiben et al., 1999; Freisleben, 2000; Francois and Lavergne, 2001; Czarn et al., 2004; Ramos et al., 2005; Samples et al., 2005; Bartz-Beielstein, 2006; de Landgraaf et al., 2007; Preuss and Bartz-Beielstein, 2007; Nannon et al., 2008; Bartz-Beielstein et al., 2009; Smit and Eiben, 2009).

## Discussion

Human-centered service robot systems have a great potential in enhancing the safety and welfare of our society in various potential applications. However, there are various issues related to dynamic resource allocation that must be addressed to leverage the benefits of the service robot systems. This research proposes to develop a unified decision framework of dynamic resource allocation that can be flexibly and effectively applied to a wide range of applications and contexts and provide a procedure that can be used in developing such a decision framework. The development of the decision framework will greatly alleviate the barriers against the introduction of complex service robot systems. The experimental and theoretical results obtained in the course of this development will be of use to practitioners as well as researchers in various profit (e.g., robot manufacturers) and nonprofit organizations (e.g., hospitals). The practitioners will be able to easily apply the decision framework to their individual problems, and the researchers can use the framework as benchmarking for their development. The decision framework is expected to have such impacts especially because it is synthesized from the decision principles that are intuitively understandable and transparent, overcoming the barriers of nonintuitive ad hoc approaches against wide dissemination.

## References

Albert R, Barabási A-L (2002). Statistical mechanics of complex networks. *Reviews of Modern Physics* 74(1): 47–97.

Anussornnitisarn P, Peralta JA, Nof SY (2002). Time-out protocol for task allocation in multi-agent systems. *Journal of intelligent Manufacturing* 13(6): 511–522.

Arras KO, Grzonka S, Luber M, Burgard W (2008). Efficient people tracking in laser range data using a multi-hypothesis leg-tracker with adaptive occlusion probabilities. *Proceedings of International Conference on Robotics & Automation*, Pasadena, CA, pp. 1710–1715.

Barrat A, Barthelemy M, Pastor-Satorras R, Vespignani A (2004). The architecture of complex weighted networks. *Proceedings of National Academy of Sciences of the United States of America* 101(11): 3747–3752.

Barthelemy M (2004). Betweenness centrality in large complex networks. *The European Physical Journal B* 38(2): 163–168.

Bartz-Beielstein T (2006). *Experimental Research in Evolutionary Computation: The New Experimentalism*, Springer, Berlin, Germany.

Bartz-Beielstein T, Chiarandini M, Paquete L, Preuss M (2009). *Empirical Methods for the Analysis of Optimization Algorithms*, Springer, Berlin, Germany.

Bellotto N, Hu H (2009). Multisensor-based human detection and tracking for mobile service robots. *IEEE Transactions on System, Man, and Cybernetics, Part C: Cybernetics* 39(1): 167–181.

Berman S, Nof SY (2011). Collaborative control theory for robotic systems with reconfigurable end effectors. *Proceedings of 21st ICPR*, Stuttgart, Germany.

Bertsimas DJ, van Ryzin G (1991). A stochastic and dynamic vehicle routing problem in the Euclidean plane. *Operations Research* 39(4): 601–615.

Burgard W, Argyros A, Hähnel D, Baltzakis H, Pfaff P, Stachniss C (2005). TOURBOT and WebFAIR: Web-operated mobile robots for tele-presence in populated exhibitions. *IEEE Robotics & Automation Magazine* 12(2): 77–89.

Ceroni JA, Nof SY (2002). A workflow model based on parallelism for distributed organizations. *Journal of Intelligent Manufacturing* 13(6): 439–461.

Ceroni JA, Nof SY (2005). Task parallelism in distributed supply organizations: A case study in the shoe industry. *Production Planning and Control* 16(5): 500–513.

Cui H, Zha H, Zhao H, Shibasaki R (2005). Tracking multiple people using laser and vision. *Proceedings of IEEE/RSJ International Conference on Intelligent Robots and Systems*, Edmonton, AB, Canada, pp. 116–2121.

Cui J, Zha H, Zhao H, Shibasaki R (2006). Laser-based interacting people tracking using multi-level observations. *Proceedings of IEEE/RSJ International Conference on Intelligent Robots and Systems*, Beijing, China.

Czarn A, MacNish C, Vijayan K, Turlach B, Gupta R (2004). Statistical exploratory analysis of genetic algorithms. *IEEE Transactions on Evolutionary Computation* 8(4): 405–421.

de Koster MBM, Le-Anh T, van der Meer JR (2004). Testing and classifying vehicle dispatching rules in three real-world settings. *Journal of Operations Management* 22(4): 369–386.

de Landgraaf WA, Eiben AE, Nannen V (2007). Parameter calibration using meta-algorithms. *Proceedings of 2007 IEEE Congress on Evolutionary Computation*, Singapore, 71–78.

Dias MB (2004). TraderBots: A new paradigm for robust and efficient multirobot coordination in dynamic environments, PhD thesis, Robotics Institute, Carnegie Mellon University, Pittsburgh, PA.

Dorogovtsev SN, Mendes JFF (2002). Evolution of networks. *Advances in Physics* 51: 1079–1187.

Egbelu PJ, Tanchoco JMA (1984). Characterization of automatic guided vehicle dispatching rules. *International Journal of Production Research* 22(3): 359–374.

Eiben AE, Hinterding R, Michalewicz Z (1999). Parameter control in evolutionary algorithms. *IEEE Transactions on Evolutionary Computation* 3(2): 124–141.

Fod A, Howard A, Matarıc M (2002). Laser-based people tracking. *Proceedings of International Conference on Robotics & Automation*, Washington, DC.

Francois O, Lavergne C (2001). Design of evolutionary algorithms—A statistical perspective. *IEEE Transactions on Evolutionary Computation* 5(2): 129–148.

Freeman LC (1977). A set of measures of centrality based on betweenness. *Sociometry* 40(1): 35–41.

Freisleben B (2000). Meta-evolutionary approaches. T. Bäck, D.B. Fogel, and Z. Michalewicz (Eds.) *Evolutionary Computation 2: Advanced Algorithms and Operators*, Taylor & Francis, pp. 212–223.

Fritsch J, Kleinehagenbrock M, Lang S, Plötz T, Fink GA, Sagerer G (2003). Multi-modal anchoring for human–robot-interaction. *Robotics and Autonomous Systems* 43(2–3): 133–147.

Gerkey BP, Matarić MJ (2002). Sold!: Auction methods for multi-robot coordination. *IEEE Transactions on Robotics and Automation* 18(2): 758–768.

Grefenstette J (1986). Optimization of control parameters for genetic algorithms. *IEEE Transactions on Systems Man and Cybernetics* 16(1): 122–128.

Grundel DA (2005). Searching for a moving target: Optimal path panning. *Proceedings of IEEE International Conference on Networking, Sensing, and Control*, Tucson, AZ, pp. 867–872.

Jeong W, Nof SY (2008). Performance evaluation of wireless sensor network protocols for industrial applications. *Journal of Intelligent Manufacturing* 19: 335–345.

Jeong W, Nof SY (2009). Design of timeout-based wireless microsensor network protocols: Energy and latency considerations. *International Journal of Sensor Networks* 5(3): 142–152.

Jones EG, Browning B, Dias MB, Argall B, Veloso M, Stentz A (2006). Dynamically formed heterogeneous robot teams performing tightly-coordinated tasks. *Proceedings of IEEE International Conference on Robotics and Automation*, Orlando, FL, pp. 570–575.

Kim M-H, Baik H, Lee S (2011). Multi-UAV search and attack task allocation for mobile targets. *Proceedings of Industrial Engineering Research Conference,* Reno, NV.

Kim M-H, Kim S-P, Lee S (2012). Social-welfare based task allocation for multi-robot systems with resource constraints. *Computers & Industrial Engineering* 63(4): 994–1002.

Kim M-H, Lee S (2010). Resource welfare based distributed task allocation scheme for multiple UAVs. *Proceedings of Industrial Engineering Research Conference,* Cancun, Mexico.

Kleinhagenbrock M, Lang S, Fritsch J, Lomker F, Fink G, Sagerer G (2002). Person tracking with a mobile robot based on multi-modal anchoring. *Proceedings of IEEE International Workshop on Robot and Human Interactive Communication*, Berlin, Germany.

Kluge B, Kohler C, Prassler E (2001). Fast and robust tracking of multiple moving objects with a laser range finder. *Proceedings of International Conference on Robotics & Automation*, Piscataway, NJ.

Lee J, Lee S (2011). Ambulance dispatching driven by parallelism, myopia, and no-reservation. *Proceedings of Industrial Engineering Research Conference*, Reno, NV.

Lee S (2011). The role of preparedness in ambulance dispatching. *Journal of the Operational Research Society* 62(10): 1888–1897.

Lee S (2012). The role of centrality in ambulance dispatching. *Decision Support Systems* 54(1): 282–291.

Lee S (2013). Centrality-based ambulance dispatching for demanding emergency situations. *Journal of the Operational Research Society* 64(4): 611–618.

Lin L, Zheng Z (2005). Combinatorial bids based multi-robot task allocation method. *Proceedings of IEEE International Conference on Robotics and Automation*, Barcelona, Spain, pp. 1157–1162.

Liu JNK, Wang M, Feng B (2005). iBotGuard: An Internet-based intelligent robot security system using invariant face recognition against intruder. *IEEE Transactions on System, Man, Cybernetics, Part C: Applications and Reviews* 35(1): 97–105.

López-Fernández Robles LG, Gonzalez-Barahona JM, Herraiz I (2006). Applying social network analysis techniques to community-driven libre software projects. *International Journal of Information Technology and Web Engineering* 1(3): 27–48.

Luber M, Tipaldi GD, Arras KO (2011). Place-dependent people tracking. *International Journal of Robotics Research* 30(3): 280–293.

Mantel RJ, Landeweerd HRA (1995). Design and operational control of an AGV system. *International Journal of Production Economics* 41: 257–266.

Mucientes M, Burgard W (2006). Multiple hypothesis tracking of clusters of people. *Proceedings of IEEE/RSJ International Conference on Intelligent Robots and Systems*, Beijing, China.

Nannen V, Smit SK, Eiben AE (2008). Costs and benefits of tuning parameters of evolutionary algorithms. *Proceedings of International Conference on Parallel Problem Solving from Nature*, Dortmund, Germany, pp. 528–538.

Newman MEJ (2003). The structure and function of complex networks. *SIAM Review* 45: 167–256.

Newman MEJ (2004). Analysis of weighted networks. *Physical Review E* 70(5): 056131.

Østergaard E, Matarić MJ, Sukhatme GS (2001). Distributed multi-robot task allocation for emergency handling. *Proceedings of IEEE/RSJ International Conference on Intelligent Robots and Systems*, Wailea, HI, pp. 821–826.

Preuss M, Bartz-Beielstein T (2007). Sequential parameter optimization applied to self-adaptation for binary-coded evolutionary algorithms. F.G. Lobo, C.F. Lima, and Z. Michalewicz (Eds.) *Parameter Setting in Evolutionary Algorithms, Studies in Computational Intelligence*, Springer, New York, pp. 91–119.

Ramos ICO, Goldbarg MC, Goldbarg EG, Neto AD (2005). Logistic regression for parameter tuning on an evolutionary algorithm. *Proceedings of 2005 IEEE Congress on Evolutionary Computation*, Edinburgh, UK, pp. 1061–1068.

Ravasz E, Somera A, Mongru D, Oltvai Z, Barabási A-L (2002). Hierarchical organization of modularity in metabolic networks. *Science* 297(5586): 1551–1555.

Sabidussi G (1996). The centrality index of a graph. *Psychometrika* 31: 581–606.

Samples ME, Daida JM, Byom M, Pizzimenti M (2005). Parameter sweeps for exploring GP parameters. *Proceedings of 2005 Conference on Genetic and Evolutionary Computation*, Ann Arbor, MI, pp. 1791–1792.

Scheutz M, McRaven J, Cserey G (2004). Fast, reliable, adaptive, bimodal people tracking for indoor environments. *Proceedings of IEEE/RSJ International Conference on Intelligent Robots and Systems* 2: 1347–1352.

Schulz D, Burgard W, Fox D, Cremers AB (2003). People tracking with mobile robots using sample-based joint probabilistic data association filters. *International Journal of Robotic Research* 22(2): 99–116.

Smit SK, Eiben AE (2009). Comparing parameter tuning methods for evolutionary algorithms. *Proceedings of 11th Conference on Evolutionary Computation*, Piscataway, NJ, pp. 399–406.

Sujit PB, Beard R (2007). Distributed sequential auctions for multiple UAV task allocation. *Proceedings of American Control Conference*, New York, pp. 3955–3960.

Sujit PB, Ghose D (2005). Multiple UAV search using agent based negotiation scheme. *Proceedings of American Control Conference*, Portland, OR, pp. 2995–3000.

Taylor G, Kleeman L (2004). A multiple hypothesis walking person tracker with switched dynamic model. *Proceedings of Australasian Conference on Robotics and Automation*, Canberra, Australia.

Tkach I, Edan Y, Nof SY (2011). A framework for automatic multi-agents collaboration in target recognition tasks. *Proceedings of the 21st ICPR*, Stuttgart, Germany.

Tkach I, Edan Y, Nof SY (2012). Security of supply chains by automatic multi-agents collaboration. *Proceedings of the INCOM*, Bucharest, Romania.

Topp E, Christensen H (2005). Tracking for following and passing persons. *Proceedings of IEEE/RSJ International Conference on Intelligent Robots and Systems*, Beijing, China.

Vig L, Adams JA (2006). Multi-robot coalition formation. *IEEE Transactions on Robotics* 22(4): 637–649.

Vig L, Adams JA (2007). Coalition formation: From software agents to robots. *Journal of Intelligent and Robotic Systems* 50(4): 85–118.

Yoon SW, Nof SY (2011). Affiliation/dissociation decisions in collaborative enterprise networks. *International Journal of Production Economics* 130(2): 135–143.

# 14

## Laser Applications in Safety and Ergonomics

Nancy Huang, Gaurav Nanda, and Mark R. Lehto

### CONTENTS

## Introduction

Lasers have become so much a part of daily life that many people may not realize how ubiquitous they are. Every home with a CD player has a laser; hardware stores are now selling a wide variety of laser-based products; many, if not most, computers, printers, and copiers are using laser technology. Laser technology is also playing an increasingly large role in a large variety of safety applications, particularly with respect to developing sophisticated warning and alerting systems. Illustrating this trend, most automobile manufacturers currently offer collision avoidance systems (CASs) as an option for higher-end vehicles. Emerging evidence, including recent simulator studies, suggests such systems can be effective (Muhrer et al., 2012). Real-world data also support this conclusion. In particular, a recent study conducted by the US insurance industry found that the frontal collision warning systems (FCWSs) offered on Acura, Mercedes-Benz, and Volvo vehicles reduced crashes by 10%–14% (IIHS news release, n.d.). The following sections provide a brief introduction to lasers and discuss the various laser applications and their effectiveness in respective industries.

The term *laser* is an acronym for *l*ight *a*mplification by *s*timulated *e*mission of *r*adiation. Lasers are devices that produce intense beams of light that are *monochromatic, coherent*, and *highly collimated*. The wavelength (color) of laser light is extremely pure (monochromatic) when compared to other sources of light, and all of the photons (energy) that make up the laser beam have a fixed phase relationship (coherence) with respect to one another. Light from a laser typically has very low divergence. It can travel over great distances or can be focused to a very small spot. Because of these properties, lasers are used in a wide variety of applications in all walks of life. The basic operating principles of the laser were put forth by Charles Townes and Arthur Schawlow from the Bell Telephone Laboratories in 1958, and the first actual laser, based on a pink ruby crystal, was demonstrated in 1960 by Theodore Maiman at Hughes Research Laboratories.

Lasers can be broadly classified into four categories: gas discharge lasers, semiconductor diode lasers, optically pumped lasers, and *others*, which can include chemical lasers, gas dynamic lasers, x-ray lasers, combustion lasers, and others developed primarily for military applications. The most common types of gas discharge lasers are helium–neon lasers, helium–cadmium lasers (a metal vapor laser), noble gas ion lasers (argon, krypton), carbon dioxide lasers, and the excimer laser family. Practical diode laser devices reach a 50% electrical-to-optical power conversion rate, at least an order of magnitude larger than most other lasers. Over the past 20 years, the trend has been one of a gradual replacement of other laser types by diode laser–based solutions, as the considerable challenges to engineering with diode lasers have been met.

Lasers have a wide range of applications in various industries. High-power lasers have long been used for cutting and welding materials. Today, the frames of automobiles are assembled using laser welding robots; complex cardboard boxes are made with laser-cut dies. Lasers are also used extensively in the scientific laboratory for a wide variety of spectroscopic and analytic tasks. One of the earliest applications of lasers in medicine was photocoagulation, using an argon ion laser to seal off ruptured blood vessels on the retina of the eye. Today, lasers are used extensively in analytical instrumentation, ophthalmology, cellular sorting, and to correct vision. Cosmetic treatment of wrinkles, moles, warts, and discolorations (birth marks) is often accomplished with near-infrared (IR) and IR lasers (Laser Sensor Applications, n.d.; List of Laser Applications—Wikipedia, the free encyclopedia, n.d.).

There are various types of laser-based noncontact measurement techniques in use today. Lasers can solve a variety of practical problems in distance measurement. The interferometric systems using the helium–neon lasers help in accurate distance measurement. Such devices are suitable for dimensional control of machine tools. The laser Doppler displacement method competes with the interferometric distance measurement. The Doppler shift of the beam reflected from a target is measured and integrated to obtain displacement. It is best suited for indoor usage at distances not more than a few hundred feet. A method used outdoor is the beam modulation telemetry. In this method, the modulated beam is transmitted to a distant object, and a reflected return signal is detected. The phase of the modulation of the return signal is compared with the phase of the outgoing signal. This method is useful for distances of hundreds of meters and is used in surveying (Ready, 1997).

As with any sensing technology, laser systems have both advantages and disadvantages. Perhaps their greatest attribute is their ability to resolve measurements below one micron at a fraction of the cost of other high-performance technologies. In addition, their measurement range is large allowing them to fulfill a variety of application requirements. The large operating distance provides sufficient standoff to reduce possible damage from contacting the moving target. Some of the disadvantages of laser-based sensors are the care involved in handling them. Laser sensors should be kept clean. Dirt or other foreign debris can affect accuracy, so frequent cleaning may be required. Because laser heads have sensitive electronic components, their operating temperature is limited and vacuum installations are not recommended without external cooling (Technology: Non-contact measurement sensors, solutions and systems—MTI instruments, n.d.).

## Laser Applications in Safety and Ergonomics

This section discusses various laser applications in the area of safety and ergonomics. Application examples include CASs, laser-based speed enforcement devices, presence sensing safety devices, intrusion alarms, in-line process monitoring, human scanning, human position tracking, remote sensing, hazardous gas leak detection, food safety, fire detection, dam safety, identity authentication systems in healthcare, and eye tracking technology in human–computer interaction (HCI). Each application is elaborated in this section.

### Collision Avoidance Systems

According to the World Health Organization, about 1.3 million people die each year as a result of traffic accidents. More than 90% of road accidents are caused by human error. Laser sensors have been widely used in CASs to improve driving safety. Lasers are one of the commonly used sensors that can quickly and accurately obtain real-time measures of position, velocity, and acceleration. The following section provides an overview of the CAS in terms of its origins, the US government's research, and the introduction to popular CAS technologies.

Laser technology is particularly good in these and a wide variety of other collision avoidance applications in which lasers can be used to quickly and accurately obtain real-time measures of position, velocity, and acceleration. Some of these other applications of collision avoidance warning systems where laser technology plays an important role include laser collision warning systems for overhead crane and hoists now offered by many different manufacturers and laser helicopter warning systems that warn pilots of power lines and other obstacles when flying at low altitudes (Schulz et al., 2002).

Laser technology is a critical element in many other safety devices including interlocks or emergency braking systems. For example, presence sensing and interlock devices are sometimes used in installations of robots to sense and react to potentially dangerous workplace conditions. Laser light curtains are often used in such applications to determine whether someone has crossed the safety boundary surrounding the perimeter of the robot. Perimeter penetration will trigger a warning signal and in some cases will cause the robot to stop. Laser and other types of end effector sensors detect the beginning of a collision and trigger emergency stops. In more sophisticated applications, computer vision and other methods of data fusion can play an important role in detecting safety problems during robot motion planning and collision avoidance (Baltzakis et al., 2003).

CASs, which are also called the *collision mitigation system*, originated in the aviation industry. An airborne CASs is an aircraft system that operates independently of ground-based equipment and air traffic control; it warns pilots of the presence of other aircraft that may present a threat of collision. Modern aircraft utilizes several types of CAS to prevent unintentional contact with other aircraft, obstacles, or the ground. For example, the *a*irborne *r*adar has been in military use since World War II to help locate bombers or detect the relative location of other aircrafts.

The technology of CAS has gradually made its way into the automotive industry. With the aim to enhance the road safety, the US Department of Transportation has performed several projects in collaboration with different research institutions since the 1990s. In 1995, the National Highway Traffic Safety Administration (NHTSA) sponsored a two-year project to launch the automotive CAS program. CAS was introduced as a promising technology to reduce the severity of vehicle crashes (Zador et al., 2000). During a two-year project from 2003 to 2005, the Federal Motor Carrier Safety Administration (FMCSA) collaborated with the trucking industry to test and evaluate several onboard safety systems, such as *c*ollision *w*arning *s*ystems and *a*utomated *c*ruise *c*ontrol, for commercial motor vehicles. The goal was twofold: to reduce the number and severity of large truck fatalities and crashes and to increase the safety and security of all roadway users (Houser et al., 2005). Another project sponsored by the NHTSA was conducted between 2005 and 2008; the University of Michigan Transportation Research Institute (UMTRI) evaluated the performance of the integrated vehicle-based safety systems, which are designed to help the driver avoid road departure, rear end, and lane change crashes by providing occasional crash alerts and advisories to enhance the driver's situation awareness (LeBlanc et al., 2008).

The CAS is comprised of one or all of three subsystems: obstacle detection system, warning system, and autonomous control system (Labayrade et al., 2005). The obstacle detection system is based on in-vehicle sensors to detect any dangers that may lie ahead on the road. Subsequently, the warning system alerts its driver to enhance their situation awareness. An advanced CAS has an autonomous control system such as adaptive cruise control (ACC) or an automatic braking system to initiate autonomous control of the host vehicle to avoid collision. The most common CAS technologies include forward or reverse collision warning systems, ACC, autonomous braking, lane departure warning or prevention, adaptive headlights, blind spot detection, and backup cameras (Collision Avoidance System, 2011).

Abdul Vahid (2011) highlights several popular applications of CAS:

1. Forward/front collision warning system (FCWS)
2. Reverse/rear end collision warning system (RCWS)
3. Adaptive cruise control (ACC)
4. Collision mitigation by braking (CMB)

FCWS attempts to improve driving safety by warning a driver of any potential dangers that lie ahead on the road. This in-vehicle electronic system monitors the roadway in front of the host vehicle and warns the driver whether a collision is imminent or not. Most warning signals are visual, audible, or both. This system is based on the sensors placed in vehicles to send and receive signals from other cars or obstacles that lie ahead on the road. FCWS, therefore, helps in dangerous situations by anticipating the distance and relative speed between the car and other cars surrounding it and providing parking assistance or curve guidance. FCWS plays a critical role in driving safety. A key technique called time to contact (TTC) is used in the FCWS to examine the speed of the host vehicle, relative speed, and relative acceleration by learning the variation in size of the image of the vehicle in front. Consequently, the driver will be warned of the dangers. Drivers often misjudge about the distance between vehicles during driving. The system warns the drivers of dangers and helps drivers more accurately measure the distance between the vehicles, allowing drivers to take proper and timely action to avoid a collision.

RCWSs have become very common in modern vehicles. In contrast to the FCWS, RCWS has built-in sensors that are normally placed on the rear bumper to detect vehicles or obstacles behind the driver. This system emits video and audio signals when the host vehicle approaches an obstacle in reverse. This system also warns the following drivers by flashing brake lights or other warning lights, so the follower can slow their vehicle down and keep a safe distance. However, the excess of warning of RCWS may potentially lead to undesired effects for drivers such as distraction and irritation. Thus, the system needs to be properly designed to avoid this problem.

ACCs are an extension of the standard cruise control system. Like a conventional cruise control system, ACC maintains a user-set speed. According to Rajamani (2012), the ACC system differs itself in the functionality of automatically maintaining its distance from preceding vehicles even when they slow down. The current system uses radar or laser-based (LIDAR) sensors mounted in the front bumper or grille to monitor the road ahead. If there are no preceding vehicles, the ACC vehicle travels at a speed set by the driver. However, if a preceding vehicle is detected, the ACC system determines whether the host vehicle can continue to travel safely at the current speed. If the distance from the preceding vehicle to the host vehicle is detected to be too close, the ACC system switches from speed control to spacing control. In the spacing control mode, the ACC vehicle controls the engine throttle and brakes in order to maintain a desired following interval.

The ACC system was first introduced in Japan and Europe in 1997 and is now available in the US market. The 2003 Mercedes S-class and E-class passenger sedans and the 2003 Lexus LS340 come with an optional radar or laser-based ACC system. An ACC system allows drivers to have a safer and more comfortable driving experience; potentially, it reduces drivers' burden by partially replacing driver operation with automated operation. This system is forward-looking for its potential benefit of reducing road accidents (Rajamani, 2012).

CMB systems, also known as *advanced emergency braking system* or *autonomous braking system*, are an evolution of the ACC system. When an accident is predicted, the CMB system alerts the host driver through audio, visual, or haptic signals to break. However, if the driver fails to react to the situation properly, this system autonomously triggers the brakes to slow the host vehicle down and prevent it from colliding with another vehicle. The European Union (EU) has announced that CMB systems will become mandatory for new heavy vehicles starting from 2013. The EU will use the United Nations Economic Commission for Europe (UNECE) regulation as a basis for the approval of such systems. This could, according to the impact assessment, ultimately prevent around 5,000 fatalities and 50,000 serious injuries per year across the EU (European Commission, 2008) (Figure 14.1).

Two commonly cited technical challenges for collision warning and avoidance systems are, namely, the development of reliable and all-weather target detection systems and the trade-off between false/nuisance alarms

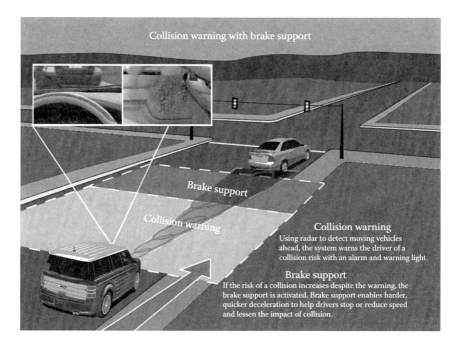

**FIGURE 14.1**
Collision warning with brake support.

(false-positive) and missed detections (false-negative) (Lee and Peng, 2005). From a short-term perspective, false alarms are merely a nuisance to the operator, but over time, false alarms create mistrust in the system and encourage people to ignore the warnings or even disable the systems (Papastavrou and Lehto, 1996; Lehto et al., 2009) The solution is to increase the quality of the information provided, by reducing the number of false alarms and misses. This can be done by either improving sensor technology or better algorithms. Simply adjusting the warning threshold to a more optimal (less conservative) setting that reduces the number of false alarms can have a large benefit by encouraging drivers to behave more rationally (Lehto et al., 2000). However, ultimately, the best solution is to build on advances in laser technology to develop systems capable of accurately detecting imminent collisions under all foreseeable conditions. Advances in sensor fusion and image recognition in which laser technology is combined with other methods of sensing environmental and operator states are, of course, a first step in developing truly intelligent warning systems with capabilities well beyond those currently available.

## Laser-Based Speed Enforcement Devices

According to the traffic safety facts published by NHTSA in 2008, 31% of all fatal crashes were speeding related, with 11,674 lives being lost. Speed not only affects the severity of a crash but is also related to the risk of being involved in a crash. Empirical studies analyzing the relationship between speed and the risk of a crash show that, at a particular road, the crash rate increases when speed increases (Aarts and Van Schagen, 2006). Laser-based enforcement devices may deter habitual violators from speeding and ultimately can help save lives. From speed enforcement to crash scene investigations, the laser-based equipments have been used for years by law enforcement agencies around the world. Laser-based devices are being used for tailgating enforcement, photo-/video-based speed enforcement, statistical data collection, and general measurements (Patent US5938717—Speed detection and image capture system for moving vehicles—Google patents, n.d.).

In one of the patented laser-based speed-measuring devices, the laser range finder fires a series of laser pulses at a selected target at known time intervals and detects reflected laser light from each pulse. A microcontroller is configured to read these and to compute from them the time of flight of laser pulse and the distance to the target and then computes the velocity of the vehicle with respect to the speed detector from the change in distance to the target divided by the known elapsed time between firing of the pulses (Patent US5521696—Laser-based speed measuring device—Google patents, n.d.). A schematic diagram of the device is shown in Figure 14.2 later.

A well-known organization called Laser Technology Inc. (LTI) introduced one of the very first laser speed devices called the LTI 20/20 Marksman. This was used by the law enforcement community to pinpoint a particular vehicle in dense traffic, and this equipment was unaffected by radar detectors (Laser technology—Traffic safety, n.d.). A study demonstrated that a system for real-time classification of vehicles in traffic situations can also be implemented with a laser range imaging system mounted over a traffic lane. A traffic monitoring system of this type offers potential for traffic controllers where the traffic control reflects the type of vehicles in the traffic pattern. It also has potential in improved highway safety because highway

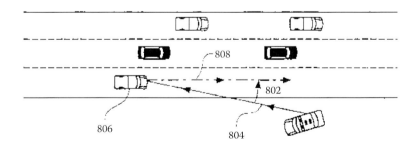

**FIGURE 14.2**
Laser-based speed measuring device. (From Patent US5521696—Laser-based speed measuring device—Google patents, Retrieved February 12, 2013, from http://www.google.com/patents? hl = en&lr = &vid = USPAT5521696&id = WJgnAAAAEBAJ&oi = fnd&dq = laser+speed+gun&pri ntsec = abstract#v = onepage&q&f = false, n.d.)

safety is related to the types of vehicles and their observance of speed limits and traffic rules (Harlow and Peng, 2001). Another study formulated a nonintrusive laser-based detection system for measurement of vehicle travel time. This system can determine the length and width of moving objects in real time with high resolution. This information can be used to differentiate similar objects and can be used later for reidentification of individual objects or object groups, providing a real measure of travel time between detection sites (Cheng et al., 2001).

## Presence Sensing Safety Devices

Presence sensing safety devices, also known as *light curtains*, are one of the main laser applications in safety engineering. These devices are designed to automatically stop the machine's operation if an interruption in the sensing field is detected. As safeguards, presence sensing devices are commonly used for power presses with an automatic feed to provide protections for operators and other people in the area. In addition to the advanced safety, presence sensing devices also increase productivity in terms of quick setup, less maintenance, and increased visibility of the work area (Presence Sensing Devices, 2012) (Figures 14.3 and 14.4).

## Intrusion Alarms

Laser scanning technology has been used for some time in a range of industrial and commercial applications. These span from detecting hands when they are close to dangerous machinery to preventing collisions at container ports and monitoring manufacturing processes. Laser scanners are noncontact measurement systems and scan their surroundings two-dimensionally.

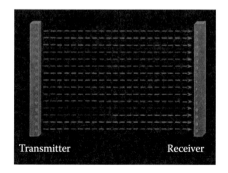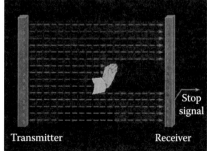

**FIGURE 14.3**
The basic operation of presence sensing. *The transmitter unit sends modulated IR light to a corresponding array of phototransistor in the receiver unit. When an opaque object interrupts one or more of the beams, the light curtain sends a stop signal to the guarded machine and initiates a stopping action of the machine.* (From Link Systems, Presence ensing devices—An introduction, n.d., Retrieved from http://www.linkelectric.com/tools_psdover.html, December 6, 2013.)

(a)

(b)

**FIGURE 14.4**

Image examples of presence sensing. (a) Point of operation light curtain. (b) Point of operation light curtain provides horizontal pass-through protection.

(c)

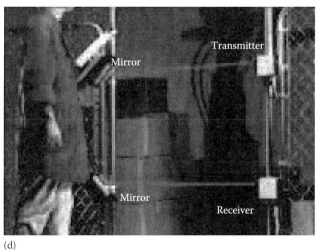

(d)

**FIGURE 14.4 (continued)**
Image examples of presence sensing. (c) Point of operation light curtain provides horizontal pass-through protection. (d) Single beam light curtain with mirrors.

*(continued)*

This describes the application of laser scanning to physical security, including modifications to such systems for their use as perimeter intrusion detection systems (PIDSs). In these systems, motion detection is provided by a pulsed laser beam by measuring the propagation time for the reflected beam to return to its source. Using this information, a contour plot of the surrounding area is built up. Thus, lasers can be used to set up an invisible fence to protect an area (Figure 14.5).

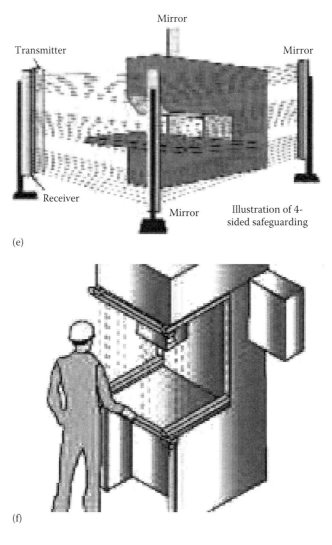

(e)

(f)

**FIGURE 14.4 (continued)**
Image examples of presence sensing. (e) Perimeter light curtain for large work envelope safeguarding. (f) Light curtain.

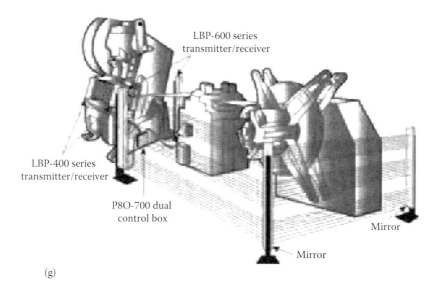

(g)

**FIGURE 14.4 (continued)**
Image examples of presence sensing. (g) Point of operation and perimeter light curtain protection for power press and feeding system. (From Machine Guarding eTool | Presses - Presence Sensing Devices, Occupational Safety & Health Administration, the U.S. Department of Labor, (n.d.). Retrieved from https://www.osha.gov/SLTC/etools/machineguarding/presses/psd.html. December 9, 2013.)

**FIGURE 14.5**
**(See color insert.)** Laser-based intrusion alert system. (From FLM—Applications.)

Once this contour information has been obtained, the detector can recognize the addition of an object to its field of view through a change in the programmed surroundings. Consequently, the size, shape, and direction of targets moving in the frame can be assessed and an alarm output given, if required. Laser scanners can be mounted in a number of different configurations, depending on the application for which they are required. They are suitable for attaching to walls, roofs, lighting poles, fences, and any other (relatively stable) object. Laser scanning is suitable for identifying the unauthorized presence of vehicles, as well as detection of personnel (Hosmer, 2004). Laser scanners of various types have been used for monitoring nuclear sites. Laser scanners are also increasingly being used for monitoring walls in galleries or museums. Some companies have developed efficient laser scanner solutions to protect exhibits, indoor areas, and corridors as well as to monitor ceilings (Lasers in security | GIT-security.com—Portal for safety and security, n.d.).

An IR laser beam in combination with an optical detector can seal a path, an area, or a volume against infiltrators. When the invisible beam is interrupted by an intruder trying to approach the protected area, it sets off a remote alarm. The laser alarm has many advantages over the conventional electric alarm. The IR beam, being invisible, cannot be spotted by the intruder. The narrowness of the beam also minimizes false alarms by the passage of birds, small animals, and objects floating in the air. Laser-based intrusion systems have also been used for monitoring the construction of swimming pools as well as detection of persons entering the pool (Patent US6278373—Laser intrusion system for detecting motion in a swimming pool—Google patents, n.d.).

## In-Line Process Monitoring

Industrial processes usually operate under highly constrained conditions due to material and energy balances, quality requirements, operational constraints, safety constraints, and control feedback (Joe Qin, 2003). Hence, efficient process monitoring is important from the viewpoint of cost, safety, and quality. The special properties of lasers enable a variety of spectroscopic methods for the chemical analysis of a substance or for the determination of the physical state, which can be used for real-time in-line process monitoring. The tunability of the laser radiation wavelength to atomic or molecular transitions offers high selectivity in analysis. The high spectral intensity of laser radiation increases the sensitivity; hence, even small trace element concentrations can be determined. For practical applications also, the laser spectroscopic methods are of special interest, as they are able to determine several species simultaneously with minimum equipment. Laser-induced breakdown spectroscopy (LIBS) can perform simultaneous multielement analysis of solid, liquid, and gaseous substances. The process control in steel production relies on the results from the chemical analysis of the slag. An analytical system developed using LIBS to analyze slag samples has been found to be two times faster than with conventional methods (Noll et al., 2001).

Online production thickness and length measurements have conventionally been made using direct contact–type measurement systems. Unfortunately, contact-type methods cause measurement problems. Not only can the material being measured be damaged but sensor wear also occurs. In addition, contact sensors are slow and may not properly track targets that may move or vibrate, making these applications ideal for laser systems. Laser displacement sensors are noncontact by design. That is, they are able to precisely measure the position or displacement of an object without touching it. Because of this, the object being measured will not be distorted or damaged and target motions will not be dampened. Noncontact sensors are ideal for measuring moving targets because they have high frequency response and do not dampen target motions by adding mass, hence making them ideal for high-speed applications such as in-line process monitoring. As shown in the following figure, laser sensors can be used to ensure if the vacuum sealing has been done properly or not on a high-speed conveyor belt (Technology: Non-contact measurement sensors, solutions and systems—MTI instruments, n.d.) (Figure 14.6).

Laser sensors can also be used to measure curved targets. As shown in the following figure, the surface profile of the manufactured objects can be measured using laser sensors for consistency and safety (Figure 14.7).

Apart from the accuracy, safety is also a concern when measuring hot or hazardous objects. For a steel pipe manufacturer, laser-based noncontact sensors were found to be very helpful for measuring the length of hot pipes

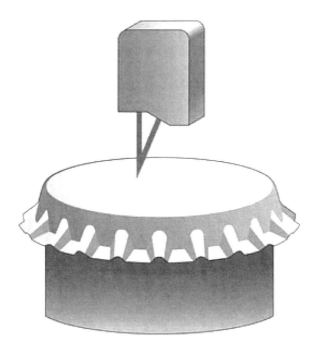

**FIGURE 14.6**
Vacuum seal integrity for canning industry. (From Technology: Non-contact measurement sensors, solutions and systems—MTI instruments, Retrieved January 31, 2013, from http://www.mtiinstruments.com/technology/triangulation.aspx, n.d.)

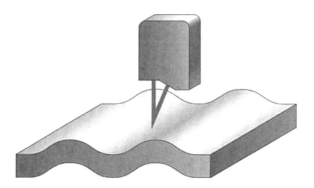

**FIGURE 14.7**
Surface profile of a wide variety of materials. (From Technology: Non-contact measurement sensors, solutions and systems—MTI instruments, Retrieved January 31, 2013, from http://www.mtiinstruments.com/technology/triangulation.aspx, n.d.)

during production. In this situation, a noncontact solution was required because these pipes are quite hot and glowing orange just after production. The laser spot was perfectly aligned to measure the edge of the pipe, on the surface between the inner and outer diameters. The range finder was installed sufficiently far away from the heavy, moving pipe. This steel pipe manufacturer planned to use this solution on all of its production lines in the future because it saves time and reduces worker exposure to the hot material (Laser Sensor Applications, n.d.).

## Human Scanning

In occupational injury prevention applications, anthropometric measurements are used to evaluate the interaction of workers with tasks, tools, machines, vehicles, and personal protective equipment (CDC—Anthropometry—NIOSH workplace safety and health topic, n.d.). Designs that are incompatible with anthropometric measurements of a workforce could result in undesired incidents. For example, inadequate length or configuration of seat belts could lead to nonuse of seat belts, which will affect postcrash survivability. Inadequate fit of personal protective equipment cannot provide workers with sufficient protection from health and injury exposures. Since different occupational groups perform different jobs, the use of inappropriate or inadequate group anthropometric data for designing workplaces, systems, and personal protective devices is a serious safety concern. Diverse workforces in many occupations, as well as new roles for women in the workforce, require detailed body size data for designing adequate workplaces, systems, and personal protective equipment. Technological development in recent years has advanced the basic science of human size and shape studies in 3D forms, and computer-generated human models are now available for anthropometric analysis

(Hsiao et al., 2002). These advancements have resulted in various safety and ergonomic improvements. Some of the projects on anthropometry research at the National Institute for Occupational Safety and Health (NIOSH) using 3D scanning include the following:

- Development of a whole-body fall-arrest-harness sizing scheme and design to control hazards during falls from elevation using 3D information for sizing scheme and developing torso/hip/thigh strap assemblies and rigging components that enhance the ability of the worker to select and use the harness. It will help the construction and other industries to reduce the risk of injury that results from poor user fit, improper selection, and the failure to don the protective equipment properly.
- Using whole-body scanners and hand scanners for various equipment safety evaluation and design applications.
- Development of a new design approach using 3D data and computer-aided face-fit methods for developing respirator fit panels and testing respirators, safety glasses, and helmets.
- Development of approaches to quantify 3D human shapes and sizing information for assessing machine and equipment accommodation level to increase the safety of farming tractor operation (CDC—Anthropometry, n.d.).

There has been an emergence of new 3D imaging devices capable of capturing the entire shape as well as the appearance of the human body. A human body scanner (HBS) is a device that generates a 3D *point cloud* from the subject's frame, that is, a constellation of 100,000–200,000 points generated by the body's surface. These data are saved into a simple digital format and can easily be converted to the most common computer-aided design formats. With the help of 3D body scanners available these days, highly accurate methods for the acquisition of body dimensions have been developed in a fully automated and contact-free manner (3D body scan laser measurements, n.d.) (Figure 14.8).

HBS has a positive impact on human engineering that relies extensively on anthropometric databases and that is also involved in custom fitting items to human surfaces. Such items are generally used by a great number of people and include protective equipment such as helmets, seat belts, desks, airplane, and car seats (Werghi, 2007). Because of the proliferation of PCs and different types of communication networks, the technology of full human shape capture is continually improving, and at the same time, its cost is going down. HBS also opens up new applications in medicine and health. Gyms can use HBS data to track the effects of diet and exercise regimens. Human shape databases will be useful for screening and survey tasks such as monitoring public health problems (e.g., obesity) and assessing child growth.

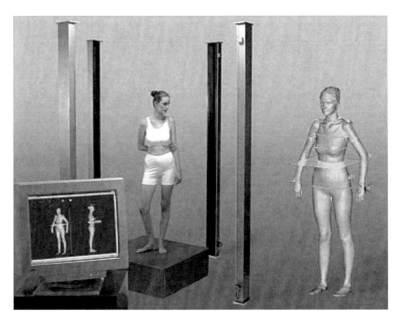

**FIGURE 14.8**
**(See color insert.)** 3D body measurements. (From 3D body scan laser measurements, Retrieved February 11, 2013, from http://www.assystbullmer.co.uk/3d_body_measurements.shtml, n.d.)

HBS would also be of significant interest particularly when the use of other standard medical tools like x-rays is precluded for safety reasons. In fact, the relatively low cost and noninvasive nature of HBS make it a promising potential complement to current medical imaging technologies used to assist medical diagnosis such as computed tomography imagery (CTI) and ultrasound imagery (USI). These 3D optical laser scanners are being used for noninvasive direct, digital capture of the outer ear, ear canal, and eardrum for developing custom hearing products. This 3D laser projection and capture method replaces the need for costly and potentially harmful silicone impressions that have an unacceptable error rate (3D laser ear measurement for custom hearing products—Tackling the most prevalent military service-connected disability, n.d.). Clothing design and human engineering are also potential beneficiaries of HBS technology. The clothing industry is presently targeting custom apparel design, commonly referred to as *apparel on demand*, which aims at producing clothing designed and fitted to an individual's size and proportions. The potential of 3D anthropometric imaging seems tremendous (Paquette, 1996).

## Human Position Tracking

Scanning laser range finders are a popular tool for human tracking and navigation applications due to their precision, effective sensing distance, and ease of use. Laser-based tracking systems are used in mobile robotics and intelligent surveillance areas for estimating human positions as well as human pose represented by human head and waist position. In comparison with cameras, laser range finders are resistant to environmental changes and simple to apply to various environments (Matsumoto et al., 2009). These applications can be used to determine the safety of people in a manufacturing shop floor and also other wide environment such as airports, train stations, and shopping malls.

Real-time human body motion estimation plays an important role in the perception for robotics nowadays, especially for the applications of human–robot interaction and service robotics. In such applications, accurate position estimation and tracking of people in the vicinity of the robot are essential, both for safety and for smooth, context-aware interaction with humans. For robots to operate within society, it will be necessary for them to detect the subtle cues of gesture and body positioning that humans naturally notice and respond to subconsciously. A study demonstrated that 3-layer laser scans can be used for real-time 3D human body motion estimation effectively (Wang et al., 2011). Existing robotic applications can track the locations, posture, and movement of humans. Researchers are developing methods of identifying a variety of human behaviors and understanding social context using data from ubiquitous sensor networks. In addition to data such as location, speed, and relative positioning within a group, pose-related information such as a person's body orientation can provide valuable insight into the dynamics of a social interaction. A study incorporated a shape model into the tracking algorithm, to estimate highly accurate estimates of torso and arm position, and showed that these position estimates can provide information valuable to the understanding of social situations (Glas et al., 2007).

## Remote Sensing

Laser remote sensing is a fast-developing area of laser spectroscopy with numerous applications (Panne, 1998). Being coherent and only diffraction limited in terms of propagation, the special features of laser light favor direct remote sensing. It can be applied to environmentally relevant subjects as global warming, stratospheric ozone depletion, urban photochemical smog (Zelinger et al., 2004), forest decline, detection of oil spill on surface waters, and depth-resolved oceanographic measurements of organic materials. Laser radiation can be produced at almost any wavelength, from the vacuum ultraviolet through to the far IR. Certain lasers can squeeze this energy into extremely short pulses. This allows them to perform a spectral analysis at a distance with excellent temporal and spatial resolution, effectively adding a new dimension to remote sensing. A short laser pulse of appropriate wavelength is directed toward a target of interest. The radiation returning to the photo detection system, mounted adjacent to the laser, provides information about the target, including its range. This is often termed as LIDAR, acronym for light detection and ranging. LIDAR observations can be undertaken from the ground or from vehicles such as trucks, ships, airplanes, helicopters, or space platforms. The scope of lasers in environmental sensing is extensive. They can undertake the following:

1. Concentration measurements of both major and minor constituents and are therefore well suited to monitor the growth of carbon dioxide and other minor constituents in the atmosphere
2. Pollution surveillance and real-time mapping of the dispersal of effluent plumes in both the atmosphere and natural bodies of water
3. Threshold detection of specific constituents for the purpose of generating an alarm when their concentration exceeds some maximum acceptable level
4. Evaluation of thermal, structural, and dynamic properties of both the atmosphere and the hydrosphere (Measures, 1989)

The sensitivity and range of laser remote sensing are very good. Although the presence of oil on natural bodies of water can be detected by a number of sensors, only the laser fluorosensor can instantly characterize the oil and unequivocally evaluate the area and depth of the extensive thin film associated with an oil spill while flying over it. Laser fluorosensors are among the most appropriate sensors for oil spill surveillance because of their ability to detect oil against a variety of backgrounds, including ice and snow (Wang and Stout, 2010). Laser fluorosensor data can be processed in real time and can reliably detect oil and hence can be the most useful remote sensing data as an input for real-time decision support systems. A study modeled the oil spill trajectory based on the available metrological and oceanic current and tidal data and hence provided various oil spill disaster products such as *oil spill location* map, oil spill risk map, oil spill–affected area map, and oil spill *emergency response* map to the oil spill responders. The emergency response system was designed in this study to be Internet-based so that users or emergency responders can access all the valuable information from anywhere in the world (Jha and Gao, 2008).

## Hazardous Gas Leak Detection

Tunable diode laser absorption spectroscopy (TDLAS) is frequently used for measurement of trace gas pollutants in the atmosphere. The TDLAS spectrometers usually work with multipass absorption cells to achieve high sensitivity. The advantage of TDLAS over other techniques for concentration measurement is its ability to achieve very low detection limits. Apart from concentration, it is also possible to determine the temperature, pressure, velocity, and mass flux of the gas under observation. TDLAS is commercially available and is rapidly becoming an accepted gas measurement and analysis technique in the traditionally conservative chemical, petrochemical, power generation, and other industries. Trace gas sensing and analysis by TDLAS has become a robust and reliable technology accepted for industrial process monitoring and control, quality assurance, environmental sensing, plant safety, and infrastructure security (Frish et al., 2005).

Gas sensors based on TDLAS enable sensing trace concentrations of many critical gases in a broad array of applications. TDLAS gas analyzers rely on well-known spectroscopic principles and sensitive detection techniques coupled with advanced diode lasers and optical fibers. Gas molecules absorb energy in narrow bands surrounding specific wavelengths in the electromagnetic spectrum. At wavelengths slightly different than these *absorption lines*, there is essentially no absorption. By transmitting a beam of light through a gas mixture sample containing a quantity of the target gas, tuning the beam's wavelength to one of the target gas's absorption lines, and accurately measuring the absorption of that beam, one can deduce the concentration of target gas molecules integrated over the beam's path length. Typically, each TDLAS system is built using a laser having a specific design wavelength chosen to optimize the sensitivity to a particular target gas while minimizing sensitivity to other gases. Figure 14.9 illustrates a commonly utilized TDLAS system architecture. The laser's fast tuning capability is exploited to rapidly and repeatedly scan the wavelength across the selected gas absorption line. When the wavelength is tuned to be off of the absorption line, the power transmitted through the gas mixture is higher than when it is on the line. The periodic power modulation, combined with well-established noise reduction techniques known as frequency

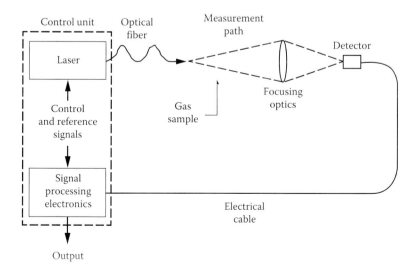

**FIGURE 14.9**
TDLAS gas detector system. (From Frish et al., 2006.)

or wavelength modulation spectroscopy (WMS) and balanced ratiometric detection (BRD), yields a precise measure of the amount of target gas along the laser path (Frish et al., 2006).

TDLAS has been widely employed in atmospheric trace gas detection and industrial control due to its high sensitivity, selectivity, and rapidity of response. TDLAS systems have been developed for monitoring large-scale methane leakage around the oil refinery (Kan et al., 2006). TDLAS systems are sometimes used in industrial environments as permanently installed open-path devices that sense releases of selected gases as they exit the processing area.

Optical tools have been developed based on TDLAS to help petrochemical refinery and chemical processing plant personnel to locate the source of a toxic or hazardous gas leak while remaining outside the perimeter of the processing area. By standing in a safe area and *shining* the laser beam emanating from this device onto suspected leak sources, operators can quickly isolate the source while minimizing their potential exposure to the hazard. This tool illuminates a distant surface with laser light and measures the amount of target gas along the line of sight transited by the laser beam. Unlike other types of portable gas detectors, this laser-based device does not need to be immersed within the gas leak. This is particularly of value in chemical and petrochemical settings that operate tanks and pipelines containing gases like hydrogen fluoride, hydrogen sulfide, ammonia, or methane. The laser-based sensor can allow plant personnel to isolate the source of a leak while remaining outside the perimeter of the processing area where the hazardous gas is highly concentrated. Thus, the risk to plant personnel is reduced, and by enhancing the speed with which leaks can be located, the risk of an incipient failure becoming a catastrophic failure is reduced significantly (Handheld laser-based sensor for remote detection of gas leaks | Research project database | NCER | ORD | US EPA, n.d.).

Another application for the handheld TDLAS gas sensor is municipal natural gas pipeline leak surveying. Traditionally, this process has been labor intensive, requiring an individual to either drive or walk over every buried natural gas pipe. A laser-based device that can rapidly survey off-road pipelines has great appeal because it would eliminate or minimize the need to walk along these pipes. The device could also be valuable for usage by first responders to determine if an area is safe for occupation or to localize the source of a leak after natural disasters causing pipeline ruptures (Werle, 1998).

Natural gas distribution companies continually survey gas pipelines to detect small leaks and correct them before they become dangerous. Some of the currently employed leak survey tools include combustible gas indicators (CGIs), flame ionization detectors (FIDs), and the relatively recent optical methane detection (OMD). All of these devices require the sensor to be physically embedded within the gas leak plume to detect the leak. Thus, to find leaks, the surveyor generally must travel directly

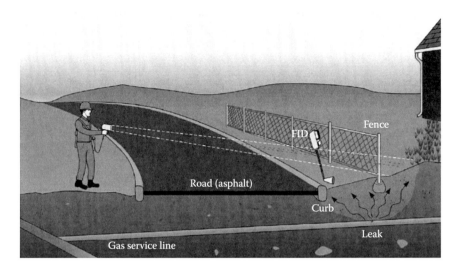

**FIGURE 14.10**
Illustration of a surveyor inspecting a gas pipeline. (From Frish, M.B. et al., Standoff sensing of natural gas leaks: Evolution of the remote methane leak detector (RMLD) (p. JThF3), Optical Society of America, retrieved from http://www.opticsinfobase.org/abstract.cfm?URI = PhAST-2005-JThF3, 2006.)

over the pipeline—a costly and at times difficult process often requiring the service person to enter a property and walk the entire length of the service line. As pictured in Figure 14.10, standoff detection by the *r*emote *m*ethane *l*eak *d*etector (RMLD) based on a laser enables detecting leaks from afar, reducing survey time and enabling more efficient use of man-power. The RMLD system includes a handheld optical transceiver and a shoulder-mounted control unit. The two sections are connected by an umbilical cable bearing an optical fiber and a few wires. The laser beam, projected from the road or sidewalk above the path of the pipeline, illumi-nates a passive surface (i.e., the ground or a building wall). By analyzing the signal embedded in the small amount of laser light reflected back to the transceiver, the RMLD deduces the presence or absence of gas along the laser path. The RMLD can also detect natural gas inside a building or confined space with clear access, as well as in difficult to reach areas, such as gas pipelines under bridges, backyard mains, and fenced-in areas (Frish et al., 2006) (Figure 14.10).

## Food Safety

Food handling, preservation, and packaging are an important aspect of great public interest and concern. The natural presence of oxygen in food products and packaging environments hastens chemical breakdown and microbial spoilage of the food products. Laser-based sensors have been found to be a powerful tool for studying foodstuffs and food packaging. Measurements on minced meat packages, bake-off bread packages, and the headspace in a milk carton illustrate the possibility to monitor the packed foodstuff, as well as the package integrity, nonintrusively. The technique named *gas* in *scattering media absorption spectroscopy* (GASMAS) is based on TDLAS and relies on the fact that free gas inside solid materials absorbs much sharper spectrally than the bulk material. Results from time-dependent measurements of molecular oxygen and water vapor in packages of minced meat, bake-off bread, and the headspace of a milk carton were presented to show that the technique allows gas measurements inside the food through the package and assessment of the integrity of the package (Lewander et al., 2008). Märta Lewander, doctor of atomic physics at Lund University in Sweden, also mentioned in a *Science Daily* (www.sciencedaily.com) article: "It will be the first non-destructive method. This means that measurements can be taken in closed packaging and the gas composition over time can be checked. This will make it possible to check a much higher number of products than at present" (Laser makes sure food is fresh, n.d.).

The automated manufacturing of food generates challenges for the automation technologies in terms of ensuring the quality and minimizing wastage. The raw products are typically natural products that can vary greatly in shape, volume, and dimension. Laser sensor–based systems provide a robust in-line inspection framework for different types of natural food products, for example, 3D measurements of size and shape of fish or ham and optimized cutting for portioning by both weight and shape. Laser-based sensors can be used to measure water content of food to determine it is not spoiled. These sensors can also be used to segregate natural products such as potatoes and fruits based on physical dimensions as shown in Figure 14.11 (Laser inline measurement of food for portioning machines—Direct industry, n.d.).

Flow cytometry is a technique used for measuring single cells. Not only is it a key research tool for cancer and immunoassay disease research, but it is also used in the food industry for monitoring natural beverage drinks for bacterial content or other disease-causing microbes. In a basic cytometer, the cells flow, one at a time, through a capillary or flow cell where they are exposed to a focused beam of laser light (see Figure 14.12). The cell then scatters the light energy onto a detector or array of detectors. The pattern and intensity of the scattered energy help to determine the cell size and shape. In many cases, the cells are tagged with a variety of fluorochromes designed to selectively adhere to cells or cell components with specific characteristics. When exposed to the laser light, only those with the tag fluoresce. This is used in many systems to assist with separation or sorting of cells or cellular components (Darzynkiewicz et al., 1999) (CVI Laser Optics and Melles Griot Home Page, n.d.).

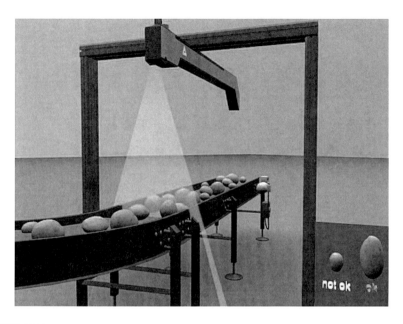

**FIGURE 14.11**
Using laser for in-line measurement. (From Laser inline measurement of food for portioning machines—Direct industry, Retrieved February 11, 2013, from http://news.directindustry.com/press/quelltech-ltd/laser-inline-measurement-of-food-for-portioning-machines-64584-378285.html, n.d.)

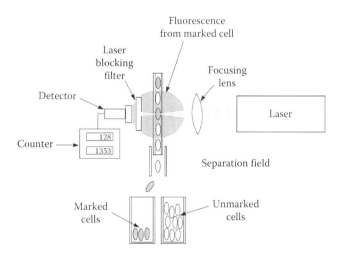

**FIGURE 14.12**
Schematic of a laser cell sorter. (From CVI Laser Optics and Melles Griot Home Page, Retrieved February 21, 2013, from http://www.cvimellesgriot.com/, n.d.)

## Fire Detection

Laser's application in fire detection is based on the principle that a laser beam is affected by hot gases emanating from a fire. A fire releases both heat and combustion products. Either or both of these changes may be used to detect a fire using lasers. All gases change their refractive index with temperature and pressure, and this means that, in a compartment, the speed of light will differ according to the condition of the atmosphere. If, therefore, the refractive index changes with height, a laser beam traversing a compartment near the ceiling will be deflected, once a fire occurs, by the rising heat and combustion products mushrooming out under the ceiling. Typically, a focused laser beam is directed across an open space near ceiling level from one side of the room to the other. It is reflected back to a photocell from a mirror fixed on the opposite wall. Any fire starting below this level will cause turbulent hot air to rise. The laser beam, normally steady, is refracted by the temperature gradients in the hot gases and is displaced from its usual position on a photocell. The deflection can be made to trigger an alarm. Alternatively, the smoke from a fire will either absorb or scatter the light in a laser beam, and this may be used for fire detection. There are thus two main possibilities for laser-based fire detectors, one operating by scattering and the other by heat. There is not much advantage in using a laser beam to detect fire by smoke absorption; ordinary light would be as effective, but there is a distinct advantage in using a laser beam to detect fires by the light scattered by smoke, because it occurs almost exclusively at low angles of scattering, that is, nearly in the direction of the laser beam, so that, by looking toward the laser nearly along the beam, any smoke particles can be clearly seen (Chen et al., 2007).

Trace gas detection systems intended to provide early warning of incipient fires have been developed based on optical absorption using near-IR diode lasers. They can be used for *high-loss* structures such as office buildings, hospitals, hotels, and shopping malls as well as airplanes and manned spacecraft where convention smoke detectors generate unacceptably high false alarm rates. Simultaneous or near-simultaneous detection of several gases, typically carbon dioxide, carbon monoxide, acetylene, and hydrogen cyanide, provides high sensitivity while reducing the chance of false alarms (Bomse et al., 2002). Results have indicated that the laser beam system is at least as fast as the most sensitive fire detection systems in use worldwide (Safety and sensor technology—Laser components USA, Inc., n.d.). Sophisticated methods such as laser sheet imaging technology have been used to develop optical, nonintrusive measuring methods to detect fire smoke (Shu et al., 2006). Lidar-based methods have also been used for the detection of smoke from sources such as power plants, factories, ships, and forest fire. Smoke contains a large number of small particles of ash or soot, leading to high backscattering efficiency and, consequently, favorable conditions for the application of LIDAR (Utkin et al., 2002).

## Dam Safety

The dam is one of the most important parts of facilities of hydraulic power station. The safety of the dam directly affects the people's live and property. The tremendous thrust and pressure of the river, lake, and sea make the dam produce micromoving and vibration. Although the dam seemed to stand firm, it is possible for the dam would be landslip because of the pressure for long time. It will lead economic losses and misfortune for people. So the dam should be measured and monitored on real time (Ioan et al., 2007). Many monitoring methods for dam safety have been proposed in recent years, but it is difficult to monitor the arch dam automatically. In the past, some traditional methods were used for dam safety monitoring. But the major disadvantages of those measuring methods cannot be used for arch dam, and they are difficult to be handled automatically. So those methods cannot be used for monitoring dam safety with network or Internet for long distance. In recent years, laser technology has been put forward for straight dam safety monitoring; it can monitor the dam safety automatically. A study developed a novel inspecting method for arch dam safety monitoring. Using this method, a large arch dam could be measured automatically. According to the temperature and humidity of the arch dam, the semiconductor diode laser is used in the measurement system, and the position-sensitive device has been used for deformation measuring. It is convenient to monitor the safety of dams from long distances. The laser safety monitoring system for large straight dam was used successfully in Fengman, Taipingshao, Gongzui, and Gezhouba dams of China. Since the main principle of the method was based on generalized three-point method, this measuring system could also be used for straight-line dam (Yue et al., 2005).

In 2007, 3D *Laser Mapping*, a UK-based company specializing in laser measurement technology and software, used laser based survey system to monitor the risk of flooding in dams due to heavy rains. Engineers trying to reinforce the embankment dam used it to undertake a detailed survey of the reservoir perimeter in order to maintain the current water level and calculate the capacity of the reservoir in advance of predicted rainfall. The team also demonstrated the use of a laser scanning system designed to measure and monitor the stability of man-made structures, rock faces, and landslips, for the ongoing observation of the dam (3D laser mapping monitors dam safety following UK floods—Directions Magazine, n.d.). A recent study formulated a measure system that included *c*harge-*c*oupled *d*evice (CCD) image sensor, microprocessor-based system, and industrial controlling computer system. The system could detect displacement of dam through CCD image sensors and data processing and communication devices. The industrial controlling computer was used to monitor, record, and store the status of the dam automatically (Ioan et al., 2007).

## Identity Authentication Systems in Healthcare

Healthcare is one of the most time-sensitive environments as it deals with numerous human lives. Every healthcare organization strives for providing patients with efficient and effective treatments. Research has been widely conducted in healthcare environments to explore any potential solutions that are able to improve the system efficiency and patient safety. Field research such as observations and interviews is often applied to better understand the healthcare workflow. The healthcare system involves a great number of information flows and HCIs due to its fundamental responsibility of managing patient medical report. A time motion study conducted in the emergency room indicates that ED doctors and nurses spend the majority of their time interacting with the computers (Zafar and Lehto, 2012). Given the nature of healthcare for the intensive involvement of HCI, health information technology (HIT) emerges as a potential solution to reduce medical errors and improve quality of care. HIT is an umbrella term that encapsulates many technology tools, processes, and policies present in healthcare today (Zafar and Lehto, 2010). Laser systems could be a promising solution to improve the efficiency of HCI in the healthcare setting. For example, healthcare staffs need to work closely with computers to access and update patient medical records in the EMR, requiring frequent login and logout actions (Zafar and Lehto, 2011). With laser-based identification systems such as fingerprint/palm print identification or face recognition in place within the system, healthcare staffs can save time from manually typing in user ID and passwords by just scanning their fingers or face. Potentially, this technology leads to better quality of care since service providers have more time available for direct patient care.

## Eye Tracking Technology in Human–Computer Interaction

According to Leigh and Zee (1999), quantitative eye movement measurements can provide invaluable information for the diagnosis and study of various neurological disorders and are an essential tool in fundamental research on the oculomotor system. The laser-based eye tracker uses a flying spot laser to selectively image landmarks on the eye and subsequently measure 3D eye movement (Irie et al., 2002). Eye tracking technology has been extensively applied in the area of HCI. The eye gaze or tracking technology can be used as a new computer input to ease the use of computers or increase interaction efficiency. Several potential benefits of such systems include ease of use, interaction speedup, maintenance free, hygienic interface, remote control, and safer interaction; it also provides more detailed information on users' activities in terms of their context awareness (Drewes, 2010). According to Poole and Ball (2005), eye trackers have been used extensively in applied human factor research to measure situation awareness in air traffic control training, to evaluate the design of cockpit controls to reduce pilot error, and to investigate and improve doctors' performance in medical procedures.

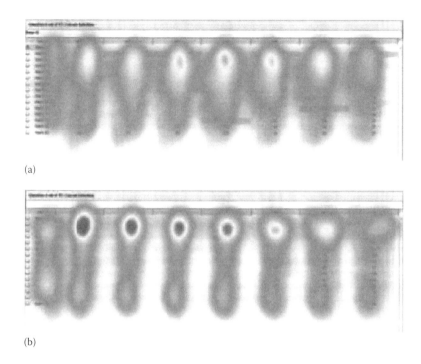

(a)

(b)

**FIGURE 14.13**
Comparing screenshots of the total aggregated fixation duration for two visual techniques; the darker part (in red on the original color-printed figure) indicates longer duration of fixations. (a) SimulSort and (b) typical sorting. (From Kim, S. et al., *IEEE Trans. Vis. Comput. Graph.*, 18(12), 2421, 2012.)

In a recent study, this technology was used to investigate research subject's underlying cognitive processes by documenting how people visually scan the spreadsheets which have been organized in different ways (Kim et al., 2012) (Figure 14.13).

In addition, marketing research and advertising testing are also a popular application area for eye tracking. The eye trackers are used to explore how customers perceive a product and how much influence the characteristic change (such as color or design) a product or web page can be noticed by customers (see the Figure 14.14).

## Conclusions

Laser technology has been widely applied and integrated to a variety of industries and our daily lives. Laser-based applications often provide

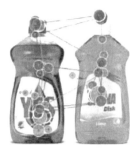

**FIGURE 14.14**
**(See color insert.)** Advertisement for Tobii eye trackers showing a possible application for their product (scanned from a Tobii flyer).

better solutions than conventional ones for the benefits of speed and precision. With the advancement of laser technology, laser-based devices have become cheaper and safer to use. This book chapter highlights several laser applications that are commonly used in practices from the safety and ergonomics perspective. In any event, laser technology has clearly given the researcher and practitioner a new tool to make life better for all mankind. However, laser technology is not without its debates and it still has many safety and accessibility issues. More future research is desired in order to further popularize this technology. To sum up this chapter, a summary table for the highlighted applications is presented in the following (Table 14.1).

As discussed in this chapter, lasers have a variety of applications in the area of ergonomics and safety. Some of the particularly useful applications from human factor perspective include sensing and identity authentication. Further work in this area and methods of actually identifying people to reduce the time of password, especially in the healthcare area, can be very beneficial. With advancements in software, laser technology can be used for a lot of sophisticated tasks like identifying chemicals or weapons in a crowded area. Laser technology has been used for preventive applications in automobiles and can be extended to other areas as well. There can be intelligent vehicles designed not only to identify the driver but also to determine the state and what is the driver looking for and respond intelligently with the advanced software. The human factors and usability aspects should also be taken care of when designing these smart systems. These systems have to be compatible with people and should work in a nonintrusive manner. As these systems can be designed to be more aware, they need to be more sensitive to the actual need of the people using them because if the system is not accepted by end users due to usability issues, it would not be used.

**TABLE 14.1**

Summary Table for Laser Applications

| Applications | Descriptions |
|---|---|
| CASs | Laser technology has been widely used in the collision avoidance applications, specifically the obstacle detection system, to quickly and accurately obtain real-time measures of position, velocity, and acceleration. The most common CAS technologies include forward or reverse collision warning systems, ACC, autonomous braking, lane departure warning or prevention, adaptive headlights, blind spot detection, and backup cameras. |
| Speed enforcement | Speed enforcement devices have utilized laser sensors for speed detection and image capture in order to deter habitual violators from speeding and improve road safety. |
| Presence sensing | Presence sensing safety devices are commonly used for the workstations that have power presses with an automatic feed for workplace safety. This system would also increase productivity in terms of quick setup, less maintenance, and increased visibility of the work area. |
| Intrusion alarms | Laser-based intrusion alarms can be used to set up an invisible fence to protect an area that is unable to be spotted by the intruder. The narrowness of the beam minimizes false alarms by the passage of birds, small animals, and objects floating in the air. |
| In-line process monitoring | Lasers have been utilized to improve industrial process monitoring by conducting spectroscopic methods for the chemical analysis of a substance or for the determination of the physical state. |
| Human scanning | Laser sensors are applied to occupational injury prevention as well. Laser-based body scanners are able to measure the human body in three dimensions precisely for designing adequate workplaces, systems, and personal protective equipment. |
| Human position tracking | Laser-based human tracking and navigation have been applied to improve the safety in manufacturing shop floors and other public areas such as airports, train stations, and shopping malls. In addition, this technology has benefited the development of robotics and social studies by providing valuable insights into the dynamics of a social interaction. |
| Remote sensing | The scope of lasers in environmental sensing is extensive. Examples include concentration measurements of both major and minor constituents to monitor the growth of carbon dioxide and other minor constituents in the atmosphere and pollution surveillance in both the atmosphere and natural bodies of water. |
| Hazardous gas leak detection | Lasers are frequently used for measuring gas pollutants in the atmosphere. TDLAS has less detection limits and is able to determine the temperature, pressure, velocity, and mass flux of the gas under observation. |
| Food safety | For food safety, laser-based systems provide a robust in-line inspection framework for different types of natural food products. Laser-based sensors can be used to measure water content of food to determine it is not spoiled. These sensors can also be used to segregate natural products such as potatoes and fruits based on physical dimensions and cut food products to predetermined weight or shape. |

*(continued)*

**TABLE 14.1 (continued)**

Summary Table for Laser Applications

| Applications | Descriptions |
|---|---|
| Fire detection | A laser-based fire detection device is based on the principle that a laser beam is affected by hot gases emanating from a fire. It has also been used for the detection of smoke from sources such as power plants, factories, ships, and forest fire. This laser beam system has been viewed as the fastest and the most sensitive fire detection system to date. |
| Dam safety | Laser technology has been used to monitor the safety of dams from long distance to overcome the limited accessibility of the network and Internet. The laser safety monitoring system for large straight dam was used successfully in Fengman, Taipingshao, Gongzui, and Gezhouba dams of China. |
| Identity authentication systems in healthcare | Laser systems could be a promising solution to improve the efficiency of HCI in the healthcare setting. With laser-based identification systems such as fingerprint/palm print identification or face recognition in place within the system, healthcare staffs can save time from manually typing in user ID and passwords by just scanning their fingers or face. |
| Eye tracking in HCI | Eye tracking technology has been used extensively in applied human factors and marketing researches as it provides valuable information on users' activities in terms of their context awareness. It has been used to measure situation awareness in air traffic control training, to evaluate the design of cockpit controls to reduce pilot error, and to explore how customers perceive a product and respond to changes of product characteristics. |

# References

3D body scan laser measurements. (n.d.). Retrieved February 11, 2013, from http://www.assystbullmer.co.uk/3d_body_measurements.shtml.

3D laser ear measurement for custom hearing products—Tackling the most prevalent military service-connected disability. (n.d.). Retrieved February 11, 2013, from http://mhsquadrupleaimchallenge.challenge.gov/submissions/13262.

3D laser mapping monitors dam safety following UK floods—*Directions Magazine*. (n.d.). Retrieved February 12, 2013, from http://www.directionsmag.com/pressreleases/3d-laser-mapping-monitors-dam-safety-following-uk-floods/112107.

Aarts, L. and Van Schagen, I. (2006). Driving speed and the risk of road crashes: A review. *Accident Analysis and Prevention*, *38*(2), 215–224. doi:10.1016/j.aap.2005.07.004.

Abdu l Vahid V. (2011). Current trends in collision avoidance systems. *Automotto Retrieved*, February 11, 2013, from http://www.automotto.com/entry/current-trends-in-collision-avoidance-systems.

Baltzakis, H., Argyros, A., and Trahanias, P. (2003). Fusion of laser and visual data for robot motion planning and collision avoidance. *Machine Vision and Applications*, *15*(2), 92–100. doi:10.1007/s00138-003-0133-2.

Bomse , D. S., Hovde, D. C., Chen, S.-J., and Silver, J. A. (2002). Early fire sensing using near-IR diode laser spectroscopy. *Proceedings—SPIE the International Society for Optical Engineering*, Diode Lasers and Applications in Atmospheric Sensing, 4817, 73–81. doi:10.1117/12.452097.

CDC—Anthropometry—NIOSH workplace safety and health topic. (n.d.). Retrieved February 11, 2013, from http://www.cdc.gov/niosh/topics/anthropometry/.

CDC—Anthropometry: Research at NIOSH—NIOSH workplace safety and health topic. (n.d.). Retrieved February 11, 2013, from http://www.cdc.gov/niosh/topics/anthropometry/projects.html.

Chen, S.-J., Hovde, D. C., Peterson, K. A., and Marshall, A. W. (2007). Fire detection using smoke and gas sensors. *Fire Safety Journal*, *42*(8), 507–515. doi:10.1016/j.firesaf.2007.01.006.

Cheng, H. H., Shaw, B. D., Palen, J., Larson, J. E., Hu, X., and Van katwyk, K. (2001). A real-time laser-based detection system for measurement of delineations of moving vehicles. *IEEE/ASME Transactions on Mechatronics*, *6*(2), 170–187. doi:10.1109/3516.928732.

Collision Avoidance System. (2011). Wikipedia, The Free Encyclopedia. Retrieved February 21, 2013, from http://en.wikipedia.org/wiki/Collision_avoidance_system.

CVI Laser Optics and Melles Griot Home Page. (n.d.). Retrieved February 21, 2013, from http://www.cvimellesgriot.com/.

Darzynkiewicz, Z., Bedner, E., Li, X., Gorczyca, W., and Melamed, M. R. (1999). Laser-scanning cytometry: A new instrumentation with many applications. *Experimental Cell Research*, *249*(1), 1–12. doi:10.1006/excr.1999.4477.

Drewes, H. (2010). *Eye Gaze Tracking for Human Computer Interaction.* Munich, Germany, LMU München.

European Commission. (2008). Proposal for a regulation of the European Parliament and of the Council Concerning Type—Approval requirements for the general safety of motor vehicles (Procedure number: COD(2008)0100) Belgium.

Frish, M. B., Wainner, R. T., Green, B. D., Laderer, M. C., and Allen, M. G. (2005). Standoff gas leak detectors based on tunable diode laser absorption spectroscopy. *Proceedings—SPIE the International Society for Optical Engineering*, 6010, 60100D–60100D. doi:10.1117/12.630599.

Frish, M. B., Wainner, R. T., Laderer, M. C., Green, B. D. and Allen, M. G. High-Altitude Airborne Standoff Sensing of Natural Gas Leaks, in Conference on Lasers and Electro-Optics/Quantum Electronics and Laser Science Conference and Photonic Applications Systems Technologies, Technical Digest (CD) (Optical Society of America, 2006), paper PThA2.

Glas, D. F., Miyashita, T., Ishiguro, H., and Hagita, N. (2007). Laser tracking of human body motion using adaptive shape modeling. *Proceedings of IEEE/RSJ International Conference Intelligent Robots and System.* New York, IEEE, pp. 602–608. doi:10.1109/IROS.2007.4399383.

Handheld laser-based sensor for remote detection of gas leaks | Research project database | NCER | ORD | US EPA. (n.d.). Retrieved February 4, 2013, from http://cfpub.epa.gov/ncer_abstracts/index.cfm/fuseaction/display.abstractDetail/abstract/1350/report/0.

Harlow, C. and Peng, S. (2001). Automatic vehicle classification system with range sensors. *Transportation Research Part C: Emerging Technologies*, 9(4), 231–247. doi:10.1016/S0968-090X(00)00034-6.

Hosmer, P. (2004). Use of laser scanning technology for perimeter protection. *IEEE Aerospace and Electronic Systems Magazine*, 19(8), 13–17. doi:10.1109/MAES.2004.1346890.

Houser, A., Pierowicz, J., and McClellan, R. (2005). *Concept of Operations and Voluntary Operational Requirements for Automated Cruise Control/Collision Warning Systems (ACC/CWS) On-board Commercial Motor Vehicles*. Washington, DC, U.S. Department of Transportation, Federal Motor Carrier Safety Administration.

Hsiao, H., Long, D., and Snyder, K. (2002). Anthropometric differences among occupational groups. *Ergonomics*, 45(2), 136–152. doi:10.1080/00140130110115372.

IIHS news release. (n.d.). Retrieved February 21, 2013, from http://www.iihs.org/news/rss/pr070312.html.

Ioan, C., Donciu, O., and Septimiu, P. (2007). Laser based displacement measurements using image processing. IEEE Proceedings of 30th International Spring Seminar on Electronics Technology, 372–376. doi:10.1109/ISSE.2007.4432882.

Irie, K., Wilson, B. A, Jones, R. D., Bones, P. J., and Anderson, T. J. (2002). A laser-based eye-tracking system. *Behavior Research Methods, Instruments, & Computers: A Journal of the Psychonomic Society, Inc.*, 34(4), 561–572. Retrieved from http://www.ncbi.nlm.nih.gov/pubmed/12564560.

Jha, M. N. and Gao, Y. (2008). Oil spill contingency planning using laser fluorosensors and web-based GIS. *Proceedings of MTS/IEEE Oceans Conference.* Quebec City, Quebec, Canada, pp. 1–8. doi:10.1109/OCEANS.2008.5151877.

Joe Qin, S. (2003). Statistical process monitoring: Basics and beyond. *Journal of Chemometrics*, 17(8–9), 480–502. doi:10.1002/cem.800.

Kan, R., Liu, W., Zhang, Y., Liu, J., Wang, M., Chen, D. et al. (2006). Large scale gas leakage monitoring with tunable diode laser absorption spectroscopy. *Chinese Optics Letters*, 4(2), 116–118.

Kim, S., Dong, Z., Xian, H., Upatising, B., Yi, J. S., and Member, I. (2012). Does an eye tracker tell the truth about visualizations?: Findings while investigating visualizations for decision making. *IEEE Transactions on Visualization and Computer Graphics*, 18(12), 2421–2430.

Labayrade, R., Royere, C., and Aubert, D. (2005). A collision mitigation system using laser scanner and stereovision fusion and its assessment. *Proceedings of the IEEE Intelligent Vehicles Symposium 2005.* Las Vegas, NV, IEEE pp. 441–446. doi:10.1109/IVS.2005.1505143.

Laser inline measurement of food for portioning machines—Direct industry. (n.d.). Retrieved February 11, 2013, from http://news.directindustry.com/press/quelltech-ltd/laser-inline-measurement-of-food-for-portioning-machines-64584-378285.html.

Laser makes sure food is fresh. (n.d.). Retrieved February 4, 2013, from http://www.sciencedaily.com/releases/2011/10/111020084821.htm.

Laser sensor applications « 2/3 « sharing experiences integrating acuity laser sensors and scanners. (n.d.). Retrieved January 31, 2013, from http://www.lasersensorapplications.com/page/2/.

Laser Technology—Traffic Safety. (n.d.). Retrieved February 1, 2013, from http://www.lasertech.com/Traffic-Safety.aspx.

Lasers in security | GIT-security.com—Portal for safety and security. (n.d.). Retrieved February 1, 2013, from http://www.git-security.com/topstories/security/lasers-security.

LeBlanc, D., Bezzina, D., Tiernan, T., Gabel, M., and Pomerleau, D. (2008). *Functional Requirements for Integrated Vehicle-Based Safety Systems (IVBSS)—Light Vehicle Platform.* Ann Arbor, MI, University of Michigan.

Lee, K. and Peng, H. (2005). Evaluation of automotive forward collision warning and collision avoidance algorithms. *Vehicle System Dynamics*, *43*(10), 735–751. doi:10.1080/00423110412331282850.

Lehto, M., Papastavrou, J., Ranney, T., and Simmons, L. (2000). An experimental comparison of conservative versus optimal collision avoidance warning system thresholds. *Safety Science*, *36*(3), 185–209. doi:10.1016/S0925-7535(00)00043-6.

Lehto, M. R., Lesch, M. F., and Horrey, W. J. (2009). Safety warnings for automation. In S. Y. Nof (Ed.), *Springer Handbook of Automation.* New York, Springer, pp. 671–695.

Leigh, R. J. and Zee, D. S. (1999). *The Neurology of Eye Movements,* 3rd edn. New York, Oxford University Press.

Lewander, M., Guan, Z. G., Persson, L., Olsson, A., & Svanberg, S. (2008). Food monitoring based on diode laser gas spectroscopy. *Applied Physics B*, 93(2-3), 619–625. doi:10.1007/s00340-008-3192-2.

LINK Systems-Presence Sensing Devices-An Introduction. (n.d.). Retrieved December 6, 2013, from http://www.linkelectric.com/tools_psdover.html.

List of Laser Applications—Wikipedia, the free encyclopedia. (n.d.). Retrieved February 4, 2013, from http://en.wikipedia.org/wiki/List_of_applications_for_lasers.

Matsumoto, T., Shimosaka, M., Noguchi, H., Sato, T., and Mori, T. (2009). Pose estimation of multiple people using contour features from multiple laser range finders. *The IEEE/RSJ International Conference on Intelligent Robots and Systems*, Piscataway, NJ, IEEE, pp. 2190–2196. doi:10.1109/IROS.2009.5354135.

Measures, R. M. (1989). Laser remote sensing: Searchlight on the environment. *Endeavour*, *13*(3), 108–116. doi:10.1016/0160-9327(89)90084-7.

Muhrer, E., Reinprecht, K., and Vollrath, M. (2012). Driving with a partially autonomous forward collision warning system: How do drivers react? *Human Factors: The Journal of the Human Factors and Ergonomics Society*, *54*(5), 698–708. doi:10.1177/0018720812439712.

Noll, R., Bette, H., Brysch, A., Kraushaar, M., Mönch, I., Peter, L., and Sturm, V. (2001). Laser-induced breakdown spectrometry—Applications for production control and quality assurance in the steel industry. *Spectrochimica Acta Part B: Atomic Spectroscopy*, *56*(6), 637–649. doi:10.1016/S0584-8547(01)00214-2.

Panne, U. (1998). Laser remote sensing. *TrAC Trends in Analytical Chemistry*, *17*(8–9), 491–500. doi:10.1016/S0165-9936(98)00054-5.

Papastavrou, J. D. and Lehto, M. R. (1996). Improving the effectiveness of warnings by increasing the appropriateness of their information content: Some hypotheses about human compliance. *Safety Science*, *21*(3), 175–189. doi:10.1016/0925-7535(95)00060-7.

Paquette, S. (1996). 3D scanning in apparel design and human engineering. *IEEE Computer Graphics and Applications*, *16*(5), 11–15. doi:10.1109/38.536269.

Patent US5521696—Laser-based speed measuring device—Google patents. (n.d.). Retrieved February 12, 2013, from http://www.google.com/patents?hl = en&lr = &vid = USPAT5521696&id = WJgnAAAAEBAJ&oi = fnd&dq = laser+speed+gun&printsec = abstract#v = onepage&q&f = false.

Patent US5938717—Speed detection and image capture system for moving vehicles—Google patents. (n.d.). Retrieved February 12, 2013, from http://www.google.com/patents?hl = en&lr = &vid = USPAT5938717&id = 9jIYAAAAEBAJ&oi = fnd&dq = laser+speed+gun&printsec = abstract#v = onepage&q = laser      speed gun&f = false.

Patent US6278373—Laser intrusion system for detecting motion in a swimming pool—Google patents. (n.d.). Retrieved February 11, 2013, from http://www.google.com/patents?hl = en&lr = &vid = USPAT6278373&id = jxsIAAAAEBAJ&oi = fnd&dq = laser+intrusion+alert&printsec = abstract#v = onepage&q = laser intrusion alert&f = false.

Poole, A. and Ball, L. J. (2005). Eye tracking in human-computer interaction and usability research: Current status and future prospects. In C. Ghaoui (Ed.), *Encyclopedia of Human-Computer Interaction.* Pennsylvania, PA, Idea Group, Inc., pp. 211–219.

Presence Sensing Devices. Occupational Safety & Health Administration, the U.S. Department of Labor. (n.d.). Retrieved December 9, 2013, from https://www.osha.gov/SLTC/etools/machineguarding/presses/psd.html.

Rajamani, R. (2012). *Vehicle Dynamics and Control.* Boston, MA, Springer US, pp. 141–170. doi:10.1007/978-1-4614-1433-9.

Ready, J. F. (1997). Chapter 10—*Distance Measurement and Dimensional Control.* San Diego, CA, Academic Press, pp. 256–277. Retrieved from http://www.science-direct.com/science/article/pii/B9780125839617500120.

Safety & Sensor Technology—Laser Components USA, Inc. (n.d.). Retrieved February 4, 2013, from http://www.lascomponents.com/us/applications/safety-sensor-technology/.

Schulz, K. R., Scherbarth, S., and Fabry, U. (2002). Hellas: Obstacle warning system for helicopters. *Proceedings of the SPIE the International Society for Optical Engineering,* *4723,* 1–8. doi:10.1117/12.476398.

Shu, X.-M., Yuan, H.-Y., Su, G.-F., Fang, J., and Zhan, F. (2006). A new method of laser sheet imaging-based fire smoke detection. *Journal of Fire Sciences,* *24*(2), 95–104. doi:10.1177/0734904106055568.

Technology: Non-contact measurement sensors, solutions and systems—MTI instruments. (n.d.). Retrieved January 31, 2013, from http://www.mtiinstruments.com/technology/triangulation.aspx.

UNECE works on new standards to increase the safety of trucks and coaches. (2011). *United Nations Economic Commission for Europe.*

Utkin, A. B., Lavrov, A. V., Costa, L., Simões, F., and Vilar, R. (2002). Detection of small forest fires by lidar. *Applied Physics B,* *74*(1), 77–83. doi:10.1007/s003400100772.

Wang, W., Brscic, D., He, Z., Hirche, S., and Kuhnlenz, K. (2011). Real-time human body motion estimation based on multi-layer laser scans. *IEEE International Conference on Ubiquitous Robots and Ambient Intelligence.* Piscataway, NJ, IEEE, pp. 297–302. doi:10.1109/URAI.2011.6145980.

Wang, Z. and Stout, S. (2010). *Oil Spill Environmental Forensics: Fingerprinting and Source Identification.* New York, Academic Press, p. 618. Retrieved from http://books.google.com/books?id = lWeqTPaHR7kC.

Werghi, N. (2007). Segmentation and modeling of full human body shape from 3-D scan data: A survey. *IEEE Transactions on Systems, Man, and Cybernetics, Part C: Applications and Reviews,* *37*(6), 1122–1136. doi:10.1109/TSMCC.2007.905808.

Werle, P. (1998). A review of recent advances in semiconductor laser based gas monitors. *Spectrochimica Acta Part A: Molecular and Biomolecular Spectroscopy*, *54*(2), 197–236. doi:10.1016/S1386-1425(97)00227-8.

Yue, K. D., Zhou, X., and Gao, J. (2005). Automatic laser monitoring for dam safety. *Key Engineering Materials*, *295–296*, 239–244.

Zador, P. L., Krawchuk, S. A., and Voas, R. B. (2000). Final report—Automotive Collision Avoidance Systems (ACAS) Program. Washington, DC, U.S. Department of Transportation, National Highway Traffic Safety Administration.

Zafar, A. and Lehto, M. (2010). First do no harm: The unintended consequences of embedding technology in healthcare. In V. G. Duffy (Ed.), *Advances in Human Factors and Ergonomics in Healthcare*. Boca Raton, FL, CRC Press, pp. 211–218.

Zafar, A. and Lehto, M. R. (2011). Cultural factors in the adoption and implementation of health information technology. In R. W. Proctor, S. Y. Nof, and Y. Yih (Eds.), *Cultural Factors in Systems Design: Decision Making and Action*. Boca Raton, FL, CRC Press, pp. 239–264.

Zafar, A. and Lehto, M. (2012). Workflow characterization in busy urban primary care and emergency room settings: Implications for clinical information systems design. In V. G. Duffy (Ed.), *Advances in Human Aspects of Healthcare*. Boca Raton, FL, CRC Press, pp. 359–368.

Zelinger, Z., Střižík, M., Kubát, P., Jaňour, Z., Berger, P., Černý, A., and Engst, P. (2004). Laser remote sensing and photoacoustic spectrometry applied in air pollution investigation. *Optics and Lasers in Engineering*, *42*(4), 403–412. doi:10.1016/j.optlaseng.2004.03.005

# 15

## Lasers in Our Life and Implications to Education

Juan D. Velasquez, Shimon Y. Nof, and Cesar Reynaga

**CONTENTS**

## Introduction

In the introduction of this book, three unique features that are enabled by laser and photonic systems were identified, adding that they promise further great advancements and innovations. They are as follows:

1. Processing at multiple scales and dimensions, from nano- and micro- to large-scale objects, from local to remote subjects, and from 0D to 3D dimensions
2. Processing and delivery of laser fields at ultrafast speeds and frequencies and the ability to shape and reshape them
3. The ability to bring closer (almost together) the process and its process control by significantly faster sensing and communication

The previous chapters explain in detail and demonstrate these three features of laser and photonic systems, in various fields and applications, from theory to development, design, integration, and innovation. The purpose of this concluding chapter is threefold: Review some additional laser and photonic systems influencing our daily life; describe educational programs, recognizing their critical role for future advancements in this area; and conclude with emerging directions in the advancement of laser and photonic systems design and integration.

## Lasers in Our Daily Lives

In the twenty-first century, laser technologies are designed for use in several industries and activities and are part of our daily lives. From storing data in a compact disc (CD) to creating laser images for laser shows, the amount of application related to nonmilitary laser activities has grown over the last years. The following section describes some of the application of lasers in five areas: electronics, automobiles, food processes, aviation, and entertainment.

### Recent Advances in Electronics

#### *Optical Storage Devices: CD, DVD, and Blu-Ray*

In 1969, the history of optical storage devices started when Phillips and Sony research teams explored the idea of storing data on rings around a disc that would have pits or lands that would correspond to data (Schouhamer Immink, 1998). Through the years, the storage devices have been decreasing in size and increasing in storage due to a reduction in the wavelength of the

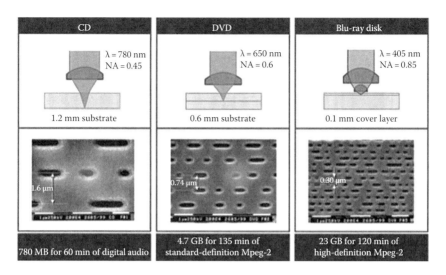

**FIGURE 15.1**
**(See color insert.)** Laser beams of the CD, DVD, and Blue-ray. (From Bergh, A.A., Commercial applications of optoelectronics, February 2006, Retrieved April 1, 2013, from Photonics Spectra: http://www.photonics.com/Article.aspx?AID=24317.)

recording beam. In CD and digital videodisc (DVD) players, an infrared (IR) laser is beamed onto a spinning disc, where the players read microscopic pits that represent ones and zeros of digital data (Bergh, 2006). DVDs have a smaller wavelength so they can store much more data than a CD. DVDs quickly became the desired entertainment medium due to their capacity, speed, and ability to provide high-quality video and sound at a low cost. The latest technology for digital media storage device has become Blu-ray. It stores data into a disc by using blue lasers that have a shorter wavelength than a red laser so they can store more data into the disc. Figure 15.1 shows the specifications for CD, DVD, and Blu-ray discs in terms of wavelength and storage capacity.

The advent of CDs, DVDs, Blu-rays, and the laser technology behind it revolutionized the entertainment industry. No longer do moviegoers have to physically go to the movie theater to experience high-definition movies or *the movie experience*. The television manufacturing industry has seen some of its best years not only because of their own technological advances but also because of the home theater experience that now most customers seek. New businesses like Netflix, Redbox, and Blockbuster have thrived by providing high-quality video and sound movies at a fraction of the cost of going to the movies.

## *Laser Printing*

In 1969, Gary Starkweather created the laser printer at Xerox (Halperin, 2010). Laser printers differ from analog printers in that direct scanning of a laser beam across the printer's photoreceptor produces the images in laser printers. Since then, laser printers have evolved and become some of the most used printers by industry and individuals; they are often held as the standard for high-quality printing at a reasonable price. One of the reasons why laser printers have become popular is their lower cost compared to inkjet and dot matrix printers; however, it is often the case that the replacement toners are more expensive than those for inkjet and dot matrix printers.

Laser printers are mechanically more complex compared to inkjet or dot matrix printers. A laser printer has several components that must work together to print properly and involve a sequence of seven steps that include the following: raster image processing, charging, exposing, developing, transferring, fusing, and cleaning. Several of the steps described previously have to occur at once, and that adds to the complexity of laser printers. When the printer receives the signal from a computer, a laser beam traces the images and text on a rotating drum that is sensitive to light. When the laser hits the drum, the area touched becomes charged with static electricity. Then, a toner is adhered to the traced image.

To print an image, the laser printer uses its feed rollers to pull the paper into the printer. The second step is to pass the paper over the corona that is charged with static electricity—opposite charge of the drum. Then, the paper rolls over the drum where it transfers the toner to the paper (Figure 15.2). Finally, the paper continues to a hot fuser where the toner is melted into fibers of the paper, and then the feed rollers send the paper out of the printer (J.D. Power and Associates, 2012).

Laser printing technology continues to evolve. For instance, Ueki and Mukoyama (2013), by using vertical-cavity surface-emitting laser (VCSEL) technology, have been able to improve quality performance to the resolution of printing that is 2400 dots per inch (dpi), which is the highest industry level. New applications of laser printing technology are also evolving and now include, for example, laser printing for nanoparticles. In the work by Polsen et al. (2013), the authors use laser printing for high-speed micropatterning of nanoparticle-containing thin films for a variety of nanostructures using engineered toners containing the desired nanoparticles (Figure 15.3a).

Laser printing is also instrumental in the emerging 3D material printers. 3D printing, also known as *additive manufacturing*, converts digital design files into physical object. The process is called additive, since it uses several technologies adding material only as necessary to fit exactly the digitally designed object. Besides prototyping (see example in Figure 15.3b), 3D printing has been increasingly applied for one-of-a-kind fabrication of medical and dental parts and devices; manufacturing and engineering tools; automotive, agricultural, and aerospace replacement components; and other fields (Anderson, 2012; Lipson and Kurman, 2013).

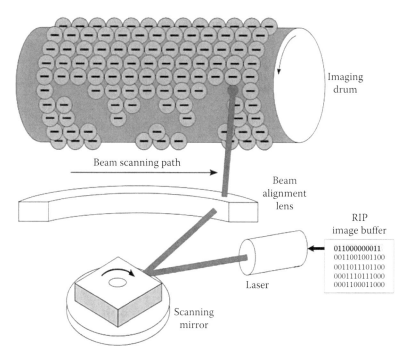

**FIGURE 15.2**
Image being written onto photoreceptive drum in a laser printer. (Courtesy of Dale Mahaiko, Wikimedia Commons. http://en.wikipedia.org/wiki/File:Laser_printer-Writing.svg)

Lasers are used in several key 3D printing technologies, such as direct metal laser sintering, and selective laser sintering, both adding tiny layers of granular materials that are solidified together under precise heating. Another key technology is light polymerization of photopolymers. This additive technology applies stereolithography and digital light processing to print the objects. Nano features can be fabricated by a technique of multiphoton polymerization. Other advantages associated with 3D printings are its availability for distributed, mobile, and remote manufacturing and providing parts and devices rapidly under emergency requirements, where and when they are needed.

### Laser Barcode Readers

A barcode reader is a device used to read barcodes imprinted in a multitude of products. Currently, there are several technologies in existence and one of those uses a laser to read the codes called laser barcode reader (Figure 15.4). The laser barcode reader consists of laser beam and photo diodes that are next to each other and work together. To read the barcode, a laser beam

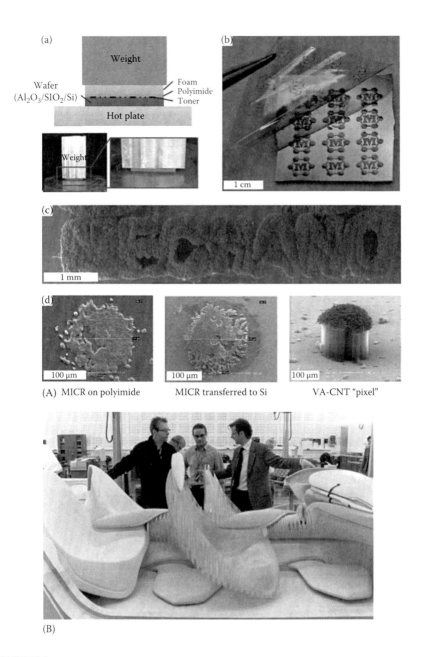

**FIGURE 15.3**
(A) Thin-film transfer process and resulting micropatterns. (a) Schematic and photograph of transfer setup; (b) removal of Kapton substrate leaving nanoparticle patterns behind; (c) top view of *Mechano* after nanogrowth, where the pattern was created by laser printing of 2-point font; (d) individual 2-point characters at each stage process. (From Polsen, E.S. et al., *ACS: Appl. Mater. Interfaces*, 5(9), 3656, 2013.) (B) Designers discuss an experimental prototype that was digitally designed and 3D-printed. (Courtesy of Materialise Co. Leuven, Belgium.)

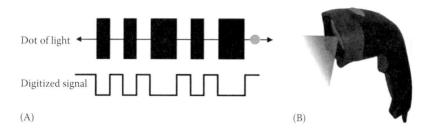

Dot of light

Digitized signal

(A)          (B)

**FIGURE 15.4**
(A) Laser barcode reader reading from a barcode. (From TAL Tech., How a barcode reader works, July 11, 2011, Retrieved April 1, 2013, from TAL Tech.: http://www.taltech.com/support/entry/how_barcode_reader_works.) (B) Laser barcode reader. (Courtesy of Adesso, Inc. California, US)

moves back and forth over the barcode by using a reciprocating mirror or a rotating prism, and the photo diode measures the intensity of light reflected back from the laser beam and generates a waveform that is used to measure the widths of the bars and spaces in the barcode (Marshall and Stutz, 2011).

> During this process, the dark bars in the barcode absorb light and white spaces reflect light so that the voltage waveform generated by the photo diode is an exact duplicate of the bar and space pattern in the barcode. The scanner decodes this waveform in a similar way as the Morse code dots and dashes are decoded. (TAL Tech, 2011)

Laser barcode scanners have become so commonplace that a majority of medium- and large-sized businesses, from coffee shops to grocery stores, now use them not only to scan products but also to track customers' shopping habits, preferences, and frequency of purchases. Membership cards to stores, like those provided by Starbucks™, enable a customer to not have to carry cash but rather pay with the membership card that has a built-in barcode. Once the barcode is scanned at the local coffee shop, the information is transmitted and stored in a corporate database that collects all customer information and provides the company with a robust customer database of shopping habits and preferences and the customer a myriad of benefits from the business. Laser barcode scanning has become so prevalent in our lives that technologies that do not include lasers have been developed in order to provide the same service that laser barcode has for years. The proliferation of mobile devices (e.g., phones, tablets, netbooks, iPads) has taken advantage of this new technology by making use of cameras rather than the laser emission for barcode reading. Applications like RedLaser (www.redlaser.com) enable anyone with a mobile device to not only scan a barcode as if they were using a laser scanner but the application also goes out and searches the databases of businesses that sell the product in order to provide the customer the lowest-priced item as well as the location and store where the item can be purchased.

## Laser Rangefinder

Laser rangefinders are devices that determine the distance of an object from the device. To calculate the distance, a rangefinder emits a laser to an object and calculates the amount of time that it takes to travel from the device to the object and back (Byren, 1993). The device was created for military purposes (e.g., shooting of snipers, artillery, military reconciliation, and engineering) and operates in ranges of 2 km to up to 25 km, typically combined with binoculars or monoculars and often equipped with night vision, wireless interfaces, etc. However, nowadays, civilians for many common day activities like golf (Figure 15.5A), construction (Figure 15.5B), hunting (Figure 15.5C), and real estate use it. Laser rangefinders are also used extensively in 3D recognition, modeling, and computer vision–related fields. For instance, autonomous mobile robots employ laser rangefinders to navigate terrains and accomplish given tasks using algorithms that enable 3D recognition of objects (Surmann et al., 2003).

(A)

(B)

(C)

**FIGURE 15.5**
Laser rangefinder applications. (A) Golf. (From IceGolf, Golf laser rangefinder reviews, September 7, 2012, Retrieved April 1, 2013, from IceGolf: http://icegolf.org/golf-laser-range-finder-reviews.html/.) (B) Construction. (From Williams, E., How to aim a handheld laser rangefinder, n.d., Retrieved April 1, 2013, from eHow: http://www.ehow.com/how_7668397_aim-handheld-laser-rangefinder.html.) (C) Hunting. (From 6mmbr.com, Laser rangefinders, 2007, Retrieved April 1, 2013, from 6mmbr.com: http://www.6mmbr.com/rangefinders.html.)

Another use of rangefinders, but in a greater scale, is for satellite laser ranging (SLR) and lunar laser ranging (LLR) that uses short-pulse lasers, optical receivers, and timing electronics to measure the two-way time of flight from ground stations to retroreflector arrays on satellites orbiting the Earth and the moon. Precise geometric positions and motions of ground stations, satellite orbits, Earth orientation parameters (OEPs), and information about the internal structure of the moon have all been possible due to the advances in SLR and LLR.

### Ring Laser Gyroscopes

The ring laser gyroscope (RLG) is a laser-based device used for measuring rotation that operates on the principle of the Sagnac effect (Figure 15.6). The instrument has mirrors arranged in the shape of a triangle or square, creating a close path (*ring*). This arrangement allows two laser beams to move around the ring. One beam travels clockwise and the other counterclockwise. For example, when it is rotated counterclockwise, the laser beam that travels counterclockwise will take more time to reach the starting position than the clockwise one. This effect causes that the beams have different frequencies. The difference is used to calculate a rotation rate of the instrument from its initial axis. So, it is possible to determine the orientation of the system at all times. RLGs are used in aircrafts, submarines, guided missiles, ships, and

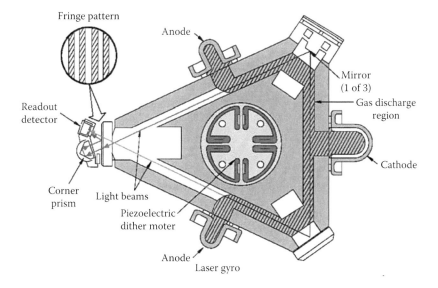

**FIGURE 15.6**
**(See color insert.)** Diagram of an RLG. (From Makris, K., Ring laser gyro, December 22, 2012, Retrieved March 4, 2013, from Kostas Makris website: http://www.k-makris.gr/AircraftComponents/Laser_Gyro/laser_gyro.htm.)

spacecrafts for navigation purposes due to their high level of accuracy and mean time between failures.

Recently, the application of RLGs has also extended to new areas like seismology (Schreiber et al., 2006) due to their high sensitivity, high resolution, good stability, and wide dynamic range, which in turn led to the development of the GEO sensor project (Schreiber et al., 2006) to measure rotations during an earthquake.

## Recent Advances in Automobiles

### *Lasers in Self-Driving Car Systems*

A self-driving car is a vehicle that does not require the assistance of a human driver. Scientists at the University of Oxford have created a self-driving car that makes use of lasers on the front to scan the surroundings and compare them to stored data to navigate the vehicle (Arthur, 2013). This technology allows the vehicle to identify its location and, for this reason, have a better understanding of the world (Figure 15.7). The main feature of the system is to recognize the objects captured by the laser by comparing them based on appearance, shape features, and geometric consistency to a database as seen in Figure 15.8 (Newman et al., 2012a,b). For the vehicle to properly operate, it needs to have the route stored before the self-driving option can be followed. During movement, the laser system scans and recognizes objects such as humans and, if necessary, stops the car if there are objects in the path of the vehicle.

The future of this technology is to create routes' databases created by the lasers of passing cars, which can be downloaded by every vehicle via 3G

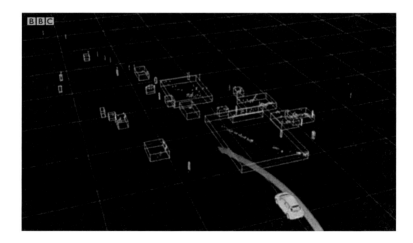

**FIGURE 15.7**
Database route scanned with the laser. (From Westcott, R., Self-driving car given test run at Oxford University, February 14, 2013, Retrieved February 28, 2013, from *BBC News*: http://www.bbc.co.uk/news/technology-21465042.)

**FIGURE 15.8**
**(See color insert.)** Comparison of the live route and the database route. (From Westcott, R., Self-driving car given test run at Oxford University, February 14, 2013, Retrieved February 28, 2013, from *BBC News*: http://www.bbc.co.uk/news/technology-21465042.)

or 4G connections; this system would provide the driver an updated route depending on the last time a car passed through the route required.

Google has been working on a driverless car since 2005 when the vehicle designed by the Stanford Artificial Intelligence Laboratory won the DARPA Grand Challenge and its $2 million prize. Google's test cars have over $150,000 in equipment including its key technology a $70,000 light detection and ranging or laser detection and ranging (LIDAR) that measures the distance to, or the properties of, targets by illuminating a target with laser light and analyzing the backscattered light (Figure 15.9).

### Speed Measurement with Lasers

Vehicles' speed measurements are controls performed by police officers to reduce the amount of traffic accidents and monitor traffic flow. For years, officers have used laser technologies to perform these measurements due to their high level of accuracy. Research by Charleston's Country Sheriff Office in South Carolina showed that laser speed devices captured more speeding vehicles than the radar speed devices. Most of the speeding vehicles were detected by the laser devices and not by the radar devices (Teed and Lund, 1993). To determine the speed, the laser device measures the round-trip time for light to reach the vehicle and return to the device (Figure 15.10).

### Lasers in Vehicles' Headlights

A growing number of vehicles on the roads today are equipped with light-emitting diode (LED) lights; however, it will not be long before lasers in

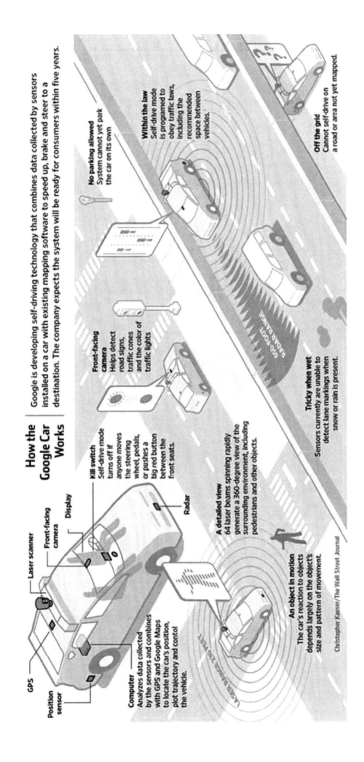

**FIGURE 15.9**
Technology that drives Google's driveless car (From Google. California, US.)

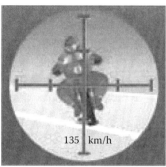

**FIGURE 15.10**
Speed measurement with a laser. (From Federal Office of Metrology METAS of Switzerland, Traffic measurement technology. Federal Office of Metrology METAS of Switzerland, Federal Department of Justice and Police of Switzerland, Bern-Wabern, Switzerland, Federal Office Metrology METAS, 2010.)

**FIGURE 15.11**
Illustration of the BMW laser headlights in the i8 Concept. (From BMW, Laser light. Leading the way to the future, n.d., Retrieved February 27, 2013, from BMW i8 Concept: http://www. bmw-i-usa.com/en_us/bmw-i8/#laser-light-leading-the-way-to-the-future.)

vehicles' headlights become the norm. BMW engineers introduced this technology as early as 2011 with their i8 Concept (Figure 15.11), and soon Mercedes Benz and Audi released their next generation of vehicles equipped with laser headlights as well. According to BMW, the lasers they use have intensity 1000 times stronger than what LED headlights can produce as a result of the monochromatic nature of lasers. Moreover, the laser headlights are capable of producing 170 lumens per watt compared to the 100 lumens per watt that the LED headlights can produce. The lower energy consumed by lasers enhances headlights and has the potential to significantly reduce power consumption and fuel usage (Photonics, 2011). Safety concerns have been taken into consideration by BMW engineers to guarantee that the high intensity of laser lights do not cause damage to humans, animals, or wildlife due to the conversion of the bluish light into a pure white light.

### Laser Ignition

Since its development in the 1970s, the laser ignition system has been an important research topic. In the laser ignition system, the laser splits into several beams that improve the chance of a complete fuel burn (Figure 15.12). According to researchers from the University of Liverpool and Japan's National Institute of Natural Sciences, experiments reveal that the system increases the fuel efficiency and reduces harmful emission levels. In the work by Lackner et al. (2004), the authors examined laser-induced ignition of hydrogen/air and biogas/air mixtures in static combustion to address the effects of pollution and start addressing the sustainability of such ignition systems.

According to Dr. Dustin McIntyre, research chemist at the US Department of Energy's National Energy Technology Laboratory in Morgantown, W. Va, there are several factors that need to be considered including cost of manufacturing, reliability and durability of the lasers due to their continuous and prolonged use, safety, and lack of knowledge by users of lasers (Marshall, 2012). However, progress continues, and in April 2013, the first laser ignition conference will be collocated with the Optics and Photonics International Congress in Yokohama, Japan.

## Recent Advances in Food Processes

### Headspace Gas Measurement by Laser Spectroscopy

Vacuum-sealed packaging is crucial to maintaining and securing the quality, freshness, and durability of food products, particularly those that are

**FIGURE 15.12**
Laser spark emitted at the end of the National Energy Technology Laboratory laser spark plug. (Image provided by the US Department of Energy National Energy Technology Laboratory.)

chilled. The process is of key importance for fresh and chilled foods due to the quality deterioration that can be caused by the presence of oxygen as well as time and temperature. To reduce the speed of deterioration, a modified atmosphere is filled into the package headspace to reduce the levels of oxygen. For this reason, to maintain the specifications of the package, it is important to monitor the amount of oxygen stored in the headspace (Lundin et al., 2012). Current techniques use intrusive methods such as waste samples as a means by which to assess the amount of oxygen in the package, but researchers are working in nonintrusive methods. Scientists at Lund University have developed GASMAS (gas in scattering media absorption spectroscopy), a method to analyze the atmosphere inside solid and liquid food packages using laser spectroscopy (Lewander et al., 2011). Laser spectroscopy is a method that shines a laser beam into the package and evaluates the returning light that contains the composition of the atmosphere. The authors are already working on techniques that can make use of their methodology on *in-line* quality control of packed items in a food packaging supply chain.

### Laser Ablation

Laser ablation is the process of using a focused laser beam to vaporize (remove through pulses) small amounts of material from a solid object (Lawrence Berkeley National Laboratory, n.d.). In food technology, the process is being studied for peeling fruits and vegetables. Through the procedure, the laser beam used does not damage the freshness, hardness, taste, and color of the food tested, yet it allows a user to obtain pectin (a gelling/coagulating agent and stabilizer found in foods like yogurt and dessert creams) and aroma from the material, which in turn can be used in further industrial processes (Panchev et al., 2011).

## Recent Advances in Aviation

### Advanced Visual Docking Guidance System

Airports around the world have been introducing advanced visual docking guidance systems (A-VDGSs) to aide airplane pilots parking aircrafts (Figure 15.13). The system enhances the efficiency and safety of and airports' ground operations by eliminating delays from unmet flights and reducing delays in irregular conditions (Safe Gate Group, 2010). The technology consists of a 3D laser scanner that guides pilots within 10 cm of the stop position. It provides the correct alignment of the aircraft for safe docking process (Figure 15.14). Additionally, the system analyzes the adjacent aircrafts to prevent collisions prior the start of the process. According to Flight Safety Foundation, aircraft ramp accidents cost at least $10 billion dollars annually worldwide (Lacagnina, 2007). Additionally, 48% of the 82% of ramp accidents

**FIGURE 15.13**
Illustration of the Advanced visual docking guided system (A-VDGS). (From Rosenkrans, W., Graceful arrivals. *Aero Safety World*, May 2007, pp. 42–45.)

occur during the arrival process. In London's Gatwick Airport, after the installation of the A-VDGS, the accidents dropped by 80% (Phippen, 2011).

## Recent Advances in Entertainment

### Laser Shows

Laser shows are common of large events such as the Super Bowl, Olympic Games ceremonies, planetariums, concerts, and corporate meetings (Daukantas, 2010). Laser shows rely on simple optical equipment and principles: moving mirrors and the persistent vision effects (afterimage produced when a point of light moves faster that the eye) (Figure 15.15). Currently, laser projectors are used in the industry to produce these images. Additionally, the industry uses galvanometers to represent images in screens or walls by sending electric signals to make small mirrors vibrate over a 2D plane. These mirrors reflect the beam path very fast to create the shape on a surface. The process is called *scanning*.

Other laser projectors use different techniques such as *chopping* and *blanking*. *Chopping* is a technique that consists in turning the laser beam on and off at a constant rate to create a dotted line effect. It is obtained by moving the mirror to the side to avoid the beam to exit the project (Laser F/X International and LaserFX.com, 2008). On the other hand, *blanking*, similar to the *chopping* effect, turns the laser beam on and off, but it produces blanks on the shapes of images. Today's laser artists use several graphic software packages to reproduce the shapes, pictures, and logos that they want. These technologies allow them to move the laser beams at a rate of 15–30 Hz due to the persistence of vision of 25 Hz.

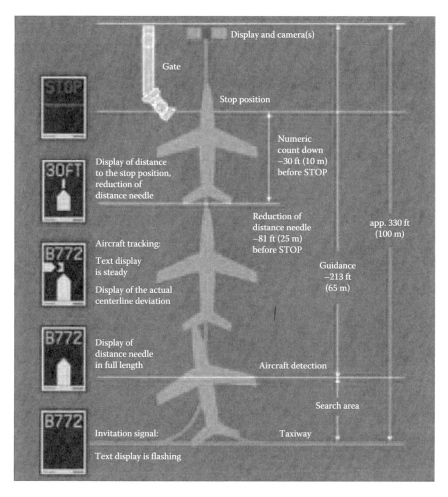

**FIGURE 15.14**
A-VDGS. (From Rosenkrans, W., Graceful arrivals. *Aero Safety World*, May 2007, pp. 42–45.)

## Building the Pipeline of Laser- and Photonics-Educated Workforce

### Lasers and Photonic Systems in P-12

To better prepare students for college and address the constant calls from experts to graduate more students pursuing degrees in science and engineering, high schools like the Regents School of Austin (www.regents-austin.com) in Austin, TX, are exposing them to cutting-edge research and

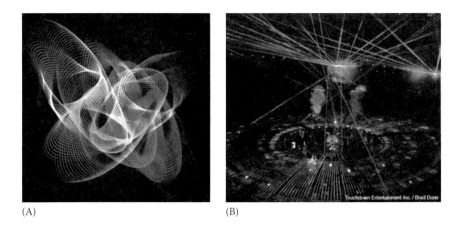

(A)　　　　　　　　　　　　　　(B)

**FIGURE 15.15**
**(See color insert.)** (A) Illustration of a laser beam art. (Cross, L., Inventing the laser light show, 2005, Retrieved March 4, 2013, from Lowell Cross: http://www.lowellcross.com/home/.). (B) Laser show at the February 2010 Super Bowl LVIV halftime show in Miami, FL. (From Daukantas, P., *Opt. Photon. News*, 21(5), 42, 2010.)

technologies in the areas of lasers and photonic systems at an early age, for instance. Curriculums are being updated to make learning science subjects like chemistry, biology, and physics relevant, affordable, accessible, but most importantly intellectually motivating and imagination driven. For schools like Regents and others across Texas, Florida, and many other US states, the limitations used to be the lack of funds, expertise, and experience, but partnerships with local, federal governments, as well as corporations and alumni donations have made possible the establishment of labs like the lasers lab in Regents.

The Laser Optics Lab efforts at Regents constitute one example of a school-specific and school-driven effort to enhance the teaching and learning of science, technology, engineering, and mathematics (STEM) focused, in their case, to connecting the existing curriculum to laser education. More comprehensive programs and initiatives do exist across the United States. One such effort is the New York State Science Initiative (www.p12.nysed.gov/ciai/mst/sci/initiative.html). The state of New York through this initiative is attempting to address the needs for improving student performance in science, particularly on physical science as identified in assessment reports such as the ones provided by the National Assessment of Educational Progress (NAEP), the Third International Mathematics and Science Study (TIMMS), and other local reports. Comprehensive initiatives like the one in New York lay the foundation for curriculum reform, professional development, and alignment of administrative and financial support to better engage students in the learning process. By exposing students at an early age to an inquiry-centered curriculum, like the one New York has worked on, it is possible to influence the way they learn and in turn encourage them to become lifelong learners and awake their appetite for exploration and discovery in the sciences. In Missouri, a 3-year high school photonics program was established to address the lack of laser electro-optical technicians in the local manufacturing industry. The goals behind the program were to increase student awareness of job opportunities in photonics and optics, teach skills needed, and foster connections with industry and postsecondary institutions (John and Shanks, 2002).

One of the most influential forces revolutionizing the science curriculum and making it more engaging and accessible to students in P-12 has been the Lego Mindstorms kits and the FIRST Robotics Competition for students in grades 9–12. The use of the Lego Mindstorms kits in the classroom reinforces traditional classroom instruction while at the same time nurturing the creativity of the students. Although the kits do not currently include lasers or photonics per se, when visiting the Lego Mindstorms website (http://us.mindstorms.lego.com), one discovers the creativity of the students to solve problems and design robots that address specific tasks that can benefit from using lasers or photonic systems. The kits typically include touch, light, sound, ultrasonic, compass, color, and accelerometer sensors that test the creativity of students to perform series of tasks.

Figure 15.16A shows a plotter constructed with one of the Lego Mindstorms kits that the developer enhanced with an IR laser to construct a high-power laser that can write on surfaces (http://us.mindstorms.lego.com/en-us/community/nxtlog/displayproject.aspx?id=1ede3811-e056-461c-8f2c-b58a9a58c108). Besides the novelty of the laser plotter, the description by the developer is perhaps what best captures the relevance and influence of the Lego Mindstorms kits in supporting creativity in today's youth:

> It looks like an ordinary plotter, like hundreds of plotters on the internet, but it is original in one way—it uses a high power LASER to write on stuff … The laser is from a broken DVD burner - this one is the IR laser that was actually used to burn CDs (I made a pointer that uses the DVD laser, so I have only CD laser for this project). It is an infrared laser with power of 100—not enough to burn a white paper, like on the pictures. It burns only black paper and black plastic foil. The laser is connected to the NXT via a simple interface. This interface uses a light on/off trigger used in Light sensor, but there is an optoelectronic relay connected instead of a LED. This is connected to a DDL circuit that drives the laser diode. The laser needs an external power supply. The laser on/off switching is very simple—it uses a Light on/off command in Light sensor block.

In Figure 15.16B, a computer numerical control (CNC) machine was constructed using a similar kit and a 1200 mW laser to cut and engrave on almost all material surfaces (http://us.mindstorms.lego.com/en-us/Community/

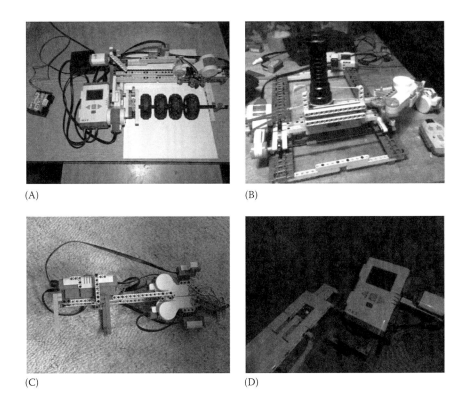

**FIGURE 15.16**
(A) NXT-LP laser plotter. (B) CNC machine constructed using Lego Mindstorms kit.
(C) Multipurpose robotic arm. (D) Self-checkout scanner.

NXTLog/DisplayProjectList.aspx?SearchText=laser). Figure 15.16C shows a multipurpose robotic arm with a built-in radar system, a flashlight, a claw, and a laser (http://us.mindstorms.lego.com/en-us/community/nxtlog/displayproject.aspx?id=df4818a7-a433-4a83-aa8e-cf3acfe908fa). In Figure 15.16D, a self-checkout scanner with an imitation laser barcode scanner and an imitation credit card reader is showcased.

Exposing students at an early age to science and mathematics topics through engaging mediums like the Lego Mindstorms kits awakens their inquisitive minds and allows them to learn the subjects in a more explorative, problem-focused, product-oriented manner (Pike, 2002; Mataric et al., 2007). Preliminary studies on the educational benefits and challenges of using the Mindstorms kits have supported their use as an engaging and educational tool (Marulcu and Barnett, 2012; Williams et al., 2012); however, more educational research work is needed in order to effectively and efficiently incorporate them into the curriculum and excite students to pursue studies in science, engineering, and mathematics.

Since 1989, the FIRST Robotics programs and the competitions they support have inspired youngsters to be science and technology leaders by engaging them in mentor-based programs that build science, technology, and engineering skills that inspire innovation that foster self-confidence, communication, and leadership. Programs like FIRST continue to build and expand the pipeline of students that pursue an education and career in STEM fields.

## Lasers and Photonic Systems in Higher Education

Community colleges in the United States have played and will continue to play an instrumental role in the development of laser and photonics curriculum to meet the needs of the industry. Community colleges are more nimble to react to changes in the marketplace and workforce and, therefore, are better equipped to prepare the technician labor force of the twenty-first century, which, as highlighted from the previous sections, will continue to grow specially in lasers and photonics as they become more prevalent in our daily lives. The Indian Hills Community College in Iowa offers, for instance, a 21-month electronic technician diploma program and an associate or applied science degree. Some of the courses that are part of the degree include PLTW digital electronics, direct current (DC) circuit analysis, photonics concepts, introductions to photonics, laser components, optical devices, geometric optics, and physical optics (http://www.indianhills.edu/courses/tech/laser.html).

The Central Carolina Community College also offers a program on laser and photonics technology that focuses on the application of electronic, fiber optic, photonic, and laser principles for industrial, medical, and business settings. Among the higher-order learning objectives of the program are the following: (1) construct, analyze, test, measure, troubleshoot, and assist in the design of optical systems; (2) describe or explain the concept, structure, and operation of gas lasers, semiconductor lasers, and solid-state lasers and be able to safely operate, test, measure, and align, where appropriate, these lasers using optical test equipment and procedures; (3) describe requirements for laser safety, specifically major dangers, the different laser class levels, and protection requirements of different class levels, and be able to calculate and specify parameters for laser systems, such as nominal hazard zone (NHZ), maximum permissible exposure (MPE), and minimum optical density (OD) (http://www.cccc.edu/curriculum/majors/lasersphotonics/).

Almost all states in the United States that have a community college infrastructure have similar programs that address the needs for a labor force educated in lasers among them: Hawaii, Illinois (13-16 credit hour certificate program), North Carolina (74 credit hour associate degree), Iowa (81 credit hour associate degree), Connecticut (68 credit hour associate degree), Delaware (34 credit hour diploma), New Mexico (20-21 credit hour certificate), and many more. As lasers expand their presence in new applications, it is expected that programs (certificates, diplomas, associate degrees) offered by community colleges will continue to grow and in some cases evolve to address new industry needs.

Four-year public and private colleges/universities in the United States and universities around the world have a similar role in educating the laser and optics workforce for the twenty-first century, but their role goes beyond education as they are also at the forefront of advancing the science of lasers and photonics. Table 15.1 showcases a few of those advanced programs (masters and PhD) both in the Unites States and around the world. The table is not meant to be comprehensive but rather is a reflection of the educational programs offered in colleges and universities with an emphasis on lasers and photonics. In the "Laser and Photonic Systems Research Centers and Labs from around the World" section, the role of these colleges as drivers of research and innovation on lasers and photonics will be addressed in more detail.

## Laser and Photonic Systems Research Centers and Labs from around the World

Solving societal problems like those put forward by the National Academy of Engineering in its Grand Challenges for Engineering report (NAE, 2008) require that the investigation and education that researchers, educators, and students are involved in crosses all boundaries. Interdisciplinary labs and centers, like the majority in photonics and lasers, are at the heart of solving many of the problems identified in the NAE report, the European Commission Principles for Innovative Doctoral Training (2011), and other reports like it. The interdisciplinary nature of these labs and centers cuts not only across disciplines but also across departments, schools/colleges, and countries and in turn exposes students to new disciplines, research experiences, ways of thinking, and problem-solving methodologies, therefore enriching the experiences of both undergraduate and graduate students. Graduates of these labs are better equipped with the necessary technical, communication, leadership, and working skills to be contributors to the new global economy. Programs on photonics and lasers are a perfect example of the direction that higher education in a global context continues to evolve toward.

The Lund Laser Centre in Lund University (Sweden), one of the members of the Laserlab-Europe, a constellation of 26 laser research facilities in 16 European Union countries, for example, has embraced lasers and photonics as an educational research area that students can obtain a masters or a diploma in. Some areas for diplomas or masters degrees include ultrahigh-intensity laser physics, quantum information, bio-photonics, attosecond physics, and high-order harmonic generation. A sample of the courses offered includes optics and optical design, atomic and molecular spectroscopy, lasers, multispectral imaging, photonics and optical communication, medical optics, advanced optics and lasers, fundamental combustion, and molecular physics. The diplomas/masters degree and classes that Lund offers do not reside in one discipline but rather brings together educators, researchers, and students from engineering, sciences, humanities, and medicine to provide for a more holistic educational experience.

In the United States, the majority of the centers advancing research in lasers and photonic systems are located in research universities with very high research activity (RU/VH) as defined by the Carnegie Classification of Institutions of Higher Education. RU/VH institutions provide the perfect conditions for the establishment and success of these centers due to their

- Wide range of baccalaureate programs, the majority of which offer undergraduate and graduate students in lasers and photonic systems or related areas and can therefore recruit top talent to advance the science

**TABLE 15.1**

Masters and PhD Programs with a Focus on Lasers and Photonics across the World

| Center | | Graduate Studies Programs |
| --- | --- | --- |
| Beckman Institute Laser Group | California Institute of Technology, United States | Masters, PhD in engineering. Focus in bio-optics; electromagnetics, optics, and optoelectronics; science with focus in photonics, optics, and electronics, solid-state devices |
| Center for Laser Based Manufacturing | Purdue University, United States | Masters, PhD in engineering. Focus in lasers (mechanical engineering) |
| Center for Laser Micro-Fabrication | Purdue University, United States | Masters, PhD in engineering. Focus in lasers (mechanical engineering) |
| Center for Lasers and Plasma for Advanced Manufacturing | University of Michigan, United States | Masters, PhD in engineering. Focus in plasma science and engineering, optics and photonics, energy, manufacturing (laser welding, laser micromachining) |
| Institute of Laser Engineering | Beijing University of Technology, China | Masters, PhD in optical engineering and optics |
| Laser Biomedical Research Center | MIT, United States | Masters, PhD in engineering. Focus in optics, electromagnetics, energy and electronic, photonic, and magnetic materials |
| Laser Group | University of Liverpool, England | Masters, PhD in laser engineering |
| Stanford Photonics Research Center | Stanford University, United States | Masters, PhD in engineering. Focus in photonics, EM, quantum, imaging |

- Strong organizational support to research and access to funding sources from federal, state, and private funding
- Commitment to graduate education and the granting of doctoral degrees (20 or more are needed for an RU/VH classification)

Table 15.2 summarizes centers of research in lasers and photonic systems around the world. Although the list is not comprehensive, it does provide a representative sample of the centers. The table highlights areas of research conducted at the centers, which is often connected to the funding agency and expertise of the faculty and researchers leading the investigative efforts. Also included in the table is a membership category that showcases the diversity of the centers. The majority of US institutions expose students at all levels of higher education to research in lasers and photonic systems. This comprehensive approach of exposing all students in the higher education pipeline provides centers continuous access to top talent to advance their educational and investigative endeavors. Programs like the National Science Foundation (NSF) Research Experiences for Undergraduates (REU) have enabled research centers in US colleges and universities to provide experiences to undergraduate students in areas of interest to NSF. This support from NSF is critical for motivating and exposing students early on to the educational richness of research and is often credited with increasing student retention and encouraging students to continue on to graduate school (Dahlberg et al., 2008).

## Conclusions

Systems, networks, and system-of-systems integration are important for the effective use and advancement of laser and photonic systems, and their scale-up even beyond local application impacts. Emerging progress in laser and photonic systems design and integration is focusing on the following directions (Nof et al., 2013):

1. Designing and developing collaboration and communication protocols to enable seamless, service-centric collaboration among a network of entities, including physical systems, and research, education, service, and industry participants
2. Establishing open manufacturing, communication, healthcare, and service environments that facilitate sustainable development and application of all processes including laser- and photonics-based manufacturing, processing, and services
3. Enhancing virtual design for manufacturing, medical, and other services to further shorten development time and provide access to distributed laser and photonic systems

**TABLE 15.2**

Centers of Research in Laser and Photonic Systems around The World

| Center | Location | Country | Research Fields | Membership |
|---|---|---|---|---|
| Beckman Institute Laser Group | California Institute of Technology | United States | Time-resolved spectroscopy for measurements of chemical and biochemical kinetics and the elucidation of reaction mechanisms. | 1 faculty, 10 graduate students, 8 postdocs, 7 undergrad students |
| Center for Laser Based Manufacturing | Purdue University | United States | Laser-assisted machining, laser hardening, laser cladding, laser 3D deposition, laser surface alloying, laser shock peening, laser direct writing, and laser micromachining. | 18 faculty and 20 graduate and postdoc students in engineering |
| Center for Laser Micro-Fabrication | Purdue University | United States | Laser micro- and nanomachining, laser fabrication of MEMS/NEMS components, development of integrated MEMS systems. Nanoscale energy transport, nano-optics and laser-based nano-optical engineering. | 1 faculty, 12 graduate students, 4 undergraduate students |
| Center for Laser and Plasma for Advanced Manufacturing | University of Michigan | United States | Mathematical modeling of transport phenomena and phase transformation for laser processing; sensor development for model validation, process monitoring, and control using optical and acoustic techniques. | 1 faculty, 4 graduate students |
| High Intensity Laser Science Group | University of Texas at Austin | United States | Study the interaction of ultraintense laser light with matter. These laser–matter interactions can create extreme and exotic conditions. This leads to application in diverse fields such as astrophysics, fusion research, and ultrafast radiation source development. | 1 faculty, 4 researchers, 12 graduate and 9 undergraduate students |
| Institute for Quantum Science and Engineering | Harvard University | United States | Photonic crystal quantum cascade laser. | 3 researchers, 2–3 visiting scholars, 6–10 students |

*(continued)*

**TABLE 15.2 (continued)**

Centers of Research in Laser and Photonic Systems around The World

| Center | Location | Country | Research Fields | Membership |
|---|---|---|---|---|
| Institute of Laser Engineering | Beijing University of Technology | China | Laser modern manufacturing science and engineering, laser micro- and nanoprocessing and ultrafast science, energy optical electronic technology and system. | 40 faculty primarily in science and engineering |
| Laser Analyst Group | University of Cambridge | UK | Laser spectroscopic methods to visualize physics, chemistry of complex processes in applications: industrial process control to biomedical. Three focus areas: microscopy, sensor design, and reactive flow imaging. | 3 faculty, 5 postdoctoral researchers, 9 graduate students, 20 outside collaborators |
| Laser Biomedical Research Center | MIT | United States | Laser-based spectroscopic techniques for medical applications: spectral diagnosis of disease, investigation of biophysical and biochemical properties of cells and tissues, and development of novel imaging techniques. | 12 faculty, 10 researchers, 3 students |
| Laser Group | University of Liverpool | England | Laser forming, laser cladding, laser scabbling, laser drilling, laser cleaning, laser welding, laser ignition in IC engines. | 20 researchers primarily in engineering |
| Laser and Optics Research Center | US Air force Academy | United States | Laser cooling and trapping, optical computing, light propagation in dense atomic vapors, optical nonlinear polymers, laser eye studies, fiber optical chemical sensors, laser and optics theory, IR laser development. | 12 researchers in science and engineering |

| | | | | |
|---|---|---|---|---|
| Laser Research Institute | Stellenbosch University | South Africa | Laser-based diagnostics (vacuum ultraviolet spectroscopy, diode laser spectroscopy), laser research and development (gas lasers, fiber lasers), ultrafast science (molecular dynamics, ultrafast electron diffraction), and laser applications in the medical and industrial field. | 15 faculty, 15 graduate students and postdocs, 10 students (MS and undergraduate) |
| Laser Spectroscopy and Applied Material Group | Lawrence Berkeley National Laboratory | United States | Chemical analysis with laser ablation inductively coupled-plasma mass spectroscopy (LA-ICP-MS), thin films synthesis by pulse laser deposition (PLD), laser nanotechnology, and laser ultrasonics. | 14 researchers |
| Lund Laser Center | Lund University | Sweden | High-intensity radiation/matter interactions, atomic and molecular laser spectroscopy, plasma spectroscopy, ultrafast spectroscopy of molecules and condensed matter, spectroscopic imaging, laser diagnostics in combustion, flows and hostile environments, environmental spectroscopy and monitoring, medical/biological laser spectroscopy. | 100 researchers in engineering, science, humanities, and medicine |
| Stanford Photonics Research Center | Stanford University | United States | Automotive photonics, information technology, integrated photonics, lasers and nonlinear optics, microscopy, nanophotonics, neuroscience, solar cell, quantum information science, telecommunications, ultrafast lasers. | 40 faculty, 200 graduate and postdoc students |

4. Enabling human-in-the-loop control through human-centered design of laser and photonic systems, new concepts of interactive work cells for operators, and streamlined organization structure and work flow

5. Modeling and optimizing laser technology and photonic systems and their potential for improving quality, effectiveness, efficiency, and robustness in implementation and integration

6. Providing digital service and digital manufacturing to integrate and optimize finite global resources and meet the increasing demand for laser technology and photonic systems without the limit of time and space

7. Promoting policies that balance scientific exploration and discovery, technology advancement, service sustainability, and economic considerations

Most of the current and emerging research advancements in laser- and photonics-based processes for traditional and nanoscales have focused on the synthesis and processing techniques for building structures and delivering outcomes of varying complexities. More efforts are needed to integrate these processes with interface and networking approaches addressing interaction, collaboration, and nanoinformatics for modeling and control purposes. A noticeable growth has been observed in the development and deployment of in situ sensing methods to gather information spread over multiple scales of nanomanufacturing and other nanoscale processes (e.g., NSF Report, 2009). It is anticipated that such developments in and deployment of the sensing methods would be precursors to the advancements in nanoinformatics methods for in situ process and quality control. In conjunction, the development of associated advanced statistical models and interaction protocols, and the integration of engineering knowledge will become part of the future, largely laser-, photonics-; and energy-based nanosystems, where closer, ultrafast sensing and decision making are applied closer to the process itself.

## References

6mmbr.com. (2007). Laser rangefinders. Retrieved April 1, 2013, from 6mmbr.com: http://www.6mmbr.com/rangefinders.html.

Adesso. (2009, March 25). Barcode scanners. Retrieved April 1, 2013, from ADESSO: http://www.adesso-shop.com/index.php?main_page=product_info &products_id=19.

Anderson, C. (2012). *Makers: The New Industrial Revolution*, New York: Crown Business.

Arthur, C. (2013, February 14). New self-driving car system tested on UK roads. Retrieved February 28, 2013, from *The Guardian*: http://www.guardian.co.uk/technology/2013/feb/14/self-driving-car-system-uk.

Bergh, A. A. (2006, February). Commercial applications of optoelectronics. Retrieved April 1, 2013, from Photonics Spectra: http://www.photonics.com/Article.aspx?AID=24317.

BMW. (n.d.). Laser light. Leading the way to the future. Retrieved February 27, 2013, from BMW i8 Concept: http://www.bmw-i-usa.com/en_us/bmw-i8/#laser-light-leading-the-way-to-the-future.

Byren, R. W. (1993). Laser rangefinders, in *Active Electro-Optical Systems*, F. S. Clifton (Ed.), Vol. 6, pp. 77–114, Ann Arbor, MI: Environmental Research Institute of Michigan and SPIE Optimal Engineering Press.

Cross, L. (2005). Inventing the laser light show. Retrieved March 4, 2013, from Lowell Cross: http://www.lowellcross.com/home/.

Dahlberg, T., Barnes, T., Rorrer, A., Powell, E., and Cairco, L. (2008). Improving retention and graduate recruitment through immersive research experiences for undergraduates, *Proceedings of the 39th SIGCSE Technical Symposium on Computer Science Education*, New York: ACM Press, pp. 466–470.

Daukantas, P. (2010). A short history of laser light shows, *Optics & Photonics News*, 21(5), 42–47.

European Commission Directorate-General For Research and Innovation. (2011). Report of mapping exercise on doctoral training in Europe "Towards a common approach." http://ec.europa.eu/euraxess/pdf/research_policies/Principles_for_Innovative_Doctoral_Training.pdf.

Federal Office of Metrology METAS of Switzerland. (2010). Traffic measurement technology, Federal Office of Metrology METAS of Switzerland, Federal Department of Justice and Police of Switzerland, Bern-Wabern, Switzerland.

Halperin, A. (2010, May 8). Laser printer's history. Retrieved April 1, 2013, from Laser Printer Resource: http://cityscoop.us/walnutcreekca-laserprinters/2010/05/08/laser-printers-history/.

IceGolf. (2012, September 7). Golf laser rangefinder reviews. Retrieved April 1, 2013, from IceGolf: http://icegolf.org/golf-laser-rangefinder-reviews.html/.

J.D. Power and Associates. (2012, February 24). The science inside a laser printer. Retrieved April 1, 2012, from J.D. Power and Associates: http://www.jdpower.com/content/detail.htm?jdpaArticleId=216.

John, P. and Shanks, R. (2002). Photonics classes in high school, *Seventh International Conference on Education and Training in Optics and Photonics*, Vol. 4588, pp. 89–102.

Lacagnina, M. (2007, May). Diffusing the ramp. *AeroSafety World*, pp. 20–24.

Lackner, M., Winter, F., Graf, J., Geringer, H., Kopecek, M., Weinrotter, E., Wintner, J., Klausner, J., and Herdin, G. (2004). Laser ignition in internal combustion engines—A contribution to a sustainable environment, *14th IFRF Members Conference*, Noordwijkerhout, the Netherlands.

Laser F/X International and LaserFX.com. (2008). Glossary of terminology. Retrieved March 4, 2012, from LaserFX: http://www.laserfx.com/Science/Science3.html.

Lawrence Berkeley National Laboratory. (n.d.). Laser ablation technology for chemical analysis. Retrieved March 1, 2013, from Lawrence Berkeley National Laboratory: http://eetd.lbl.gov/l2m2/laser.html.

Lewander, M., Lundin, P., Svensson, T., Svanberg, S., and Olsson, A. (2011). Non-intrusive measurements of headspace gas composition in liquid food packages made of translucent materials, *Packaging Technology and Science*, 24(5), 271–280.

Lipson, H. and Kurman, M. (2013). *Fabricated: The New World of 3D Printing*, Indianapolis, IN: Wiley.

Lundin, P., Cocola, L., Lewander, M., Olsson, A., and Svanberg, S. (2012). Non-intrusive headspace gas measurement by laser spectroscopy—Performance validation by a reference sensor, *Journal of Food Engineering*, 111(4), 612–617.

Makris, K. (2012, December 22). Ring laser gyro. Retrieved March 4, 2013, from Kostas Makris website: http://www.k-makris.gr/AircraftComponents/Laser_Gyro/laser_gyro.htm.

Marshall, L. (2012, September). Laser car ignition dream sparks multiple approaches. Retrieved March 21, 2013, from Photonics Spectra: http://www.photonics.com/Article.aspx?AID=51731.

Marshall, G. F. and Stutz, G. E. (2011). *Handbook of Optical and Laser Scanning*, 2nd edn., Hoboken, NJ: CRC Press, 788pp.

Marulcu, I. and Barnett, M. (2012). Fifth graders' learning about simple machines through engineering design-based instruction using LEGO materials, *Research in Science Education*, 43(5), 1825–1850.

Mataric, M. J., Koenig, N., and Feil-Seifer, D. (2007). Materials for enabling hands-on robotics and STEM education, *AAAI Spring Symposium on Robots and Robot Venues: Resources for AI Education*, Menlo Park, CA: AAAI Press.

National Academy of Engineering. (2008). Grand challenges for engineering. http://www.engineeringchallenges.org.

National Assessment of Educational Progress (NAEP). nces.ed.gov/nations reportcard/science.

Newman, P., Posner, I., and Wang, D. Z. (2012a). What could move? Finding cars, pedestrians and bicyclists in 3D laser data, *Proceedings of the IEEE International Conference on Robotics and Automation* (ICRA), St. Paul, MN, pp. 1–5.

Newman, P., Rus, D., Rohan, P., and Triebel, R. (2012b). Parsing outdoor scenes from streamed 3d laser data using online clustering and incremental belief updates, *AAAI Conference on Artificial Intelligence*, Toronto, Ontario, Canada, pp. 1–8.

Nof, S. Y., Cheng, G. J., Weiner, A. M., Chen, X. W., Bechar, A., Jones, M. G., Reed, C. B. et al. (2013). Laser and photonic systems integration: Emerging innovations and framework for research and education, *Human Factors and Ergonomics in Manufacturing & Service Industries*.

NSF-Report. (2009). *Research Challenges for Integrated Systems Nanomanufacturing*, J.D. Morse et al. (Eds.), http://eprints.internano.org/49/1/NMS_Workshop_Report.pdf.

Panchev, I., Kirtchev, N., and Dimitrov, D. (2011). Possibilities for application of laser ablation in food technologies, *Innovative Food Science & Emerging Technologies*, 12(3), 369–374.

Phippen, T. (2011, January 11). Docking guidance systems: Tools for ramp management. Retrieved March 01, 2013, from AviationPros.com: http://www.aviationpros.com/article/10218389/docking-guidance-systems-tools-for-ramp-management.

Photonics. (2011, September 14). BMW to replace LED headlights with lasers. Retrieved February 27, 2013, from Photonics: http://www.photonics.com/Article.aspx?AID=48335.

Pike, C. (2002). Exploring the conceptual space of LEGO: Teaching and learning the psychology of creativity, *Psychology Learning & Teaching*, 2(2), 87–94.

Polsen, E. S., Stevens, A. G., and Hart, A. J. (2013). Laser printing of nanoparticle toner enables digital control of micropatterned carbon nanotube growth, *ACS: Applied Materials and Interfaces*, 5(9), 3656–3662.

Rosenkrans, W. (2007, May). Graceful arrivals. *Aero Safety World*, pp. 42–45.

Safe Gate Group. (2010, October 11). Safegate Group announces new contract with American Airlines. Retrieved March 1, 2013, from Safe Gate Group: http://www.safegate.com/press-releases/2010-10-11-safegate-group-announces-new-contract-with-american-airlines?selmenuid=3440.

Schreiber, K. U., Stedman, G. E., Igel, H., and Flaws, A. (2006). Ring laser gyroscopes as rotation sensors for seismic wave studies, in *Earthquake Source Asymmetry, Structural Media and Rotation Effects*, R. Teisseyre, M. Takeo, and E. Majewski (Eds.), Berlin, Germany: Springer, pp. 377–390.

Surmann, H., Nuchter, A., and Hertzberg, J. (2003). An autonomous mobile robot with a 3D laser range finder for 3D exploration and digitalization of indoor environments, *Robotics and Autonomous Systems*, 45(3–4), 181–198.

TAL Tech. (2011, July 11). How a barcode reader works. Retrieved April 1, 2013, from TAL Tech: http://www.taltech.com/support/entry/how_barcode_reader_works.

Teed, N. and Lund, A. K. (1993). The effect of laser speed-measuring devices on speed limit law enforcement in Charleston, South Carolina, *Accident Analysis & Prevention*, 25, 459–463.

Ueki, N. and Mukoyama, N. (2013). VCSEL-based laser printing system, in *VCSELs Fundamentals, Technology and Applications of Vertical-Cavity Surface-Emitting Lasers*, R. Michalzik (Ed.) Heidelberg, Germany: Springer, pp. 539–548.

Westcott, R. (2013, February 14). Self-driving car given test run at Oxford University. Retrieved February 28, 2013, from *BBC News*: http://www.bbc.co.uk/news/technology-21465042.

Williams, E. (n.d.). How to aim a handheld laser rangefinder. Retrieved April 1, 2013, from eHow: http://www.ehow.com/how_7668397_aim-handheld-laser-range finder.html.

Williams, K., Igel, I., Poveda, R., and Kapila, V. (2012). Enriching K-12 science and mathematics education using LEGOs, *Advances in Engineering Education*, 3(2), 27pp.

# Index

Printed and bound by CPI Group (UK) Ltd, Croydon, CR0 4YY

18/10/2024

01776270-0008